Famous Mathematical Proofs

" Detailed Solutions "

Edited by Paul F. Kisak

Contents

1 Mathematical proof   1
    1.1 History and etymology 2
    1.2 Nature and purpose 2
    1.3 Methods 3
       1.3.1 Direct proof 3
       1.3.2 Proof by mathematical induction 3
       1.3.3 Proof by contraposition 4
       1.3.4 Proof by contradiction 4
       1.3.5 Proof by construction 4
       1.3.6 Proof by exhaustion 4
       1.3.7 Probabilistic proof 5
       1.3.8 Combinatorial proof 5
       1.3.9 Nonconstructive proof 5
       1.3.10 Statistical proofs in pure mathematics 5
       1.3.11 Computer-assisted proofs 5
    1.4 Undecidable statements 6
    1.5 Heuristic mathematics and experimental mathematics 6
    1.6 Related concepts 6
       1.6.1 Visual proof 6
       1.6.2 Elementary proof 7
       1.6.3 Two-column proof 7
       1.6.4 Colloquial use of "mathematical proof" 8
       1.6.5 Statistical proof using data 8
       1.6.6 Inductive logic proofs and Bayesian analysis 8
       1.6.7 Proofs as mental objects 8
       1.6.8 Influence of mathematical proof methods outside mathematics 8
    1.7 Ending a proof 9
    1.8 See also 9
    1.9 References 9

1.10 Sources 10

1.11 External links 10

2 List of mathematical proofs   **11**

2.1 Theorems of which articles are primarily devoted to proving them 11

2.2 Articles devoted to theorems of which a (sketch of a) proof is given 11

2.3 Articles devoted to algorithms in which their correctness is proven 13

2.4 Articles where example statements are proven 14

2.5 Other articles containing proofs 14

2.6 Articles which mention dependencies of theorems 16

2.7 Articles giving mathematical proofs within a physical model 16

2.8 See also 16

3 Bertrand's postulate   **17**

3.1 Sylvester's theorem 17

3.2 Erdős's theorems 18

3.3 Better results 18

3.4 Consequences 19

3.5 See also 19

3.6 Notes 19

3.7 Bibliography 19

3.8 External links 20

4 Proof of Bertrand's postulate   **21**

4.1 Lemmas and computation 21

4.1.1 Lemma 1: A lower bound on the central binomial coefficients 21

4.1.2 Lemma 2: An upper bound on prime powers dividing central binomial coefficients 22

4.1.3 Lemma 3: The exact power of a large prime in a central binomial coefficient 22

4.1.4 Lemma 4: An upper bound on the primorial 22

4.2 Proof of Bertrand's Postulate 23

4.2.1 Proof by Shigenori Tochiori 24

4.3 References 25

4.4 External links 25

5 Estimation of covariance matrices   **26**

5.1 Estimation in a general context 26

5.2 Maximum-likelihood estimation for the multivariate normal distribution 27

5.2.1 First steps 27

5.2.2 The trace of a 1×1 matrix 28

5.2.3 Using the spectral theorem 29

		5.2.4 Concluding steps	29

        5.2.4   Concluding steps 29
        5.2.5   Alternative derivation 30
    5.3 Intrinsic covariance matrix estimation 31
        5.3.1   Intrinsic expectation 31
        5.3.2   Bias of the sample covariance matrix 31
    5.4 Shrinkage estimation 32
    5.5 See also 32
    5.6 References 32

6 Fermat's little theorem 34

    6.1 History 34
        6.1.1   Further history 36
    6.2 Proofs 36
    6.3 Generalizations 36
    6.4 Converse 37
    6.5 Pseudoprimes 37
    6.6 See also 37
    6.7 Notes 38
    6.8 References 38
    6.9 Further reading 38
    6.10 External links 38

7 Proofs of Fermat's little theorem 39

    7.1 Simplifications 39
    7.2 Combinatorial proofs 39
        7.2.1   Proof by counting necklaces 39
        7.2.2   Proof using dynamical systems 42
        7.2.3   Multinomial proofs 46
        7.2.4   Proof using power product expansions 48
    7.3 Proofs using modular arithmetic 48
        7.3.1   An example 49
        7.3.2   The cancellation law 50
        7.3.3   The rearrangement property 50
        7.3.4   Applications to Euler's theorem 51
    7.4 Proof using group theory 51
        7.4.1   The invertibility property 51
    7.5 Notes 52

8 Gödel's completeness theorem 53

	8.1	Statement of the theorem	53
		8.1.1 Preliminaries	53
		8.1.2 Gödel's original formulation	53
		8.1.3 Model existence theorem	54
		8.1.4 More general form	54
		8.1.5 As a theorem of arithmetic	54
	8.2	Consequences	54
	8.3	Relationship to the incompleteness theorem	55
	8.4	Relationship to the compactness theorem	55
	8.5	Completeness in other logics	55
	8.6	Proofs	56
	8.7	See also	56
	8.8	Further reading	56
	8.9	External links	56
9	**Original proof of Gödel's completeness theorem**		**57**
	9.1	Definitions and assumptions	57
	9.2	Statement of the theorem and its proof	57
		9.2.1 Theorem 1. Every valid formula (true in all structures) is provable.	58
		9.2.2 Theorem 2. Every formula φ is either refutable or satisfiable in some structure.	58
		9.2.3 Equivalence of both theorems	58
		9.2.4 Proof of theorem 2: first step	58
		9.2.5 Reducing the theorem to formulas of degree 1	59
		9.2.6 Proving the theorem for formulas of degree 1	60
	9.3	Extensions	61
		9.3.1 Extension to first-order predicate calculus with equality	62
		9.3.2 Extension to countable sets of formulas	62
		9.3.3 Extension to arbitrary sets of formulas	62
	9.4	References	62
	9.5	External links	62
10	**Mathematical induction**		**63**
	10.1	History	64
	10.2	Description	64
	10.3	Example	64
	10.4	Axiom of induction	65
		10.4.1 Characterizing the structure of N by the induction axiom	66
	10.5	Variants	66
		10.5.1 Induction basis other than 0 or 1	66

		10.5.2 Induction basis equal to 2	67

- 10.5.2 Induction basis equal to 2 . . . 67
- 10.5.3 Induction on more than one counter . . . 67
- 10.5.4 Infinite descent . . . 68
- 10.5.5 Prefix induction . . . 68
- 10.5.6 Complete induction . . . 69
- 10.6 Equivalence with the well-ordering principle . . . 70
- 10.7 Example of error in the inductive step . . . 71
- 10.8 See also . . . 71
- 10.9 Notes . . . 71
- 10.10 References . . . 72
  - 10.10.1 Introduction . . . 72
  - 10.10.2 History . . . 72

11 0.999...   74

- 11.1 Algebraic proofs . . . 74
  - 11.1.1 Fractions and long division . . . 75
  - 11.1.2 Digit manipulation . . . 75
  - 11.1.3 Discussion . . . 75
- 11.2 Analytic proofs . . . 76
  - 11.2.1 Infinite series and sequences . . . 76
  - 11.2.2 Nested intervals and least upper bounds . . . 77
- 11.3 Proofs from the construction of the real numbers . . . 78
  - 11.3.1 Dedekind cuts . . . 79
  - 11.3.2 Cauchy sequences . . . 79
  - 11.3.3 Infinite decimal representation . . . 80
- 11.4 Generalizations . . . 80
  - 11.4.1 Impossibility of unique representation . . . 81
- 11.5 Applications . . . 81
- 11.6 Skepticism in education . . . 82
- 11.7 In popular culture . . . 83
- 11.8 In alternative number systems . . . 83
  - 11.8.1 Infinitesimals . . . 84
  - 11.8.2 Hackenbush . . . 85
  - 11.8.3 Revisiting subtraction . . . 85
  - 11.8.4 p-adic numbers . . . 85
- 11.9 Related questions . . . 86
- 11.10 See also . . . 86
- 11.11 Notes . . . 86
- 11.12 References . . . 89

11.13 Further reading 93

11.14 External links 94

12 Proof that 22/7 exceeds π   97

12.1 Background 97

12.2 The proof 97

12.3 Details of evaluation of the integral 98

12.4 Quick upper and lower bounds 98

12.5 Proof that 355/113 exceeds π 99

12.6 Extensions 99

12.7 See also 100

12.8 References 101

12.9 External links 101

13 Proof that e is irrational   102

13.1 Euler's proof 102

13.2 Fourier's proof 102

13.3 Alternate proofs 103

13.4 Generalizations 104

13.5 See also 104

13.6 References 104

14 Proof that π is irrational   105

14.1 Lambert's proof 105

14.2 Hermite's proof 106

14.3 Cartwright's proof 108

14.4 Niven's proof 109

14.5 Bourbaki's proof 111

14.6 Laczkovich's proof 112

14.7 See also 113

14.8 References 114

15 Divergence of the sum of the reciprocals of the primes   115

15.1 The harmonic series 116

15.2 Proofs 116

15.2.1 First 116

15.2.2 Second 117

15.2.3 Third 119

15.2.4 Fourth 120

15.3 Partial sums 120

15.4 See also 121

   15.5 References 121

   15.6 References 121

   15.7 External links 121

16 Gödel's ontological proof   **122**

   16.1 History of Gödel's proof 122

   16.2 Outline of Gödel's proof 123

   16.3 See also 124

   16.4 Notes 124

   16.5 References 125

   16.6 External links 125

17 Proofs involving the addition of natural numbers   **126**

   17.1 Definitions 126

   17.2 Proof of associativity 126

   17.3 Proof of identity element 127

   17.4 Proof of commutativity 127

   17.5 See also 127

   17.6 References 127

18 Analyticity of holomorphic functions   **128**

   18.1 Proof 128

   18.2 Remarks 130

   18.3 External links 130

19 Proofs involving covariant derivatives   **131**

   19.1 Contracted Bianchi identities 131

         19.1.1 Proof 131

   19.2 The covariant divergence of the Einstein tensor vanishes 132

         19.2.1 Proof 132

   19.3 See also 132

   19.4 References 132

   19.5 Books 133

20 Derivation of the Cartesian form for an ellipse   **134**

21 Derivation of the Routh array   **137**

   21.1 The Cauchy index 137

   21.2 The Routh criterion 139

   21.3 Sturm's theorem 140

	21.4 References 142

22 Deriving the Schwarzschild solution   143

    22.1 Assumptions and notation 143

    22.2 Diagonalising the metric 143

    22.3 Simplifying the components 144

    22.4 Calculating the Christoffel symbols 145

    22.5 Using the field equations to find A(r) and B(r) 145

    22.6 Using the Weak-Field Approximation to find K and S 146

    22.7 Alternative form in isotropic coordinates 147

    22.8 Dispensing with the static assumption - Birkhoff's theorem 147

    22.9 See also 147

    22.10 References 147

23 Dual of BCH is an independent source   148

    23.1 Lemma 148

    23.2 Proof of Lemma 148

    23.3 Corollary 148

    23.4 Proof of Corollary 149

    23.5 References 149

24 Proofs of Fermat's theorem on sums of two squares   150

    24.1 Euler's proof by infinite descent 150

    24.2 Lagrange's proof through quadratic forms 152

    24.3 Dedekind's two proofs using Gaussian integers 153

    24.4 Zagier's "one-sentence proof" 154

    24.5 References 154

    24.6 Notes 154

    24.7 External links 155

25 Furstenberg's proof of the infinitude of primes   156

    25.1 Furstenberg's proof 156

    25.2 Notes 157

    25.3 References 157

    25.4 External links 157

26 Proofs involving the Moore–Penrose pseudoinverse   158

    26.1 Useful lemmas 158

        26.1.1 Lemma 1: $A*A = 0 \Rightarrow A = 0$ 158

        26.1.2 Lemma 2: $A*AB = 0 \Rightarrow AB = 0$ 159

26.1.3 Lemma 3: $ABB^* = 0 \Rightarrow AB = 0$ 159

26.2 Existence and uniqueness 159

    26.2.1 Proof of uniqueness 159

    26.2.2 Proof of existence 159

26.3 Basic properties 160

    26.3.1 $A^{*+}=A^{+*}$ 160

    26.3.2 Identities 160

26.4 Reduction to the Hermitian case 161

    26.4.1 $A^+ = A^* (A A^*)^+$ 161

    26.4.2 $A^+ = (A^* A)^+ A^*$ 161

26.5 Products 161

    26.5.1 A has orthonormal columns 161

    26.5.2 B has orthonormal rows 161

    26.5.3 A has full column rank and B has full row rank 162

    26.5.4 Conjugate transpose 162

26.6 Projectors and subspaces 163

26.7 Additional properties 163

    26.7.1 Least-squares minimization 163

    26.7.2 Minimum-norm solution to a linear system 164

26.8 References 164

27 Sharp-P-completeness of 01-permanent   165

27.1 Significance 165

27.2 Ben-Dor and Halevi's proof 166

    27.2.1 Overview 166

    27.2.2 Constructing the integer graph 166

    27.2.3 01-Matrix 168

27.3 Aaronson's proof 171

27.4 References 171

28 Proof of Fermat's Last Theorem for specific exponents   172

28.1 Mathematical preliminaries 172

    28.1.1 Primitive solutions 172

    28.1.2 Even and odd 173

    28.1.3 Prime factorization 173

    28.1.4 Two cases 173

28.2 $n = 4$ 174

    28.2.1 Application to right triangles 174

    28.2.2 Proof for Case A 174

		28.2.3 Proof for Case B	175
28.3	$n = 3$		176
		28.3.1 Proof for Case A	176
		28.3.2 Proof for Case B	177
28.4	$n = 5$		178
		28.4.1 Proof for Case A	178
		28.4.2 Proof for Case B	179
28.5	$n = 7$		179
28.6	$n = 6, 10$, and 14		179
28.7	Notes		179
28.8	References		182
28.9	Further reading		182
28.10	External links		183

29 Proof of the Euler product formula for the Riemann zeta function — 188

- 29.1 The Euler product formula — 188
- 29.2 Proof of the Euler product formula — 188
- 29.3 The case $s = 1$ — 190
- 29.4 Another proof — 190
- 29.5 References — 191
- 29.6 Notes — 191

30 Proofs involving ordinary least squares — 192

- 30.1 Least squares estimator for β — 192
- 30.2 Unbiasedness and Variance of $\hat{\beta}$ — 192
- 30.3 Expected value of $\hat{\sigma}^2$ — 193
- 30.4 Consistency and asymptotic normality of $\hat{\beta}$ — 194
- 30.5 Maximum likelihood approach — 194
  - 30.5.1 Finite sample distribution — 195

31 Proofs involving the Laplace–Beltrami operator — 196

- 31.1 −div is adjoint to d — 196
- 31.2 Laplace–de Rham operator — 196
- 31.3 Properties — 197
  - 31.3.1 Proof — 197

32 Proofs of convergence of random variables — 198

- 32.1 Convergence almost surely implies convergence in probability — 198
- 32.2 Convergence in probability does not imply almost sure convergence in the discrete case — 199
- 32.3 Convergence in probability implies convergence in distribution — 199

CONTENTS

    32.3.1 Proof for the case of scalar random variables 199

    32.3.2 Proof for the generic case 200

32.4 Convergence in distribution to a constant implies convergence in probability 200

32.5 Convergence in probability to a sequence converging in distribution implies convergence to the same distribution 200

32.6 Convergence of one sequence in distribution and another to a constant implies joint convergence in distribution 201

32.7 Convergence of two sequences in probability implies joint convergence in probability 201

32.8 See also 202

32.9 References 202

33 Proofs related to chi-squared distribution   203

33.1 Derivations of the pdf 203

    33.1.1 Derivation of the pdf for one degree of freedom 203

    33.1.2 Derivation of the pdf for two degrees of freedom 204

    33.1.3 Derivation of the pdf for k degrees of freedom 205

34 Proofs of quadratic reciprocity   207

34.1 Proofs that are accessible 207

34.2 Eisenstein's proof 207

    34.2.1 Proof of Eisenstein's lemma 209

34.3 Proof using algebraic number theory 210

    34.3.1 Cyclotomic field setup 210

    34.3.2 The Frobenius automorphism 211

    34.3.3 Completing the proof 212

34.4 References 214

34.5 External links 214

35 Proof of Stein's example   215

35.1 Sketched proof 215

36 Proofs of trigonometric identities   218

36.1 Elementary trigonometric identities 218

    36.1.1 Definitions 218

    36.1.2 Ratio identities 218

    36.1.3 Complementary angle identities 220

    36.1.4 Pythagorean identities 220

    36.1.5 Angle sum identities 221

    36.1.6 Double-angle identities 224

    36.1.7 Half-angle identities 225

- 36.1.8 Miscellaneous -- the triple tangent identity 226
- 36.1.9 Miscellaneous -- the triple cotangent identity 227
- 36.1.10 Prosthaphaeresis identities 227
- 36.1.11 Inequalities 228
- 36.2 Identities involving calculus 230
  - 36.2.1 Preliminaries 230
  - 36.2.2 Sine and angle ratio identity 230
  - 36.2.3 Cosine and angle ratio identity 231
  - 36.2.4 Cosine and square of angle ratio identity 231
  - 36.2.5 Proof of Compositions of trig and inverse trig functions 231
- 36.3 See also 232
- 36.4 Notes 232
- 36.5 References 232

37 Union of two regular languages 233
- 37.1 Theorem 233
- 37.2 References 235
- 37.3 Text and image sources, contributors, and licenses 236
  - 37.3.1 Text 236
  - 37.3.2 Images 241
  - 37.3.3 Content license 243

Chapter 1

Mathematical proof

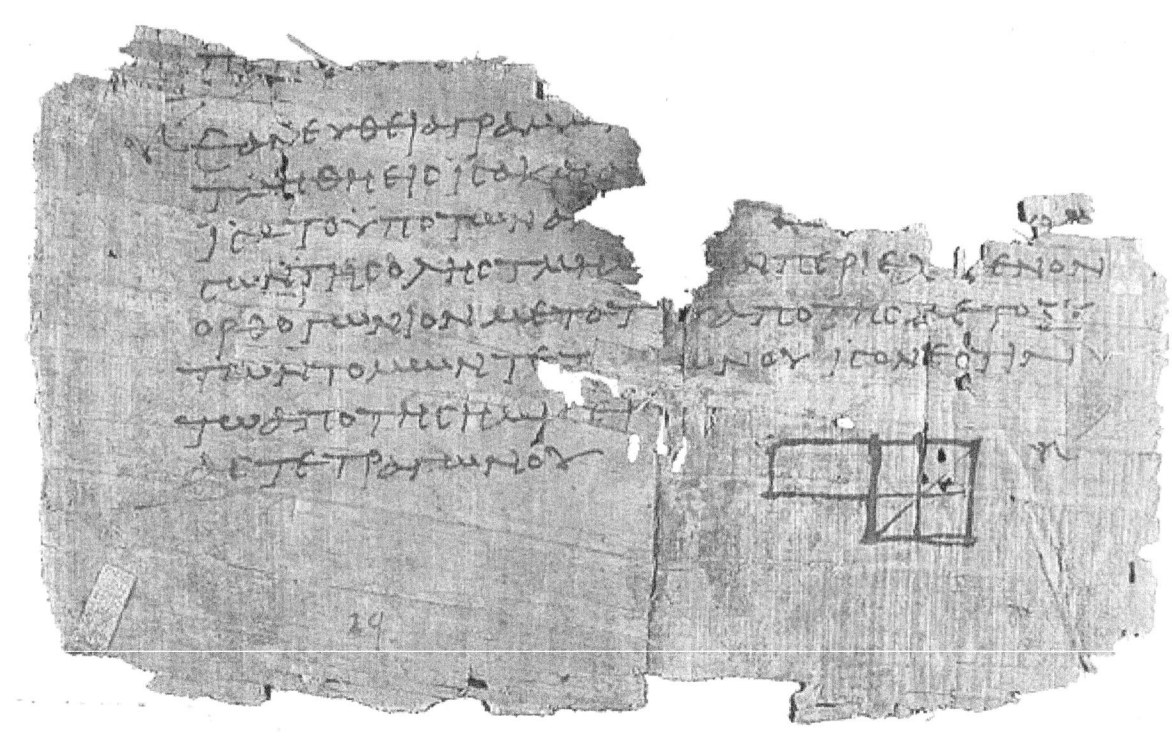

One of the oldest surviving fragments of Euclid's Elements, *a textbook used for millennia to teach proof-writing techniques. The diagram accompanies Book II, Proposition 5.*[1]

In mathematics, a **proof** is a deductive argument for a mathematical statement. In the argument, other previously established statements, such as theorems, can be used. In principle, a proof can be traced back to self-evident or assumed statements, known as axioms.[2][3][4] Proofs are examples of deductive reasoning and are distinguished from inductive or empirical arguments; a proof must demonstrate that a statement is always true (occasionally by listing *all* possible cases and showing that it holds in each), rather than enumerate many confirmatory cases. An unproved proposition that is believed true is known as a conjecture.

Proofs employ logic but usually include some amount of natural language which usually admits some ambiguity. In fact, the vast majority of proofs in written mathematics can be considered as applications of rigorous informal logic. Purely formal proofs, written in symbolic language instead of natural language, are considered in proof theory. The distinction between formal and informal proofs has led to much examination of current and historical mathematical practice, quasi-empiricism in mathematics, and so-called folk mathematics (in both senses of that term). The philosophy of mathematics is concerned with the role of language and logic in proofs, and mathematics as a language.

1.1 History and etymology

See also: History of logic

The word "proof" comes from the Latin *probare* meaning "to test". Related modern words are the English "probe", "probation", and "probability", the Spanish *probar* (to smell or taste, or (lesser use) touch or test),[5] Italian *provare* (to try), and the German *probieren* (to try). The early use of "probity" was in the presentation of legal evidence. A person of authority, such as a nobleman, was said to have probity, whereby the evidence was by his relative authority, which outweighed empirical testimony.[6]

Plausibility arguments using heuristic devices such as pictures and analogies preceded strict mathematical proof.[7] It is likely that the idea of demonstrating a conclusion first arose in connection with geometry, which originally meant the same as "land measurement".[8] The development of mathematical proof is primarily the product of ancient Greek mathematics, and one of its greatest achievements. Thales (624–546 BCE) proved some theorems in geometry. Eudoxus (408–355 BCE) and Theaetetus (417–369 BCE) formulated theorems but did not prove them. Aristotle (384–322 BCE) said definitions should describe the concept being defined in terms of other concepts already known. Mathematical proofs were revolutionized by Euclid (300 BCE), who introduced the axiomatic method still in use today, starting with undefined terms and axioms (propositions regarding the undefined terms assumed to be self-evidently true from the Greek "axios" meaning "something worthy"), and used these to prove theorems using deductive logic. His book, the *Elements*, was read by anyone who was considered educated in the West until the middle of the 20th century.[9] In addition to the familiar theorems of geometry, such as the Pythagorean theorem, the *Elements* includes a proof that the square root of two is irrational and that there are infinitely many prime numbers.

Further advances took place in medieval Islamic mathematics. While earlier Greek proofs were largely geometric demonstrations, the development of arithmetic and algebra by Islamic mathematicians allowed more general proofs that no longer depended on geometry. In the 10th century CE, the Iraqi mathematician Al-Hashimi provided general proofs for numbers (rather than geometric demonstrations) as he considered multiplication, division, etc. for "lines." He used this method to provide a proof of the existence of irrational numbers.[10] An inductive proof for arithmetic sequences was introduced in the *Al-Fakhri* (1000) by Al-Karaji, who used it to prove the binomial theorem and properties of Pascal's triangle. Alhazen also developed the method of proof by contradiction, as the first attempt at proving the Euclidean parallel postulate.[11]

Modern proof theory treats proofs as inductively defined data structures. There is no longer an assumption that axioms are "true" in any sense; this allows for parallel mathematical theories built on alternate sets of axioms (see Axiomatic set theory and Non-Euclidean geometry for examples).

1.2 Nature and purpose

As practiced, a proof is expressed in natural language and is a rigorous argument intended to convince the audience of the truth of a statement. The standard of rigor is not absolute and has varied throughout history. A proof can be presented differently depending on the intended audience. In order to gain acceptance, a proof has to meet communal statements of rigor; an argument considered vague or incomplete may be rejected.

The concept of a proof is formalized in the field of mathematical logic.[12] A formal proof is written in a formal language instead of a natural language. A formal proof is defined as sequence of formulas in a formal language, in which each formula is a logical consequence of preceding formulas. Having a definition of formal proof makes the concept of proof amenable to study. Indeed, the field of proof theory studies formal proofs and their properties, for example, the property that a statement has a formal proof. An application of proof theory is to show that certain undecidable statements are not provable.

The definition of a formal proof is intended to capture the concept of proofs as written in the practice of mathematics. The soundness of this definition amounts to the belief that a published proof can, in principle, be converted into a formal proof. However, outside the field of automated proof assistants, this is rarely done in practice. A classic question in philosophy asks whether mathematical proofs are analytic or synthetic. Kant, who introduced the analytic-synthetic distinction, believed mathematical proofs are synthetic.

Proofs may be viewed as aesthetic objects, admired for their mathematical beauty. The mathematician Paul Erdős was known for describing proofs he found particularly elegant as coming from "The Book", a hypothetical tome containing the most beautiful method(s) of proving each theorem. The book *Proofs from THE BOOK*, published in 2003, is devoted to presenting 32 proofs its editors find particularly pleasing.

1.3 Methods

1.3.1 Direct proof

Main article: Direct proof

In direct proof, the conclusion is established by logically combining the axioms, definitions, and earlier theorems.[13] For example, direct proof can be used to establish that the sum of two even integers is always even:

> Consider two even integers x and y. Since they are even, they can be written as $x = 2a$ and $y = 2b$, respectively, for integers a and b. Then the sum $x + y = 2a + 2b = 2(a+b)$. Therefore $x+y$ has 2 as a factor and, by definition, is even. Hence the sum of any two even integers is even.

This proof uses the definition of even integers, the integer properties of closure under addition and multiplication, and distributivity.

1.3.2 Proof by mathematical induction

Main article: Mathematical induction

Mathematical induction is not a form of inductive reasoning. In proof by mathematical induction, a single "base case" is proved, and an "induction rule" is proved, which establishes that a certain case implies the next case. Applying the induction rule repeatedly, starting from the independently proved base case, proves many, often infinitely many, other cases.[14] Since the base case is true, the infinity of other cases must also be true, even if all of them cannot be proved directly because of their infinite number. A subset of induction is infinite descent. Infinite descent can be used to prove the irrationality of the square root of two.

A common application of proof by mathematical induction is to prove that a property known to hold for one number holds for all natural numbers:[15] Let $\mathbf{N} = \{1,2,3,4,...\}$ be the set of natural numbers, and $P(n)$ be a mathematical statement involving the natural number n belonging to $\mathbf{N}$ such that

- **(i)** $P(1)$ is true, i.e., $P(n)$ is true for $n = 1$.

- **(ii)** $P(n+1)$ is true whenever $P(n)$ is true, i.e., $P(n)$ is true implies that $P(n+1)$ is true.

- **Then $P(n)$ is true for all natural numbers n.**

For example, we can prove by induction that all integers of the form $2n + 1$ are odd:

> **(i)** For $n = 1$, $2n + 1 = 2(1) + 1 = 3$, and 3 is odd. Thus $P(1)$ is true.
>
> **(ii)** For $2n + 1$ for some n, $2(n+1) + 1 = (2n+1) + 2$. If $2n + 1$ is odd, then $(2n+1) + 2$ must also be odd, because adding 2 to an odd number results in an odd number. So $P(n+1)$ is true if $P(n)$ is true.
>
> **Thus** $2n + 1$ is odd, for all natural numbers n.

It is common for the phrase "proof by induction" to be used for a "proof by mathematical induction".[16]

1.3.3 Proof by contraposition

Main article: Contraposition

Proof by contraposition infers the conclusion "if p then q" from the premise "if *not q* then *not p*". The statement "if *not q* then *not p*" is called the contrapositive of the statement "if p then q". For example, contraposition can be used to establish that, given an integer x, if x^2 is even, then x is even:

> Suppose x is not even. Then x is odd. The product of two odd numbers is odd, hence $x^2 = x \cdot x$ is odd. Thus x^2 is not even.

1.3.4 Proof by contradiction

Main article: Proof by contradiction

In proof by contradiction (also known as *reductio ad absurdum*, Latin for "by reduction to the absurd"), it is shown that if some statement were true, a logical contradiction occurs, hence the statement must be false. A famous example of proof by contradiction shows that $\sqrt{2}$ is an irrational number:

> Suppose that $\sqrt{2}$ were a rational number, so by definition $\sqrt{2} = \frac{a}{b}$ where a and b are non-zero integers with no common factor. Thus, $b\sqrt{2} = a$. Squaring both sides yields $2b^2 = a^2$. Since 2 divides the left hand side, 2 must also divide the right hand side (as they are equal and both integers). So a^2 is even, which implies that a must also be even. So we can write $a = 2c$, where c is also an integer. Substitution into the original equation yields $2b^2 = (2c)^2 = 4c^2$. Dividing both sides by 2 yields $b^2 = 2c^2$. But then, by the same argument as before, 2 divides b^2, so b must be even. However, if a and b are both even, they share a factor, namely 2. This contradicts our assumption, so we are forced to conclude that $\sqrt{2}$ is an irrational number.

1.3.5 Proof by construction

Main article: Proof by construction

Proof by construction, or proof by example, is the construction of a concrete example with a property to show that something having that property exists. Joseph Liouville, for instance, proved the existence of transcendental numbers by constructing an explicit example. It can also be used to construct a counterexample to disprove a proposition that all elements have a certain property.

1.3.6 Proof by exhaustion

Main article: Proof by exhaustion

In proof by exhaustion, the conclusion is established by dividing it into a finite number of cases and proving each one separately. The number of cases sometimes can become very large. For example, the first proof of the four color theorem was a proof by exhaustion with 1,936 cases. This proof was controversial because the majority of the cases were checked by a computer program, not by hand. The shortest known proof of the four color theorem as of 2011 still has over 600 cases.

1.3.7 Probabilistic proof

Main article: Probabilistic method

A probabilistic proof is one in which an example is shown to exist, with certainty, by using methods of probability theory. Probabilistic proof, like proof by construction, is one of many ways to show existence theorems.

This is not to be confused with an argument that a theorem is 'probably' true, a 'plausibility argument'. The work on the Collatz conjecture shows how far plausibility is from genuine proof.[17]

1.3.8 Combinatorial proof

Main article: Combinatorial proof

A combinatorial proof establishes the equivalence of different expressions by showing that they count the same object in different ways. Often a bijection between two sets is used to show that the expressions for their two sizes are equal. Alternatively, a double counting argument provides two different expressions for the size of a single set, again showing that the two expressions are equal.

1.3.9 Nonconstructive proof

Main article: Nonconstructive proof

A nonconstructive proof establishes that a mathematical object with a certain property exists without explaining how such an object can be found. Often, this takes the form of a proof by contradiction in which the nonexistence of the object is proved to be impossible. In contrast, a constructive proof establishes that a particular object exists by providing a method of finding it. A famous example of a nonconstructive proof shows that there exist two irrational numbers a and b such that a^b is a rational number:

Either $\sqrt{2}^{\sqrt{2}}$ is a rational number and we are done (take $a = b = \sqrt{2}$), or $\sqrt{2}^{\sqrt{2}}$ is irrational so we can write $a = \sqrt{2}^{\sqrt{2}}$ and $b = \sqrt{2}$. This then gives $\left(\sqrt{2}^{\sqrt{2}}\right)^{\sqrt{2}} = \sqrt{2}^2 = 2$, which is thus a rational of the form a^b.

1.3.10 Statistical proofs in pure mathematics

Main article: Statistical proof

The expression "statistical proof" may be used technically or colloquially in areas of pure mathematics, such as involving cryptography, chaotic series, and probabilistic or analytic number theory.[18][19][20] It is less commonly used to refer to a mathematical proof in the branch of mathematics known as mathematical statistics. See also "Statistical proof using data" section below.

1.3.11 Computer-assisted proofs

Main article: Computer-assisted proof

Until the twentieth century it was assumed that any proof could, in principle, be checked by a competent mathematician to confirm its validity.[7] However, computers are now used both to prove theorems and to carry out calculations that are too long for any human or team of humans to check; the first proof of the four color theorem is an example of a computer-assisted proof. Some mathematicians are concerned that the possibility of an error in a computer program or a run-time error in its calculations calls the validity of such computer-assisted proofs into question. In practice, the chances of an error invalidating a computer-assisted proof can be reduced by incorporating redundancy and self-checks into calculations, and by developing multiple independent approaches and programs. Errors can never be completely ruled out in case of verification of a proof by humans either, especially if the proof contains natural language and requires deep mathematical insight.

1.4 Undecidable statements

A statement that is neither provable nor disprovable from a set of axioms is called undecidable (from those axioms). One example is the parallel postulate, which is neither provable nor refutable from the remaining axioms of Euclidean geometry.

Mathematicians have shown there are many statements that are neither provable nor disprovable in Zermelo-Fraenkel set theory with the axiom of choice (ZFC), the standard system of set theory in mathematics (assuming that ZFC is consistent); see list of statements undecidable in ZFC.

Gödel's (first) incompleteness theorem shows that many axiom systems of mathematical interest will have undecidable statements.

1.5 Heuristic mathematics and experimental mathematics

Main article: Experimental mathematics

While early mathematicians such as Eudoxus of Cnidus did not use proofs, from Euclid to the foundational mathematics developments of the late 19th and 20th centuries, proofs were an essential part of mathematics.[21] With the increase in computing power in the 1960s, significant work began to be done investigating mathematical objects outside of the proof-theorem framework,[22] in experimental mathematics. Early pioneers of these methods intended the work ultimately to be embedded in a classical proof-theorem framework, e.g. the early development of fractal geometry,[23] which was ultimately so embedded.

1.6 Related concepts

1.6.1 Visual proof

Although not a formal proof, a visual demonstration of a mathematical theorem is sometimes called a "proof without words". The left-hand picture below is an example of a historic visual proof of the Pythagorean theorem in the case of the (3,4,5) triangle.

- Visual proof for the (3, 4, 5) triangle as in the Chou Pei Suan Ching 500–200 BC.
- Animated visual proof for the Pythagorean theorem by rearrangement.
- A second animated proof of the Pythagorean theorem.

Some illusory visual proofs, such as the missing square puzzle, can be constructed in a way which appear to prove a supposed mathematical fact but only do so under the presence of tiny errors (for example, supposedly straight lines which actually bend slightly) which are unnoticeable until the entire picture is closely examined, with lengths and angles precisely measured or calculated.

1.6.2 Elementary proof

Main article: Elementary proof

An elementary proof is a proof which only uses basic techniques. More specifically, the term is used in number theory to refer to proofs that make no use of complex analysis. For some time it was thought that certain theorems, like the prime number theorem, could only be proved using "higher" mathematics. However, over time, many of these results have been reproved using only elementary techniques.

1.6.3 Two-column proof

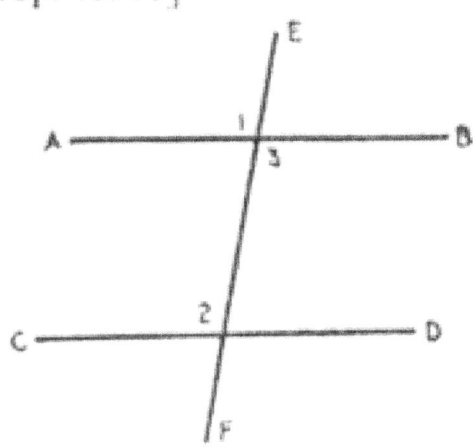

A two-column proof published in 1913

A particular way of organising a proof using two parallel columns is often used in elementary geometry classes in the United States.[24] The proof is written as a series of lines in two columns. In each line, the left-hand column contains a proposition, while the right-hand column contains a brief explanation of how the corresponding proposition in the left-hand column is either an axiom, a hypothesis, or can be logically derived from previous propositions. The left-hand column is typically headed "Statements" and the right-hand column is typically headed "Reasons".[25]

1.6.4 Colloquial use of "mathematical proof"

The expression "mathematical proof" is used by lay people to refer to using mathematical methods or arguing with mathematical objects, such as numbers, to demonstrate something about everyday life, or when data used in an argument is numerical. It is sometimes also used to mean a "statistical proof" (below), especially when used to argue from data.

1.6.5 Statistical proof using data

Main article: Statistical proof

"Statistical proof" from data refers to the application of statistics, data analysis, or Bayesian analysis to infer propositions regarding the probability of data. While *using* mathematical proof to establish theorems in statistics, it is usually not a mathematical proof in that the *assumptions* from which probability statements are derived require empirical evidence from outside mathematics to verify. In physics, in addition to statistical methods, "statistical proof" can refer to the specialized *mathematical methods of physics* applied to analyze data in a particle physics experiment or observational study in cosmology. "Statistical proof" may also refer to raw data or a convincing diagram involving data, such as scatter plots, when the data or diagram is adequately convincing without further analysis.

1.6.6 Inductive logic proofs and Bayesian analysis

Main articles: Inductive logic and Bayesian analysis

Proofs using inductive logic, while considered mathematical in nature, seek to establish propositions with a degree of certainty, which acts in a similar manner to probability, and may be less than one certainty. Bayesian analysis establishes assertions as to the degree of a person's subjective belief. Inductive logic should not be confused with mathematical induction.

1.6.7 Proofs as mental objects

Main articles: Psychologism and Language of thought

Psychologism views mathematical proofs as psychological or mental objects. Mathematician philosophers, such as Leibniz, Frege, and Carnap have attempted to develop a semantics for what they considered to be the language of thought, whereby standards of mathematical proof might be applied to empirical science.

1.6.8 Influence of mathematical proof methods outside mathematics

Philosopher-mathematicians such as Spinoza have attempted to formulate philosophical arguments in an axiomatic manner, whereby mathematical proof standards could be applied to argumentation in general philosophy. Other mathematician-philosophers have tried to use standards of mathematical proof and reason, without empiricism, to arrive at statements outside of mathematics, but having the certainty of propositions deduced in a mathematical proof, such as Descarte's *cogito* argument.

1.7 Ending a proof

Main article: Q.E.D.

Sometimes, the abbreviation *"Q.E.D."* is written to indicate the end of a proof. This abbreviation stands for *"Quod Erat Demonstrandum"*, which is Latin for *"that which was to be demonstrated"*. A more common alternative is to use a square or a rectangle, such as □ or ∎, known as a "tombstone" or "halmos" after its eponym Paul Halmos. Often, "which was to be shown" is verbally stated when writing "QED", "□", or "∎" in an oral presentation on a board.

1.8 See also

- Automated theorem proving
- Invalid proof
- List of incomplete proofs
- List of long proofs
- List of mathematical proofs
- Nonconstructive proof
- Proof by intimidation
- Termination analysis
- *What the Tortoise Said to Achilles*

1.9 References

[1] Bill Casselman. "One of the Oldest Extant Diagrams from Euclid". University of British Columbia. Retrieved 2008-09-26.

[2] Clapham, C. and Nicholson, JN. *The Concise Oxford Dictionary of Mathematics, Fourth edition*. A statement whose truth is either to be taken as self-evident or to be assumed. Certain areas of mathematics involve choosing a set of axioms and discovering what results can be derived from them, providing proofs for the theorems that are obtained.

[3] Cupillari, Antonella. *The Nuts and Bolts of Proofs*. Academic Press, 2001. Page 3.

[4] Gossett, Eric. *Discrete Mathematics with Proof*. John Wiley and Sons, 2009. Definition 3.1 page 86. ISBN 0-470-45793-7

[5] New Shorter Oxford English Dictionary, 1993, OUP, Oxford.

[6] The Emergence of Probability, Ian Hacking

[7] The History and Concept of Mathematical Proof, Steven G. Krantz. 1. February 5, 2007

[8] Kneale, p. 2

[9] Howard Eves, *An Introduction to the History of Mathematics*, Saunders, 1990, ISBN 0-03-029558-0 p. 141: "No work, except The Bible, has been more widely used...."

[10] Matvievskaya, Galina (1987), "The Theory of Quadratic Irrationals in Medieval Oriental Mathematics", *Annals of the New York Academy of Sciences* **500**: 253–277 [260], doi:10.1111/j.1749-6632.1987.tb37206.x

[11] Eder, Michelle (2000), *Views of Euclid's Parallel Postulate in Ancient Greece and in Medieval Islam*, Rutgers University, retrieved 2008-01-23

[12] Buss, 1997, p. 3

[13] Cupillari, page 20.

[14] Cupillari, page 46.

[15] Examples of simple proofs by mathematical induction for all natural numbers

[16] Proof by induction, University of Warwick Glossary of Mathematical Terminology

[17] While most mathematicians do not think that probabilistic evidence ever counts as a genuine mathematical proof, a few mathematicians and philosophers have argued that at least some types of probabilistic evidence (such as Rabin's probabilistic algorithm for testing primality) are as good as genuine mathematical proofs. See, for example, Davis, Philip J. (1972), "Fidelity in Mathematical Discourse: Is One and One Really Two?" *American Mathematical Monthly* 79:252-63. Fallis, Don (1997), "The Epistemic Status of Probabilistic Proof." *Journal of Philosophy* 94:165-86.

[18] "in number theory and commutative algebra... in particular the statistical proof of the lemma."

[19] "Whether constant π (i.e., pi) is normal is a confusing problem without any strict theoretical demonstration except for some **statistical** proof"" (Derogatory use.)

[20] "these observations suggest a statistical proof of Goldbach's conjecture with very quickly vanishing probability of failure for large E"

[21] "*What to do with the pictures? Two thoughts surfaced: the first was that they were unpublishable in the standard way, there were no theorems only very suggestive pictures. They furnished convincing evidence for many conjectures and lures to further exploration, but theorems were coins of the realm ant the conventions of that day dictated that journals only published theorems*", David Mumford, Caroline Series and David Wright, Indra's Pearls, 2002

[22] "*Mandelbrot, working at the IBM Research Laboratory, did some computer simulations for these sets on the reasonable assumption that, if you wanted to prove something, it might be helpful to know the answer ahead of time.*" A Note on the History of Fractals,

[23] "*... brought home again to Benoit [Mandelbrot] that there was a 'mathematics of the eye', that visualization of a problem was as valid a method as any for finding a solution. Amazingly, he found himself alone with this conjecture. The teaching of mathematics in France was dominated by a handful of dogmatic mathematicians hiding behind the pseudonym 'Bourbaki'...* ", Introducing Fractal Geometry, Nigel Lesmoir-Gordon

[24] Patricio G. Herbst, Establishing a Custom of Proving in American School Geometry: Evolution of the Two-Column Proof in the Early Twentieth Century, Educational Studies in Mathematics, Vol. 49, No. 3 (2002), pp. 283-312,

[25] Introduction to the Two-Column Proof, Carol Fisher

1.10 Sources

- Pólya, G. (1954), *Mathematics and Plausible Reasoning*, Princeton University Press.
- Fallis, Don (2002), "What Do Mathematicians Want? Probabilistic Proofs and the Epistemic Goals of Mathematicians", *Logique et Analyse* **45**: 373–388.
- Franklin, J.; Daoud, A. (2011), *Proof in Mathematics: An Introduction*, Kew Books, ISBN 0-646-54509-4.
- Solow, D. (2004), *How to Read and Do Proofs: An Introduction to Mathematical Thought Processes*, Wiley, ISBN 0-471-68058-3.
- Velleman, D. (2006), *How to Prove It: A Structured Approach*, Cambridge University Press, ISBN 0-521-67599-5.

1.11 External links

- Proofs in Mathematics: Simple, Charming and Fallacious
- A lesson about proofs, in a course from Wikiversity

Chapter 2

List of mathematical proofs

A list of articles with mathematical proofs:

2.1 Theorems of which articles are primarily devoted to proving them

See also: Category:Article proofs

- Bertrand's postulate and a proof
- Estimation of covariance matrices
- Fermat's little theorem and some proofs
- Gödel's completeness theorem and its original proof
- Mathematical induction and a proof
- Proof that 0.999... equals 1
- Proof that 22/7 exceeds π
- Proof that e is irrational
- Proof that π is irrational
- Proof that the sum of the reciprocals of the primes diverges

2.2 Articles devoted to theorems of which a (sketch of a) proof is given

See also: Category:Articles containing proofs

- Banach fixed point theorem
- Banach–Tarski paradox
- Basel problem
- Bolzano–Weierstrass theorem

- Brouwer fixed point theorem
- Buckingham π theorem (proof in progress)
- Burnside's lemma
- Cantor's theorem
- Cantor–Bernstein–Schroeder theorem
- Cayley's formula
- Cayley's theorem
- Clique problem (to do)
- Compactness theorem (very compact proof)
- Erdős–Ko–Rado theorem
- Euler's formula
- Euler's four-square identity
- Euler's theorem
- Five color theorem
- Five lemma
- Fundamental theorem of arithmetic
- Gauss–Markov theorem (brief pointer to proof)
- Gödel's incompleteness theorem
    - Gödel's first incompleteness theorem
    - Gödel's second incompleteness theorem
- Goodstein's theorem
- Green's theorem (to do)
    - Green's theorem when D is a simple region
- Heine–Borel theorem
- Intermediate value theorem
- Itō's lemma
- König's lemma
- König's theorem (set theory)
- König's theorem (graph theory)
- Lagrange's theorem
- Liouville's theorem (brief pointer to proof)
- Markov's inequality (proof of a generalization)
- Mean value theorem

- Multivariate normal distribution (to do)
- Holomorphic functions are analytic
- Pythagorean theorem
- Quadratic equation
- Quotient rule
- Ramsey's theorem
- Rao–Blackwell theorem
- Rice's theorem
- Rolle's theorem
- Splitting lemma
- squeeze theorem
- Sum rule in differentiation
- Sum rule in integration
- Sylow theorems
- Transcendence of e and π (as corollaries of Lindemann–Weierstrass)
- Tychonoff's theorem (to do)
- Ultrafilter lemma
- Ultraparallel theorem
- Urysohn's lemma
- Van der Waerden's theorem
- Wilson's theorem
- Zorn's lemma

2.3 Articles devoted to algorithms in which their correctness is proven

- Bellman–Ford algorithm (to do)
- Euclidean algorithm
- Kruskal's algorithm
- Gale–Shapley algorithm
- Prim's algorithm
- Shor's algorithm (incomplete)

2.4 Articles where example statements are proven

See also: Category:Articles containing proofs

- Basis (linear algebra)
- Burrows–Abadi–Needham logic
- Direct proof
- Generating a vector space
- Linear independence
- Polynomial
- Proof
- Pumping lemma
- Simpson's rule

2.5 Other articles containing proofs

See also: Category:Articles containing proofs

- Addition in N
  - associativity of addition in N
  - commutativity of addition in N
  - uniqueness of addition in N
- Algorithmic information theory
- Boolean ring
  - commutativity of a boolean ring
- Boolean satisfiability problem
  - NP-completeness of the Boolean satisfiability problem
- Cantor's diagonal argument
  - set is smaller than its power set
  - uncountability of the real numbers
- Cantor's first uncountability proof
  - uncountability of the real numbers
- Combinatorics
- Combinatory logic
- Co-NP

- Coset
- Countable
  - countability of a subset of a countable set (to do)
- Angle of parallelism
- Galois group
  - Fundamental theorem of Galois theory (to do)
- Gödel number
  - Gödel's incompleteness theorem
- Group (mathematics)
- Halting problem
  - insolubility of the halting problem
- Harmonic series (mathematics)
  - divergence of the (standard) harmonic series
- Highly composite number
- Area of hyperbolic sector, basis of hyperbolic angle
- Infinite series
  - convergence of the geometric series with first term 1 and ratio 1/2
- Integer partition
- Irrational number
  - irrationality of $\log_2 3$
  - irrationality of the square root of 2
- Limit point
- Mathematical induction
  - sum identity
- Power rule
  - differential of x^n
- Product and Quotiont Rules
- Derivation of Product and Quotient rules for differentiating.
- Prime number
  - Infinitude of the prime numbers
- Primitive recursive function
- Principle of bivalence
  - no propositions are neither true nor false in intuitionistic logic

- Recursion
- Relational algebra (to do)
- Solvable group
- Square root of 2
- Tetris
- Algebra of sets
    - idempotent laws for set union and intersection

2.6 Articles which mention dependencies of theorems

- Cauchy's integral formula
- Cauchy integral theorem
- Computational geometry
- Fundamental theorem of algebra
- Lambda calculus
- Invariance of domain
- Minkowski inequality
- Nash embedding theorem
- Open mapping theorem (functional analysis)
- Product topology
- Riemann integral
- Time hierarchy theorem
    - Deterministic time hierarchy theorem

2.7 Articles giving mathematical proofs within a physical model

- No cloning theorem
- Torque

2.8 See also

- Gödel's ontological proof
- Invalid proof
- List of theorems
- List of incomplete proofs
- List of long proofs

Chapter 3

Bertrand's postulate

Bertrand's postulate is a theorem stating that for any integer $n > 3$, there always exists at least one prime number p with

$n < p < 2n - 2$

A weaker but more elegant formulation is: for every $n > 1$ there is always at least one prime p such that

$n < p < 2n$

Another formulation, where p_n is the n-th prime, is for $n \geq 1$

$p_{n+1} < 2p_n$.[1]

This statement was first conjectured in 1845 by Joseph Bertrand [2] (1822–1900). Bertrand himself verified his statement for all numbers in the interval $[2, 3 \times 10^6]$. His conjecture was completely proved by Chebyshev (1821–1894) in 1852[3] and so the postulate is also called the **Bertrand–Chebyshev theorem** or **Chebyshev's theorem**. Chebyshev's theorem can also be stated as a relationship with $\pi(x)$, where $\pi(x)$ is the prime counting function (number of primes less than or equal to x):

$\pi(x) - \pi(\frac{x}{2}) \geq 1$, for all $x \geq 2$.

In 1919, Ramanujan (1887–1920) used properties of the Gamma function to give a simpler proof,[4] from which the concept of Ramanujan primes would later arise, and Erdős (1913–1996) in 1932 published a simpler proof using binomial coefficients and the Chebyshev function ϑ, defined as:

$$\vartheta(x) = \sum_{p=2}^{x} \ln(p)$$

where $p \leq x$ runs over primes. See proof of Bertrand's postulate for the details.

3.1 Sylvester's theorem

Bertrand's postulate was proposed for applications to permutation groups. Sylvester (1814–1897) generalized the weaker statement with the statement: the product of k consecutive integers greater than k is divisible by a prime greater than

k. Bertrand's (weaker) postulate follows from this by taking $k = n$, and considering the k numbers $n+1$, $n+2$, up to and including $n+k = 2n$, where $n > 1$. According to Sylvester's generalization, one of these numbers has a prime factor greater than k. Since all these numbers are less than $2(k+1)$, the number with a prime factor greater than k has only one prime factor, and thus is a prime. Note that $2n$ is not prime, and thus indeed we now know there exists a prime p with $n < p < 2n$.

3.2 Erdős's theorems

Erdős proved in 1934 that for any positive integer k, there is a natural number N such that for all $n > N$, there are at least k primes between n and $2n$. An equivalent statement had been proved in 1919 by Ramanujan (see Ramanujan prime).

The prime number theorem (PNT) implies that the number of primes up to x is roughly $x/\ln(x)$, so if we replace x with $2x$ then we see the number of primes up to $2x$ is asymptotically twice the number of primes up to x (the terms $\ln(2x)$ and $\ln(x)$ are asymptotically equivalent). Therefore the number of primes between n and $2n$ is roughly $n/\ln(n)$ when n is large, and so in particular there are many more primes in this interval than are guaranteed by Bertrand's Postulate. So Bertrand's postulate is comparatively weaker than the PNT. But PNT is a deep theorem, while Bertrand's Postulate can be stated more memorably and proved more easily, and also makes precise claims about what happens for small values of n. (In addition, Chebyshev's theorem was proved before the PNT and so has historical interest.)

The similar and still unsolved Legendre's conjecture asks whether for every $n > 1$, there is a prime p, such that $n^2 < p < (n + 1)^2$. Again we expect that there will be not just one but many primes between n^2 and $(n + 1)^2$, but in this case the PNT doesn't help: the number of primes up to x^2 is asymptotic to $x^2/\ln(x^2)$ while the number of primes up to $(x + 1)^2$ is asymptotic to $(x + 1)^2/\ln((x + 1)^2)$, which is asymptotic to the estimate on primes up to x^2. So unlike the previous case of x and $2x$ we don't get a proof of Legendre's conjecture even for all large n. Error estimates on the PNT are not (indeed, cannot be) sufficient to prove the existence of even one prime in this interval.

3.3 Better results

It follows from the prime number theorem that for any real $\epsilon > 0$ there is a $n_0 > 0$ such that for all $n > n_0$ there is a prime p such $n < p < (1 + \epsilon)n$. It can be shown, for instance, that

$$\lim_{n \to \infty} \frac{\pi((1 + \epsilon)n) - \pi(n)}{n/\log n} = \epsilon,$$

which implies that $\pi((1 + \epsilon)n) - \pi(n)$ goes to infinity (and, in particular, is greater than 1 for sufficiently large n).[5]

Non-asymptotic bounds have also been proved. In 1952, Jitsuro Nagura proved that for $n \geq 25$, there is always a prime between n and $(1 + 1/5)n$.[6]

In 1976, Lowell Schoenfeld showed that for $n \geq 2010760$, there is always a prime between n and $(1 + 1/16597)n$.[7]

In 1998, Pierre Dusart improved the result in his doctoral thesis, showing that for $k \geq 463$, $p_{k+1} \leq (1 + 1/(\ln^2 p_k))p_k$, and in particular for $x \geq 3275$, there exists a prime number between x and $(1 + 1/(2\ln^2 x))x$.[8] In 2010 he proved, that for $x \geq 396738$ there is at least one prime between x and $(1 + 1/(25\ln^2 x))x$.[9]

Baker, Harman and Pintz proved that there is a prime in the interval $[x, x + O(x^{21/40})]$ for all large x.[10]

Generalizations of Bertrand's Postulate have also been obtained by elementary methods. (In the following, n runs through the set of positive integers.) In 2006, M. El Bachraoui proved that there exists a prime between $2n$ and $3n$.[11] In 2011, Andy Loo proved that there exists a prime between $3n$ and $4n$. Furthermore, he proved that as n tends to infinity, the number of primes between $3n$ and $4n$ also goes to infinity, thereby generalizing Erdős' and Ramanujan's results (see the section on Erdős' theorems above).[12] In 2015, Irsen Virnoy proved the same theorem for $5n$ and $6n$.[13] None of these proofs requires the use of deep analytic results.

3.4 Consequences

- The sequence of primes, along with 1, is a complete sequence; any positive integer can be written as a sum of primes (and 1) using each at most once.
- The number 1 is the only integer which is a harmonic number.

3.5 See also

- Oppermann's conjecture

3.6 Notes

[1] Ribenboim, Paulo (2004). *The Little Book of Bigger Primes*. New York: Springer-Verlag. p. 181. ISBN 0-387-20169-6.

[2] Joseph Bertrand. Mémoire sur le nombre de valeurs que peut prendre une fonction quand on y permute les lettres qu'elle renferme. Journal de l'Ecole Royale Polytechnique, Cahier 30, Vol. 18 (1845), 123-140.

[3] P. Tchebychev. Mémoire sur les nombres premiers. Journal de mathématiques pures et appliquées, Sér. 1(1852), 366-390. (Proof of the postulate: 371-382). Also see Mémoires de l'Académie Impériale des Sciences de St. Pétersbourg, vol. 7, pp.15-33, 1854

[4] Ramanujan, S. (1919). "A proof of Bertrand's postulate". *Journal of the Indian Mathematical Society* **11**: 181–182.

[5] G. H. Hardy and E. M. Wright, *An Introduction to the Theory of Numbers*, 6th ed., Oxford University Press, 2008, p. 494.

[6] Nagura, J. "On the interval containing at least one prime number." *Proceedings of the Japan Academy, Series A* **28** (1952), pp. 177–181.

[7] Lowell Schoenfeld (April 1976). "Sharper Bounds for the Chebyshev Functions $\theta(x)$ and $\psi(x)$, II". *Mathematics of Computation* **30** (134): 337–360. doi:10.2307/2005976.

[8] Dusart, Pierre (1998), *Autour de la fonction qui compte le nombre de nombres premiers* (PDF) (in French)

[9] Dusart, Pierre (2010). "Estimates of Some Functions Over Primes without R.H.". arXiv:1002.0442.

[10] Baker, R. C.; Harman, G.; Pintz, J. (2001). "The difference between consecutive primes, II". *Proceedings of the London Mathematical Society* **83** (3): 532–562. doi:10.1112/plms/83.3.532.

[11] M. El Bachraoui, Primes in the Interval $(2n, 3n)$

[12] Loo, Andy (2011), "On the Primes in the Interval $(3n, 4n)$" (PDF), *International Journal of Contemporary Mathematical Sciences* **6** (38): 1871–1882

[13] Irsen Virnoy, On the existence of at least prime number between $(5n, 6n)$

3.7 Bibliography

- P. Erdős (1934). "A Theorem of Sylvester and Schur". *Journal of the London Mathematical Society* **9** (4): 282–288. doi:10.1112/jlms/s1-9.4.282.
- Jitsuro Nagura (1952). "On the interval containing at least one prime number". *Proc. Japan Acad.* **28** (4): 177–181. doi:10.3792/pja/1195570997.
- Jonathan Sondow and Eric W. Weisstein, "Bertrand's Postulate", *MathWorld*.
- Chris Caldwell, *Bertrand's postulate* at Prime Pages glossary.

- H. Ricardo (2005). "Goldbach's Conjecture Implies Bertrand's Postulate". *Amer. Math. Monthly* **112**: 492.

- Hugh L. Montgomery; Robert C. Vaughan (2007). *Multiplicative number theory I. Classical theory*. Cambridge tracts in advanced mathematics **97**. Cambridge: Cambridge Univ. Press. p. 49. ISBN 0-521-84903-9.

- J. Sondow (2009). "Ramanujan primes and Bertrand's postulate". *Amer. Math. Monthly* **116**: 630–635. arXiv:0907.5232. doi:10.4169/193009709x458609.

3.8 External links

- A proof of the weak version in the Mizar system: http://mizar.org/version/current/html/nat_4.html#T56

Chapter 4

Proof of Bertrand's postulate

In mathematics, Bertrand's postulate (actually a theorem) states that for each $n \geq 1$ there is a prime p such that $n < p \leq 2n$. It was first proven by Pafnuty Chebyshev, and a short but advanced proof was given by Srinivasa Ramanujan.[1] The gist of the following elementary proof is due to Paul Erdős. The basic idea of the proof is to show that a certain central binomial coefficient needs to have a prime factor within the desired interval in order to be large enough. This is made possible by a careful analysis of the prime factorization of central binomial coefficients.

The main steps of the proof are as follows. First, one shows that every prime power factor p^r that enters into the prime decomposition of the central binomial coefficient $\binom{2n}{n} := \frac{(2n)!}{(n!)^2}$ is at most $2n$. In particular, every prime larger than $\sqrt{2n}$ can enter at most once into this decomposition; that is, its exponent r is at most one. The next step is to prove that $\binom{2n}{n}$ has no prime factors at all in the gap interval $\left(\frac{2n}{3}, n\right)$. As a consequence of these two bounds, the contribution to the size of $\binom{2n}{n}$ coming from all the prime factors that are at most n grows asymptotically as $O(\theta^n)$ for some $\theta < 4$. Since the asymptotic growth of the central binomial coefficient is at least $4^n/2n$, one concludes that for n large enough the binomial coefficient must have another prime factor, which can only lie between n and $2n$. Indeed, making these estimates quantitative, one obtains that this argument is valid for all $n > 468$. The remaining smaller values of n are easily settled by direct inspection, completing the proof of Bertrand's postulate.

4.1 Lemmas and computation

4.1.1 Lemma 1: A lower bound on the central binomial coefficients

Lemma: For any integer $n > 0$, we have

$$\frac{4^n}{2n} \leq \binom{2n}{n}.$$

Proof: Applying the binomial theorem,

$$4^n = (1+1)^{2n} = \sum_{k=0}^{2n} \binom{2n}{k} = 2 + \sum_{k=1}^{2n-1} \binom{2n}{k} \leq 2n \binom{2n}{n},$$

since $\binom{2n}{n}$ is the largest term in the sum in the right-hand side, and the sum has $2n$ terms (including the initial two outside the summation).

4.1.2 Lemma 2: An upper bound on prime powers dividing central binomial coefficients

For a fixed prime p, define $R(p, n)$ to be the largest natural number r such that p^r divides $\binom{2n}{n}$.

Lemma: For any prime p, $p^{R(p,n)} \leq 2n$.

Proof: The exponent of p in $n!$ is (see Factorial#Number theory):

$$\sum_{j=1}^{\infty} \left\lfloor \frac{n}{p^j} \right\rfloor,$$

so

$$R(p, n) = \sum_{j=1}^{\infty} \left\lfloor \frac{2n}{p^j} \right\rfloor - 2 \sum_{j=1}^{\infty} \left\lfloor \frac{n}{p^j} \right\rfloor = \sum_{j=1}^{\infty} \left(\left\lfloor \frac{2n}{p^j} \right\rfloor - 2 \left\lfloor \frac{n}{p^j} \right\rfloor \right).$$

But each term of the last summation can either be zero (if $n/p^j \bmod 1 < 1/2$) or 1 (if $n/p^j \bmod 1 \geq 1/2$) and all terms with $j > \log_p(2n)$ are zero. Therefore

$$R(p, n) \leq \log_p(2n),$$

and

$$p^{R(p,n)} \leq p^{\log_p 2n} = 2n.$$

This completes the proof of the lemma.

4.1.3 Lemma 3: The exact power of a large prime in a central binomial coefficient

Lemma: If p is odd and $\frac{2n}{3} < p \leq n$, then $R(p, n) = 0$.

Proof: There are exactly two factors of p in the numerator of the expression $\binom{2n}{n} = \frac{(2n)!}{(n!)^2}$, coming from the two terms p and $2p$ in $2n!$, and also two factors of p in the denominator from two copies of the term p in $n!$. These factors all cancel, leaving no factors of p in $\binom{2n}{n}$. (The bound on p in the preconditions of the lemma ensures that $3p$ is too large to be a term of the numerator, and the assumption that p is odd is needed to ensure that $2p$ contributes only one factor of p to the numerator.)

4.1.4 Lemma 4: An upper bound on the primorial

We estimate the primorial function,

$$x\# = \prod_{p \leq x} p,$$

where the product is taken over all *prime* numbers p less than or equal to the real number x.

Lemma: For all real numbers $x \geq 3$, $x\# < 2^{2x-3}$ [2]

Proof: Since $x\# = \lfloor x \rfloor \#$, it suffices to prove the result under the assumption that $x = n$ is an integer. Since $\binom{2n}{n}$ is an integer and all the primes $n + 1 \leq p \leq 2n - 1$ appear in its numerator, $(2n - 1)\#/(n)\# \leq \binom{2n}{n} < 2^{2n-2}$ must hold. The proof proceeds by mathematical induction.

- $n = 3: n\# = 6 < 8$.

- $n = 4: n\# = 6 < 32$.

- If n odd, $n\# = (2m-1)\# < 2^{2(2m-1)-3}$

- If n even, $n\# = (2m)\# < 2^{2(2m)-3}$

- $n\# < 2^{2n-3}$

Thus the lemma is proven.

4.2 Proof of Bertrand's Postulate

Assume there is a counterexample: an integer $n \geq 2$ such that there is no prime p with $n < p < 2n$.

If $2 \leq n < 468$, then p can be chosen from among the prime numbers 3, 5, 7, 13, 23, 43, 83, 163, 317, 631 (each being less than twice its predecessor) such that $n < p < 2n$. Therefore $n \geq 468$.

There are no prime factors p of $\binom{2n}{n}$ such that:

- $2n < p$, because every factor must divide $(2n)!$;

- $p = 2n$, because $2n$ is not prime;

- $n < p < 2n$, because we assumed there is no such prime number;

- $2n/3 < p \leq n$: by Lemma 3.

Therefore, every prime factor p satisfies $p \leq 2n/3$.

When $p > \sqrt{2n}$, the number $\binom{2n}{n}$ has at most one factor of p. By Lemma 2, for any prime p we have $p^{R(p,n)} \leq 2n$, so the product of the $p^{R(p,n)}$ over the primes less than or equal to $\sqrt{2n}$ is at most $(2n)^{\sqrt{2n}}$. Then, starting with Lemma 1 and decomposing the right-hand side into its prime factorization, and finally using Lemma 4, these bounds give:

$$\frac{4^n}{2n} \leq \binom{2n}{n} = \left(\prod_{p \leq \sqrt{2n}} p^{R(p,n)}\right)\left(\prod_{\sqrt{2n} < p \leq \frac{2n}{3}} p^{R(p,n)}\right) < (2n)^{\sqrt{2n}} \prod_{1 < p \leq \frac{2n}{3}} p = (2n)^{\sqrt{2n}} \left(\frac{2n}{3}\right)\# \leq (2n)^{\sqrt{2n}} 4^{2n/3}.$$

Taking logarithms yields to

$$\frac{\log 4}{3} n \leq (\sqrt{2n} + 1) \log 2n .$$

By concavity of the right-hand side as a function of n, the last inequality is necessarily verified on an interval. Since it holds true for $n=467$ and it does not for $n=468$, we obtain

$n < 468$.

But these cases have already been settled, and we conclude that no counterexample to the postulate is possible.

4.2.1 Proof by Shigenori Tochiori

Using Lemma 4, Tochiori refined Erdos's method and proved if there exists a positive integer $n \geq 5$ such that there is no prime number $n < p \leq 2n$ then $n < 64$. [3]

First, refine lemma 1 to:

Lemma 1': For any integer $n \geq 4$, we have

$$\frac{4^n}{n} < \binom{2n}{n}.$$

Proof: By induction: $\frac{4^4}{4} = 64 < 70 = \binom{8}{4}$, and assuming the truth of the lemma for $n-1$,

$$\binom{2n}{n} = 2\frac{2n-1}{n}\binom{2(n-1)}{n-1} > 2\frac{2n-1}{n}\frac{4^{n-1}}{n-1} > 2 \cdot 2\frac{4^{n-1}}{n} = \frac{4^n}{n}.$$

Then, refine the estimate of the product of all small primes via a better estimate on $\pi(x)$ (the number of primes at most n):

Lemma 5: For any natural number n, we have

$$\pi(n) \leq \frac{1}{3}n + 2.$$

Proof: Except for $p = 2, 3$, every prime number has $p \equiv 1$ or $p \equiv 5 \pmod{6}$. Thus $\pi(n)$ is upper bounded by the number of numbers with $k \equiv 1$ or $k \equiv 5 \pmod{6}$, plus one (since this counts 1 and misses $2, 3$). Thus

$$\pi(n) \leq \left\lfloor\frac{n+5}{6}\right\rfloor + \left\lfloor\frac{n+1}{6}\right\rfloor + 1 \leq \frac{n+5}{6} + \frac{n+1}{6} + 1 = \frac{1}{3}n + 2.$$

Now, calculating the binomial coefficient as in the previous section, we can use the improved bounds to get (for $n \geq 5$, which implies $\sqrt{2n} \geq 3$ so that $\sqrt{2n}\# \geq 3\# = 6$):

$$\frac{4^n}{n} \leq \binom{2n}{n}$$

$$= \prod_{p \leq \sqrt{2n}} p^{R(p,n)} \cdot \prod_{\sqrt{2n} < p \leq \frac{2n}{3}} p^{R(p,n)}$$

$$< (2n)^{\pi(\sqrt{2n})} \prod_{\sqrt{2n} < p \leq \frac{2n}{3}} p = (2n)^{\frac{1}{3}\sqrt{2n}+2}\frac{(2n/3)\#}{\sqrt{2n}\#}$$

$$< (2n)^{\frac{1}{3}\sqrt{2n}+2}\frac{2^{2 \cdot 2n/3 - 3}}{6} < (2n)^{\frac{1}{3}\sqrt{2n}+2}2^{4n/3-5}.$$

Taking logarithms to get

$$\frac{2}{3}n\log 2 < \frac{1}{3}\sqrt{2n}\log 2n + 3\log\frac{n}{2}$$

and dividing both sides by $\frac{2}{3}n$:

$$\log 2 < \sqrt{2} \cdot \frac{\log \sqrt{n}}{\sqrt{n}} + \frac{9}{4} \frac{\log \frac{n}{2}}{\frac{n}{2}} + \frac{\log 2}{\sqrt{2n}} \equiv f(n) \ .$$

Now the function $g(x) = \frac{\log x}{x}$ is decreasing for $x \geq e$, so $f(n)$ is decreasing when $n \geq e^2 > 2e$. But

$$\frac{f(2^6)}{\log 2} = \sqrt{2} \cdot \frac{3}{8} + \frac{9}{4} \cdot \frac{5}{32} + \frac{\sqrt{2}}{16} = 0.97 \cdots < 1 < \frac{f(n)}{\log 2},$$

so $n < 2^6 = 64$. The remaining cases are proven by an explicit list of primes, as above.

4.3 References

[1] Ramanujan, S. (1919), "A proof of Bertrand's postulate", *Journal of the Indian Mathematical Society* **11**: 181–182

[2] http://www.chart.co.jp/subject/sugaku/suken_tsushin/76/76-8.pdf

[3] http://www.chart.co.jp/subject/sugaku/suken_tsushin/76/76-8.pdf

- Aigner, Martin, G., Günter M. Ziegler, Karl H. Hofmann, *Proofs from THE BOOK*, Fourth edition, Springer, 2009. ISBN 978-3-642-00855-9.

4.4 External links

- Proof in the Mizar system: http://mizar.org/version/current/html/nat_4.html#T56

Chapter 5

Estimation of covariance matrices

In statistics, sometimes the covariance matrix of a multivariate random variable is not known but has to be estimated. **Estimation of covariance matrices** then deals with the question of how to approximate the actual covariance matrix on the basis of a sample from the multivariate distribution. Simple cases, where observations are complete, can be dealt with by using the sample covariance matrix. The sample covariance matrix (SCM) is an unbiased and efficient estimator of the covariance matrix if the space of covariance matrices is viewed as an extrinsic convex cone in $\mathbf{R}^{p \times p}$; however, measured using the intrinsic geometry of positive-definite matrices, the SCM is a biased and inefficient estimator.[1] In addition, if the random variable has normal distribution, the sample covariance matrix has Wishart distribution and a slightly differently scaled version of it is the maximum likelihood estimate. Cases involving missing data require deeper considerations. Another issue is the robustness to outliers:[2] "Sample covariance matrices are extremely sensitive to outliers".[3][4]

Statistical analyses of multivariate data often involve exploratory studies of the way in which the variables change in relation to one another and this may be followed up by explicit statistical models involving the covariance matrix of the variables. Thus the estimation of covariance matrices directly from observational data plays two roles:

- to provide initial estimates that can be used to study the inter-relationships;
- to provide sample estimates that can be used for model checking.

Estimates of covariance matrices are required at the initial stages of principal component analysis and factor analysis, and are also involved in versions of regression analysis that treat the dependent variables in a data-set, jointly with the independent variable as the outcome of a random sample.

5.1 Estimation in a general context

Given a sample consisting of n independent observations $x_1,..., x_n$ of a p-dimensional random vector $X \in \mathbf{R}^{p \times 1}$ (a $p \times 1$ column-vector), an unbiased estimator of the ($p \times p$) covariance matrix

$$\operatorname{cov}(X) = \operatorname{E}\left[(X - \operatorname{E}[X])(X - \operatorname{E}[X])^{\mathrm{T}}\right]$$

is the sample covariance matrix

$$\mathbf{Q} = \frac{1}{n-1} \sum_{i=1}^{n} (x_i - \overline{x})(x_i - \overline{x})^{\mathrm{T}},$$

where x_i is the i-th observation of the p-dimensional random vector, and

$$\bar{x} = \begin{bmatrix} \bar{x}_1 \\ \vdots \\ \bar{x}_p \end{bmatrix} = \frac{1}{n} \sum_{i=1}^{n} x_i$$

is the sample mean. This is true regardless of the distribution of the random variable X, provided of course that the theoretical means and covariances exist. The reason for the factor $n-1$ rather than n is essentially the same as the reason for the same factor appearing in unbiased estimates of sample variances and sample covariances, which relates to the fact that the mean is not known and is replaced by the sample mean.

In cases where the distribution of the random variable X is known to be within a certain family of distributions, other estimates may be derived on the basis of that assumption. A well-known instance is when the random variable X is normally distributed: in this case the maximum likelihood estimator of the covariance matrix is slightly different from the unbiased estimate, and is given by

$$\mathbf{Q_n} = \frac{1}{n} \sum_{i=1}^{n} (x_i - \bar{x})(x_i - \bar{x})^{\mathrm{T}}.$$

A derivation of this result is given below. Clearly, the difference between the unbiased estimator and the maximum likelihood estimator diminishes for large n.

In the general case, the unbiased estimate of the covariance matrix provides an acceptable estimate when the data vectors in the observed data set are all complete: that is they contain no missing elements. One approach to estimating the covariance matrix is to treat the estimation of each variance or pairwise covariance separately, and to use all the observations for which both variables have valid values. Assuming the missing data are missing at random this results in an estimate for the covariance matrix which is unbiased. However, for many applications this estimate may not be acceptable because the estimated covariance matrix is not guaranteed to be positive semi-definite. This could lead to estimated correlations having absolute values which are greater than one, and/or a non-invertible covariance matrix.

When estimating the cross-covariance of a pair of signals that are wide-sense stationary, missing samples do *not* need be random (e.g., sub-sampling by an arbitrary factor is valid).

5.2 Maximum-likelihood estimation for the multivariate normal distribution

Main article: Multivariate normal distribution

A random vector $X \in \mathbf{R}^p$ (a $p \times 1$ "column vector") has a multivariate normal distribution with a nonsingular covariance matrix Σ precisely if $\Sigma \in \mathbf{R}^{p \times p}$ is a positive-definite matrix and the probability density function of X is

$$f(x) = (2\pi)^{-p/2} \det(\Sigma)^{-1/2} \exp\left(-\frac{1}{2}(x-\mu)^{\mathrm{T}} \Sigma^{-1}(x-\mu)\right)$$

where $\mu \in \mathbf{R}^{p \times 1}$ is the expected value of X. The covariance matrix Σ is the multidimensional analog of what in one dimension would be the variance, and $(2\pi)^{-p/2} \det(\Sigma)^{-1/2}$ normalizes the density $f(x)$ so that it integrates to 1.

Suppose now that $X_1, ..., X_n$ are independent and identically distributed samples from the distribution above. Based on the observed values $x_1, ..., x_n$ of this sample, we wish to estimate Σ.

5.2.1 First steps

The likelihood function is:

$$\mathcal{L}(\mu, \Sigma) = (2\pi)^{-np/2} \prod_{i=1}^{n} \det(\Sigma)^{-1/2} \exp\left(-\frac{1}{2}(x_i - \mu)^T \Sigma^{-1}(x_i - \mu)\right)$$

It is fairly readily shown that the maximum-likelihood estimate of the mean vector μ is the "sample mean" vector:

$$\bar{x} = (x_1 + \cdots + x_n)/n.$$

See the section on estimation in the article on the normal distribution for details; the process here is similar.

Since the estimate $\bar{x}$ does not depend on Σ, we can just substitute it for μ in the likelihood function, getting

$$\mathcal{L}(\bar{x}, \Sigma) \propto \det(\Sigma)^{-n/2} \exp\left(-\frac{1}{2}\sum_{i=1}^{n}(x_i - \bar{x})^T \Sigma^{-1}(x_i - \bar{x})\right),$$

and then seek the value of Σ that maximizes the likelihood of the data (in practice it is easier to work with $\log \mathcal{L}$).

5.2.2 The trace of a 1 × 1 matrix

Now we come to the first surprising step: regard the scalar $(x_i - \bar{x})^T \Sigma^{-1}(x_i - \bar{x})$ as the trace of a 1×1 matrix.

This makes it possible to use the identity tr(*AB*) = tr(*BA*) whenever *A* and *B* are matrices so shaped that both products exist. We get

$$\mathcal{L}(\bar{x}, \Sigma) \propto \det(\Sigma)^{-n/2} \exp\left(-\frac{1}{2}\sum_{i=1}^{n} \operatorname{tr}((x_i - \bar{x})^T \Sigma^{-1}(x_i - \bar{x}))\right)$$

$$= \det(\Sigma)^{-n/2} \exp\left(-\frac{1}{2}\sum_{i=1}^{n} \operatorname{tr}((x_i - \bar{x})(x_i - \bar{x})^T \Sigma^{-1})\right)$$

(so now we are taking the trace of a *p*×*p* matrix)

$$= \det(\Sigma)^{-n/2} \exp\left(-\frac{1}{2} \operatorname{tr}\left(\sum_{i=1}^{n}(x_i - \bar{x})(x_i - \bar{x})^T \Sigma^{-1}\right)\right)$$

$$= \det(\Sigma)^{-n/2} \exp\left(-\frac{1}{2} \operatorname{tr}\left(S \Sigma^{-1}\right)\right)$$

where

$$S = \sum_{i=1}^{n}(x_i - \bar{x})(x_i - \bar{x})^T \in \mathbf{R}^{p \times p}.$$

S is sometimes called the scatter matrix, and is positive definite if there exists a subset of the data consisting of p linearly independent observations (which we will assume).

5.2.3 Using the spectral theorem

It follows from the spectral theorem of linear algebra that a positive-definite symmetric matrix S has a unique positive-definite symmetric square root $S^{1/2}$. We can again use the "cyclic property" of the trace to write

$$\det(\Sigma)^{-n/2} \exp\left(-\frac{1}{2} \operatorname{tr}\left(S^{1/2}\Sigma^{-1}S^{1/2}\right)\right).$$

Let $B = S^{1/2}\,\Sigma^{-1}\,S^{1/2}$. Then the expression above becomes

$$\det(S)^{-n/2} \det(B)^{n/2} \exp\left(-\frac{1}{2}\operatorname{tr}(B)\right).$$

The positive-definite matrix B can be diagonalized, and then the problem of finding the value of B that maximizes

$$\det(B)^{n/2} \exp\left(-\frac{1}{2}\operatorname{tr}(B)\right)$$

Since the trace of a square matrix equals the sum of eigen-values ("trace and eigenvalues"), the equation reduces to the problem of finding the eigen values $\lambda_1, ..., \lambda p$ that maximize

$$\lambda_i^{n/2} \exp(-\lambda_i/2).$$

This is just a calculus problem and we get $\lambda i = n$ for all i. Thus, assume Q is the matrix of eigen vectors, then

$$B = Q(nI_p)Q^{-1} = nI_p$$

i.e., n times the $p{\times}p$ identity matrix.

5.2.4 Concluding steps

Finally we get

$$\Sigma = S^{1/2}B^{-1}S^{1/2} = S^{1/2}\left(\frac{1}{n}I_p\right)S^{1/2} = \frac{S}{n},$$

i.e., the $p{\times}p$ "sample covariance matrix"

$$\frac{S}{n} = \frac{1}{n}\sum_{i=1}^{n}(X_i - \overline{X})(X_i - \overline{X})^{\mathrm{T}}$$

is the maximum-likelihood estimator of the "population covariance matrix" Σ. At this point we are using a capital X rather than a lower-case x because we are thinking of it "as an estimator rather than as an estimate", i.e., as something random whose probability distribution we could profit by knowing. The random matrix S can be shown to have a Wishart distribution with $n-1$ degrees of freedom.[5] That is:

$$\sum_{i=1}^{n}(X_i - \overline{X})(X_i - \overline{X})^{\mathrm{T}} \sim W_p(\Sigma, n-1).$$

5.2.5 Alternative derivation

An alternative derivation of the maximum likelihood estimator can be performed via matrix calculus formulae (see also differential of a determinant and differential of the inverse matrix). It also verifies the aforementioned fact about the maximum likelihood estimate of the mean. Re-write the likelihood in the log form using the trace trick:

$$\ln \mathcal{L}(\mu, \Sigma) = \text{const} - \frac{n}{2} \ln \det(\Sigma) - \frac{1}{2} \text{tr} \left[\Sigma^{-1} \sum_{i=1}^{n} (x_i - \mu)(x_i - \mu)^{\text{T}} \right].$$

The differential of this log-likelihood is

$$d \ln \mathcal{L}(\mu, \Sigma) = -\frac{n}{2} \text{tr} \left[\Sigma^{-1} \{d\Sigma\} \right]$$

$$- \frac{1}{2} \text{tr} \left[-\Sigma^{-1} \{d\Sigma\} \Sigma^{-1} \sum_{i=1}^{n} (x_i - \mu)(x_i - \mu)^{\text{T}} - 2\Sigma^{-1} \sum_{i=1}^{n} (x_i - \mu)\{d\mu\}^{\text{T}} \right].$$

It naturally breaks down into the part related to the estimation of the mean, and to the part related to the estimation of the variance. The first order condition for maximum, $d \ln \mathcal{L}(\mu, \Sigma) = 0$, is satisfied when the terms multiplying $d\mu$ and $d\Sigma$ are identically zero. Assuming (the maximum likelihood estimate of) Σ is non-singular, the first order condition for the estimate of the mean vector is

$$\sum_{i=1}^{n} (x_i - \mu) = 0,$$

which leads to the maximum likelihood estimator

$$\widehat{\mu} = \bar{X} = \frac{1}{n} \sum_{i=1}^{n} X_i.$$

This lets us simplify $\sum_{i=1}^{n}(x_i - \mu)(x_i - \mu)^{\text{T}} = \sum_{i=1}^{n}(x_i - \bar{x})(x_i - \bar{x})^{\text{T}} = S$ as defined above. Then the terms involving $d\Sigma$ in $d \ln L$ can be combined as

$$-\frac{1}{2} \text{tr} \left(\Sigma^{-1} \{d\Sigma\} \left[nI_p - \Sigma^{-1} S \right] \right).$$

The first order condition $d \ln \mathcal{L}(\mu, \Sigma) = 0$ will hold when the term in the square bracket is (matrix-valued) zero. Pre-multiplying the latter by Σ and dividing by n gives

$$\widehat{\Sigma} = \frac{1}{n} S,$$

which of course coincides with the canonical derivation given earlier.

Dwyer [6] points out that decomposition into two terms such as appears above is "unnecessary" and derives the estimator in two lines of working. Note that it may be not trivial to show that such derived estimator is the unique global maximizer for likelihood function.

5.3 Intrinsic covariance matrix estimation

5.3.1 Intrinsic expectation

Given a sample of n independent observations $x_1,..., x_n$ of a p-dimensional zero-mean Gaussian random variable X with covariance $\mathbf{R}$, the maximum likelihood estimator of $\mathbf{R}$ is given by

$$\hat{\mathbf{R}} = \frac{1}{n} \sum_{i=1}^{n} x_i x_i^\mathsf{T}.$$

The parameter $\mathbf{R}$ belongs to the set of positive-definite matrices, which is a Riemannian manifold, not a vector space, hence the usual vector-space notions of expectation, i.e. "E[$\hat{\mathbf{R}}$]", and estimator bias must be generalized to manifolds to make sense of the problem of covariance matrix estimation. This can be done by defining the expectation of an manifold-valued estimator $\hat{\mathbf{R}}$ with respect to the manifold-valued point $\mathbf{R}$ as

$$\mathrm{E}_\mathbf{R}[\hat{\mathbf{R}}] \stackrel{\mathrm{def}}{=} \exp_\mathbf{R} \mathrm{E}\left[\exp_\mathbf{R}^{-1} \hat{\mathbf{R}}\right]$$

where

$$\exp_\mathbf{R}(\hat{\mathbf{R}}) = \mathbf{R}^{\frac{1}{2}} \exp\left(\mathbf{R}^{-\frac{1}{2}} \hat{\mathbf{R}} \mathbf{R}^{-\frac{1}{2}}\right) \mathbf{R}^{\frac{1}{2}}$$

$$\exp_\mathbf{R}^{-1}(\hat{\mathbf{R}}) = \mathbf{R}^{\frac{1}{2}} \left(\log \mathbf{R}^{-\frac{1}{2}} \hat{\mathbf{R}} \mathbf{R}^{-\frac{1}{2}}\right) \mathbf{R}^{\frac{1}{2}}$$

are the exponential map and inverse exponential map, respectively, "exp" and "log" denote the ordinary matrix exponential and matrix logarithm, and E[·] is the ordinary expectation operator defined on a vector space, in this case the tangent space of the manifold.[1]

5.3.2 Bias of the sample covariance matrix

The intrinsic bias vector field of the SCM estimator $\hat{\mathbf{R}}$ is defined to be

$$\mathbf{B}(\hat{\mathbf{R}}) = \exp_\mathbf{R}^{-1} \mathrm{E}_\mathbf{R}\left[\hat{\mathbf{R}}\right] = \mathrm{E}\left[\exp_\mathbf{R}^{-1} \hat{\mathbf{R}}\right]$$

The intrinsic estimator bias is then given by $\exp_\mathbf{R} \mathbf{B}(\hat{\mathbf{R}})$.

For complex Gaussian random variables, this bias vector field can be shown[1] to equal

$$\mathbf{B}(\hat{\mathbf{R}}) = -\beta(p,n)\mathbf{R}$$

where

$$\beta(p,n) = \frac{1}{p}\left(p \log n + p - \psi(n-p+1) + (n-p+1)\psi(n-p+2) + \psi(n+1) - (n+1)\psi(n+2)\right)$$

and $\psi(\cdot)$ is the digamma function. The intrinsic bias of the sample covariance matrix equals

$$\exp_\mathbf{R} \mathbf{B}(\hat{\mathbf{R}}) = e^{-\beta(p,n)} \mathbf{R}$$

and the SCM is asymptotically unbiased as $n \to \infty$.

Similarly, the intrinsic inefficiency of the sample covariance matrix depends upon the Riemannian curvature of the space of positive-define matrices.

5.4 Shrinkage estimation

If the sample size n is small and the number of considered variables p is large, the above empirical estimators of covariance and correlation are very unstable. Specifically, it is possible to furnish estimators that improve considerably upon the maximum likelihood estimate in terms of mean squared error. Moreover, for $n < p$, the empirical estimate of the covariance matrix becomes singular, i.e. it cannot be inverted to compute the precision matrix.

As an alternative, many methods have been suggested to improve the estimation of the covariance matrix. All of these approaches rely on the concept of shrinkage. This is implicit in Bayesian methods and in penalized maximum likelihood methods and explicit in the Stein-type shrinkage approach.

A simple version of a shrinkage estimator of the covariance matrix is constructed as follows. One considers a convex combination of the empirical estimator (A) with some suitable chosen target (B), e.g., the diagonal matrix. Subsequently, the mixing parameter (δ) is selected to maximize the expected accuracy of the shrunken estimator. This can be done by cross-validation, or by using an analytic estimate of the shrinkage intensity. The resulting regularized estimator ($\delta A + (1 - \delta)B$) can be shown to outperform the maximum likelihood estimator for small samples. For large samples, the shrinkage intensity will reduce to zero, hence in this case the shrinkage estimator will be identical to the empirical estimator. Apart from increased efficiency the shrinkage estimate has the additional advantage that it is always positive definite and well conditioned.

Various shrinkage targets have been proposed:

1. the identity matrix, scaled by the average sample variance;[7]

2. the single-index model;[8]

3. the constant-correlation model, where the sample variances are preserved, but all pairwise correlation coefficients are assumed to be equal to one another;[9]

4. the two-parameter matrix, where all variances are identical, and all covariances are identical to one another (although *not* identical to the variances);[10]

5. the diagonal matrix containing sample variances on the diagonal and zeros everywhere else.[11]

A review on this topic is given, e.g., in Schäfer and Strimmer 2005.[12] The shrinkage estimator can be generalized to a multi-target shrinkage estimator that utilizes several targets simultaneously.[13] Software for computing a covariance shrinkage estimator is available in R (packages **corpcor**[14] and **ShrinkCovMat**[15]), in Python (library scikit-learn), and in MATLAB.[16]

5.5 See also

- Propagation of uncertainty
- Sample mean and sample covariance

5.6 References

[1] Smith, Steven Thomas (May 2005). "Covariance, Subspace, and Intrinsic Cramér–Rao Bounds". *IEEE Trans. Signal Processing* **53** (5): 1610–1630. doi:10.1109/TSP.2005.845428.

[2] Devlin, Susan J.; Gnanadesikan, R.; Kettenring, J. R. (1975). "Robust Estimation and Outlier Detection with Correlation Coefficients". *Biometrika* **62** (3): 531–545. doi:10.1093/biomet/62.3.531.

[3] *Robust Statistics*, Peter. J. Huber, Wiley, 1981 (republished in paperback, 2004)

5.6. REFERENCES

[4] "Modern applied statistics with S", William N. Venables, Brian D. Ripley, Springer, 2002, ISBN 0-387-95457-0, ISBN 978-0-387-95457-8, page 336

[5] K.V. Mardia, J.T. Kent, and J.M. Bibby (1979) *Multivariate Analysis*, Academic Press.

[6] Dwyer, Paul S. (June 1967). "Some applications of matrix derivatives in multivariate analysis". *Journal of the American Statistical Association* (Journal of the American Statistical Association, Vol. 62, No. 318) **62** (318): 607–625. doi:10.2307/2283988. JSTOR 2283988.

[7] O. Ledoit and M. Wolf (2004a) "A well-conditioned estimator for large-dimensional covariance matrices" *Journal of Multivariate Analysis* **88** (2): 365—411.

[8] O. Ledoit and M. Wolf (2003) "Improved estimation of the covariance matrix of stock returns with an application to portofolio selection" *Journal of Empirical Finance* **10** (5): 603—621.

[9] O. Ledoit and M. Wolf (2004b) "Honey, I shrunk the sample covariance matrix" *The Journal of Portfolio Management* **30** (4): 110—119.

[10] Appendix B.1 of O. Ledoit (1996) "Improved Covariance Matrix Estimation" Finance Working Paper No. 5-96, Anderson School of Management, University of California, Los Angeles.

[11] Appendix B.2 of O. Ledoit (1996).

[12] J. Schäfer and K. Strimmer (2005) *A Shrinkage Approach to Large-Scale Covariance Matrix Estimation and Implications for Functional Genomics*, Statistical Applications in Genetics and Molecular Biology: Vol. 4: No. 1, Article 32.

[13] T. Lancewicki and M. Aladjem (2014) "Multi-Target Shrinkage Estimation for Covariance Matrices", *IEEE Transactions on Signal Processing*, Volume: 62, Issue 24, pages: 6380-6390.

[14] *corpcor: Efficient Estimation of Covariance and (Partial) Correlation*, CRAN

[15] *ShrinkCovMat: Shrinkage Covariance Matrix Estimators*, CRAN

[16] MATLAB code for shrinkage targets: scaled identity, single-index model, constant-correlation model, two-parameter matrix, and diagonal matrix.

Chapter 6

Fermat's little theorem

For other theorems named after Pierre de Fermat, see Fermat's theorem (disambiguation).

Fermat's little theorem states that if p is a prime number, then for any integer a, the number $a^p - a$ is an integer multiple of p. In the notation of modular arithmetic, this is expressed as

$$a^p \equiv a \pmod{p}.$$

For example, if a = 2 and p = 7, 2^7 = 128, and 128 − 2 = 7 × 18 is an integer multiple of 7.

If a is not divisible by p, Fermat's little theorem is equivalent to the statement that $a^{p-1} - 1$ is an integer multiple of p, or in symbols

$$a^{p-1} \equiv 1 \pmod{p}.\ [1][2]$$

For example, if a = 2 and p = 7 then 2^6 = 64 and 64 − 1 = 63 is a multiple of 7.

Fermat's little theorem is the basis for the Fermat primality test and is one of the fundamental results of elementary number theory. The theorem is named after Pierre de Fermat, who stated it in 1640. It is called the "little theorem" to distinguish it from Fermat's last theorem.[3]

6.1 History

Pierre de Fermat first stated the theorem in a letter dated October 18, 1640, to his friend and confidant Frénicle de Bessy as the following:[3]

> If p is a prime and a is any integer not divisible by p, then $a^{p-1} - 1$ is divisible by p.

Fermat did not prove his assertion, only stating:[4]

> Et cette proposition est généralement vraie en toutes progressions et en tous nombres premiers; de quoi je vous envoierois la démonstration, si je n'appréhendois d'être trop long.

> (And this proposition is generally true for all series and for all prime numbers; the proof of which I would send to you, if I did not fear it being too long.)[5]

6.1. HISTORY

Pierre de Fermat

Euler provided the first published proof in 1736 in a paper entitled "Theorematum Quorundam ad Numeros Primos Spectantium Demonstratio" in the *Proceedings* of the St. Petersburg Academy,[6] but Leibniz had given virtually the same proof in an unpublished manuscript from sometime before 1683.[3]

The term "Fermat's Little Theorem" was probably first used in print in 1913 in *Zahlentheorie* by Kurt Hensel:

> Für jede endliche Gruppe besteht nun ein Fundamentalsatz, welcher der kleine Fermatsche Satz genannt zu werden pflegt, weil ein ganz spezieller Teil desselben zuerst von Fermat bewiesen worden ist."

> (There is a fundamental theorem holding in every finite group, usually called Fermat's little Theorem because Fermat was the first to have proved a very special part of it.)

An early use in English occurs in A.A. Albert, Modern Higher Algebra (1937), which refers to "the so-called "little" Fermat theorem" on page 206.

6.1.1 Further history

Main article: Chinese hypothesis

Some mathematicians independently made the related hypothesis (sometimes incorrectly called the Chinese Hypothesis) that $2^p \equiv 2 \pmod{p}$ if and only if p is a prime. Indeed, the "if" part is true, and is a special case of Fermat's little theorem. However, the "only if" part of this hypothesis is false: for example, $2^{341} \equiv 2 \pmod{341}$, but $341 = 11 \times 31$ is a pseudoprime. See below.

6.2 Proofs

Main article: Proofs of Fermat's little theorem

Several proofs of Fermat's little theorem are known. It is frequently proved as a corollary of Euler's theorem.

6.3 Generalizations

Fermat's little theorem is a special case of Euler's theorem: for any modulus n and any integer a coprime to n, we have

$$a^{\varphi(n)} \equiv 1 \pmod{n},$$

where φ(n) denotes Euler's totient function (which counts the integers between 1 and n that are coprime to n). Euler's theorem is indeed a generalization, because if $n = p$ is a prime number, then $\varphi(p) = p - 1$.

A slight generalization of Euler's theorem, which immediately follows from it, is: if a, n, x, y are integers with n *positive* and a and n coprime, then

If $x \equiv y \pmod{\varphi(n)}$, then $a^x \equiv a^y \pmod{n}$.

This follows as x is of the form y + φ(n)k, so

$$a^x = a^{y+\varphi(n)k} = a^y(a^{\varphi(n)})^k \equiv a^y 1^k \equiv a^y \pmod{n}.$$

In this form, the theorem finds many uses in cryptography and, in particular, underlies the computations used in the RSA public key encryption method.[7] The special case with n a prime may be considered a consequence of Fermat's little theorem.

Fermat's little theorem is also related to Carmichael's theorem, as well as to Lagrange's theorem in group theory.

The algebraic setting of Fermat's little theorem can be generalized to finite fields.

6.4 Converse

The converse of Fermat's little theorem is not generally true, as it fails for Carmichael numbers. However, a slightly stronger form of the theorem is true, and is known as Lehmer's theorem. The theorem is as follows:

If there exists an a such that

$$a^{p-1} \equiv 1 \pmod{p}$$

and for all primes q dividing p − 1

$$a^{(p-1)/q} \not\equiv 1 \pmod{p}$$

then p is prime.

This theorem forms the basis for the Lucas–Lehmer test, an important primality test.

6.5 Pseudoprimes

If a and p are coprime numbers such that $a^{p-1} - 1$ is divisible by p, then p need not be prime. If it is not, then p is called a pseudoprime to base a (or a Fermat pseudoprime). F. Sarrus in 1820 found 341 = 11 × 31 as one of the first pseudoprimes, to base 2.

A number p that is a pseudoprime to base a for every number a coprime to p is called a Carmichael number (e.g. 561). Alternately, any number p satisfying the equality

$$\gcd\left(\sum_{a=1}^{p-1} a^{p-1}, p\right) = 1$$

is either a prime or a Carmichael number.

6.6 See also

- Fractions with prime denominators: numbers with behavior relating to Fermat's little theorem
- RSA
- p-derivation
- Frobenius endomorphism
- Table of congruences

6.7 Notes

[1] Long 1972, pp. 87–88

[2] Pettofrezzo & Byrkit 1970, pp. 110–111

[3] Burton 2011, p. 514

[4] Bergeron & Zhao (2004)

[5] Ore 1988, p. 272 for the English translation

[6] Ore 1988, p. 273

[7] Trappe, Wade; Washington, Lawrence C. (2002), *Introduction to Cryptography with Coding Theory*, Prentice-Hall, p. 78, ISBN 0-13-061814-4

6.8 References

- Bergeron, Amanda; Zhao, David (2004). "Translation of letter from Fermat to de Bessy, Thursday, 18 October 1640" (PDF). Archived from the original (PDF) on 22 December 2006.

- Burton, David M. (2011), *The History of Mathematics / An Introduction* (7th ed.), McGraw-Hill, ISBN 978-0-07-338315-6

- Long, Calvin T. (1972), *Elementary Introduction to Number Theory* (2nd ed.), Lexington: D. C. Heath and Company, LCCN 77171950

- Ore, Oystein (1988) [1948], *Number Theory and Its History*, Dover, ISBN 978-0-486-65620-5

- Pettofrezzo, Anthony J.; Byrkit, Donald R. (1970), *Elements of Number Theory*, Englewood Cliffs: Prentice Hall, LCCN 71081766

6.9 Further reading

- Paulo Ribenboim (1995). *The New Book of Prime Number Records* (3rd ed.). New York: Springer-Verlag. ISBN 0-387-94457-5. pp. 22–25, 49.

6.10 External links

- János Bolyai and the pseudoprimes (in Hungarian)

- Fermat's Little Theorem at cut-the-knot

- Euler Function and Theorem at cut-the-knot

- Fermat's Little Theorem and Sophie's Proof

- Hazewinkel, Michiel, ed. (2001), "Fermat's little theorem", *Encyclopedia of Mathematics*, Springer, ISBN 978-1-55608-010-4

- Weisstein, Eric W., "Fermat's Little Theorem", *MathWorld*.

- Weisstein, Eric W., "Fermat's Little Theorem Converse", *MathWorld*.

Chapter 7

Proofs of Fermat's little theorem

This article collects together a variety of proofs of Fermat's little theorem, which states that

$$a^p \equiv a \pmod{p}$$

for every prime number p and every integer a (see modular arithmetic).

7.1 Simplifications

Some of the **proofs of Fermat's little theorem** given below depend on two simplifications.

The first is that we may assume that a is in the range $0 \leq a \leq p - 1$. This is a simple consequence of the laws of modular arithmetic; we are simply saying that we may first reduce a modulo p.

Secondly, it suffices to prove that

$$a^{p-1} \equiv 1 \pmod{p} \quad (X)$$

for a in the range $1 \leq a \leq p - 1$. Indeed, if (X) holds for such a, multiplying both sides by a yields the original form of the theorem,

$$a^p \equiv a \pmod{p}$$

On the other hand, if a equals zero, the theorem holds trivially.

7.2 Combinatorial proofs

7.2.1 Proof by counting necklaces

This is perhaps the simplest known proof, requiring the least mathematical background. It is an attractive example of a combinatorial proof (a proof that involves counting a collection of objects in two different ways).

The proof given here is an adaptation of Golomb's proof.[1]

To keep things simple, let us assume that a is a positive integer. Consider all the possible strings of p symbols, using an alphabet with a different symbols. The total number of such strings is a^p, since there are a possibilities for each of p positions (see rule of product).

For example, if $p = 5$ and $a = 2$, then we can use an alphabet with two symbols (say A and B), and there are $2^5 = 32$ strings of length five:

AAAAA, AAAAB, AAABA, AAABB, AABAA, AABAB, AABBA, AABBB,

ABAAA, ABAAB, ABABA, ABABB, ABBAA, ABBAB, ABBBA, ABBBB,

BAAAA, BAAAB, BAABA, BAABB, BABAA, BABAB, BABBA, BABBB,

BBAAA, BBAAB, BBABA, BBABB, BBBAA, BBBAB, BBBBA, BBBBB.

We will argue below that if we remove the strings consisting of a single symbol from the list (in our example, AAAAA and BBBBB), the remaining $a^p - a$ strings can be arranged into groups, each group containing exactly p strings. It follows that $a^p - a$ is divisible by p.

Necklaces

Let us think of each such string as representing a necklace. That is, we connect the two ends of the string together, and regard two strings as the same necklace if we can rotate one string to obtain the second string; in this case we will say that the two strings are **friends**. In our example, the following strings are all friends:

AAAAB, AAABA, AABAA, ABAAA, BAAAA.

Similarly, each line of the following list corresponds to a single necklace.

AAABB, AABBA, ABBAA, BBAAA, BAAAB,

AABAB, ABABA, BABAA, ABAAB, BAABA,

AABBB, ABBBA, BBBAA, BBAAB, BAABB,

ABABB, BABBA, ABBAB, BBABA, BABAB,

ABBBB, BBBBA, BBBAB, BBABB, BABBB,

AAAAA,

BBBBB.

Notice that in the above list, some necklaces are represented by five different strings, and some only by a single string, so the list shows very clearly why $32 - 2$ is divisible by 5.

One can use the following rule to work out how many friends a given string S has:

If S is built up of several copies of the string T, and T cannot itself be broken down further into repeating strings, then the number of friends of S (including S itself) is equal to the length *of T.*

For example, suppose we start with the string $S =$ "ABBABBABBABB", which is built up of several copies of the shorter string $T =$ "ABB". If we rotate it one symbol at a time, we obtain the following three strings:

ABBABBABBABB,

BBABBABBABBA,

BABBABBABBAB.

There aren't any others, because ABB is exactly three symbols long, and cannot be broken down into further repeating strings.

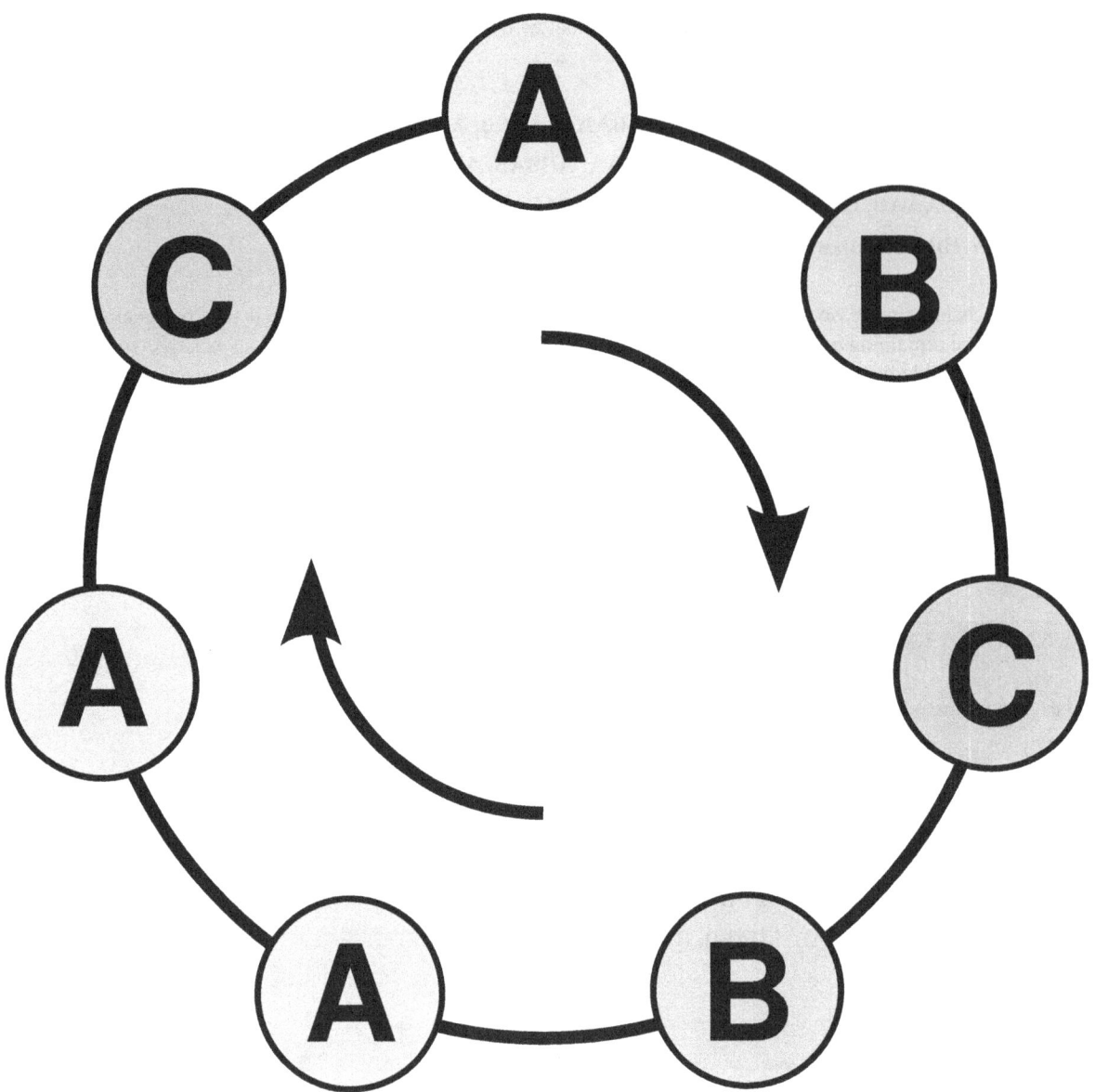

Necklace representing seven different strings (ABCBAAC, BCBAACA, CBAACAB, BAACABC, AACABCB, ACABCBA, CABCBAA)

Completing the proof

Using the above rule, we can complete the proof of Fermat's little theorem quite easily, as follows. Our starting pool of a^p strings may be split into two categories:

- Some strings contain p identical symbols. There are exactly a of these, one for each symbol in the alphabet. (In our running example, these are the strings AAAAA and BBBBB.)

- The rest of the strings use at least two distinct symbols from the alphabet. If we can break up S into repeating copies of some string T, the length of T must divide the length of S. But, since the length of S is the prime p, the only possible length for T is also p. Therefore, the above rule tells us that S has exactly p friends (including S itself).

The second category contains $a^p - a$ strings, and they may be arranged into groups of p strings, one group for each necklace. Therefore $a^p - a$ must be divisible by p, as promised.

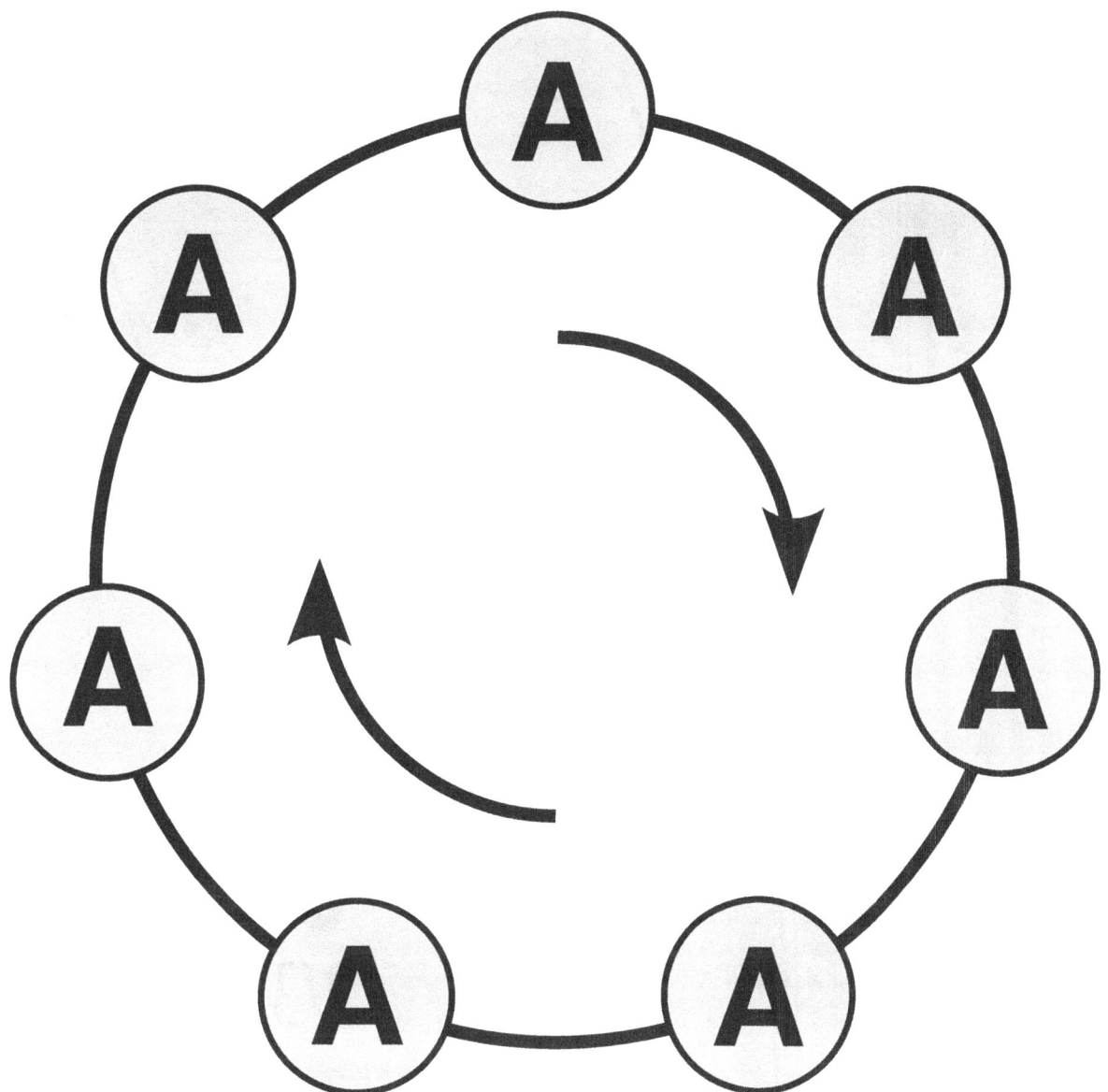

Necklace representing only one string (AAAAAAA)

7.2.2 Proof using dynamical systems

This proof uses some basic concepts from dynamical systems.

We start by considering a family of functions, $Tn(x)$, where $n \geq 2$ is an integer, mapping the interval $[0, 1]$ to itself by the formula

$$T_n(x) = \begin{cases} \{nx\} & 0 \leq x < 1, \\ 1 & x = 1, \end{cases}$$

where $\{y\}$ denotes the fractional part of y. For example, the function $T_3(x)$ is illustrated below:

A number x_0 is said to be a **fixed point** of a function $f(x)$ if $f(x_0) = x_0$; in other words, if f leaves x_0 fixed. The fixed points of a function can be easily found graphically: they are simply the x-coordinates of the points where the graph of

7.2. COMBINATORIAL PROOFS

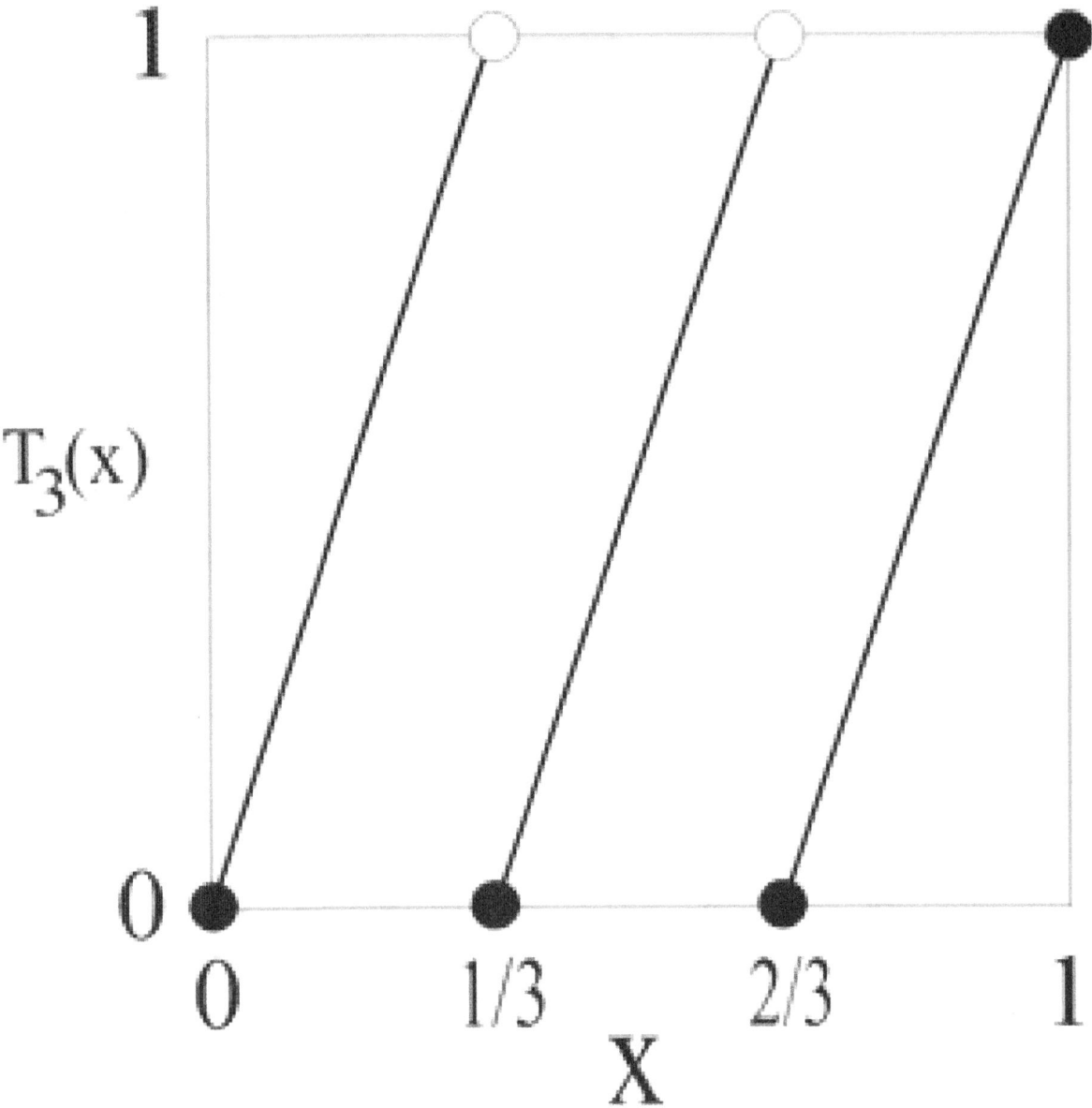

An example of a T_n function

$f(x)$ intersects the graph of the line $y = x$. For example, the fixed points of the function $T_3(x)$ are 0, 1/2, and 1; they are marked by black circles on the following diagram.

We will require the following two lemmas.

Lemma 1. For any $n \geq 2$, the function $Tn(x)$ has exactly n fixed points.

Proof. There are three fixed points in the illustration above, and the same sort geometrical argument applies for any $n \geq 2$.

Lemma 2. For any positive integers n and m, and any $0 \leq x \leq 1$,

$$T_m(T_n(x)) = T_{mn}(x).$$

In other words, $Tmn(x)$ is the composition of $Tn(x)$ and $Tm(x)$.

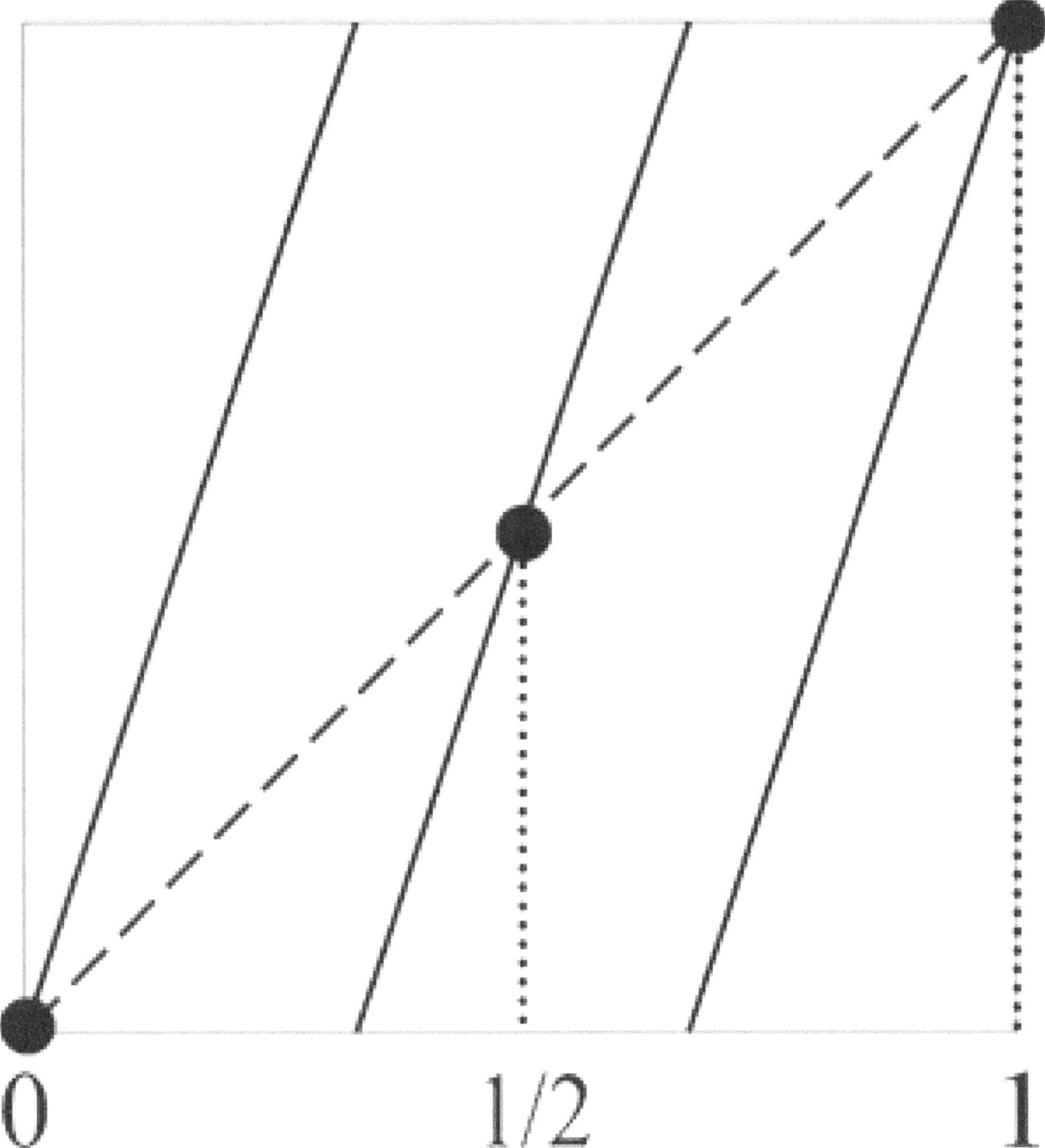

Fixed points of a T_n function

Proof. The proof of this lemma is not difficult, but we need to be slightly careful with the endpoint $x = 1$. For this point the lemma is clearly true since

$$T_m(T_n(1)) = T_m(1) = 1 = T_{mn}(1).$$

So let us assume that $0 \leq x < 1$. In this case,

$$T_n(x) = \{nx\} < 1,$$

7.2. COMBINATORIAL PROOFS

so $Tm(Tn(x))$ is given by

$$T_m(T_n(x)) = \{m\{nx\}\}.$$

Therefore, what we really need to show is that

$$\{m\{nx\}\} = \{mnx\}.$$

To do this we observe that $\{nx\} = nx - k$, where k is the integer part of nx; then

$$\{m\{nx\}\} = \{mnx - mk\} = \{mnx\}$$

since mk is an integer.

Now let us properly begin the proof of Fermat's little theorem, by studying the function $Ta^p(x)$. We will assume that a is ≥ 2. From Lemma 1, we know that it has a^p fixed points. By Lemma 2 we know that

$$T_{a^p}(x) = \underbrace{T_a(T_a(\cdots T_a(x)\cdots))}_{p \text{ times}},$$

so any fixed point of $Ta(x)$ is automatically a fixed point of $Ta^p(x)$.

We are interested in the fixed points of $Ta^p(x)$ that are *not* fixed points of $Ta(x)$. Let us call the set of such points S. There are $a^p - a$ points in S, because by Lemma 1 again, $Ta(x)$ has exactly a fixed points. The following diagram illustrates the situation for $a = 3$ and $p = 2$. The black circles are the points of S, of which there are $3^2 - 3 = 6$.

The main idea of the proof is now to split the set S up into its **orbits** under Ta. What this means is that we pick a point x_0 in S, and repeatedly apply $Ta(x)$ to it, to obtain the sequence of points

$$x_0, T_a(x_0), T_a(T_a(x_0)), T_a(T_a(T_a(x_0))), \ldots.$$

This sequence is called the orbit of x_0 under Ta. By Lemma 2, this sequence can be rewritten as

$$x_0, T_a(x_0), T_{a^2}(x_0), T_{a^3}(x_0), \ldots.$$

Since we are assuming that x_0 is a fixed point of $Ta^p(x)$, after p steps we hit $Ta^p(x_0) = x_0$, and from that point onwards the sequence repeats itself.

However, the sequence *cannot* begin repeating itself any earlier than that. If it did, the length of the repeating section would have to be a divisor of p, so it would have to be 1 (since p is prime). But this contradicts our assumption that x_0 is not a fixed point of Ta.

In other words, the orbit contains exactly p distinct points. This holds for every orbit of S. Therefore, the set S, which contains $a^p - a$ points, can be broken up into orbits, each containing p points, so $a^p - a$ is divisible by p.

(This proof is essentially the same as the necklace-counting proof given above, simply viewed through a different lens: one may think of the interval [0, 1] as given by sequences of digits in base a (our distinction between 0 and 1 corresponding to the familiar distinction between representing integers as ending in ".0000..." and ".9999..."). Ta^n amounts to shifting such a sequence by n many digits. The fixed points of this will be those sequences which are cyclic with period dividing n. In particular, the fixed points of Ta^p can be thought of as the necklaces of length p, with Ta^n corresponding to rotation of such necklaces by n many spots.

This proof could also be presented without distinguishing between 0 and 1, simply using the half-open interval [0, 1); then T_n would only have $n - 1$ many fixed points, but $Ta^p - Ta$ would still work out to $a^p - a$, as needed.)

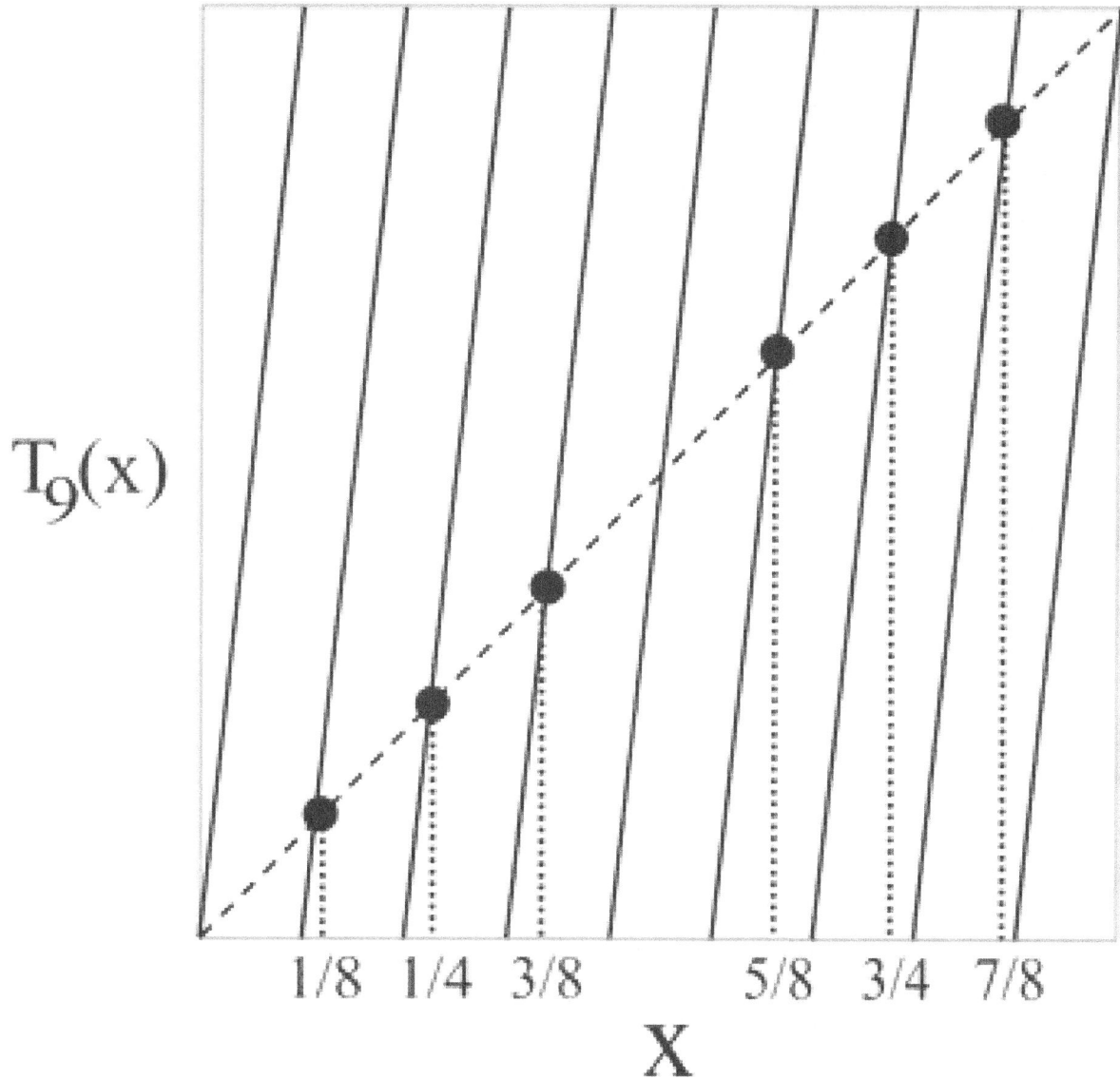

Fixed points in the set S

7.2.3 Multinomial proofs

Proof using the binomial theorem

This proof uses induction to prove the theorem for all integers $a \geq 0$.

The base step, that $0^p \equiv 0 \pmod{p}$, is true for modular arithmetic because it is true for integers. Next, we must show that if the theorem is true for $a = k$, then it is also true for $a = k+1$. For this inductive step, we need the following lemma.

Lemma. For any prime p,

$$(x + y)^p \equiv x^p + y^p \pmod{p}.$$

An alternative way of viewing this lemma is that it states that

7.2. COMBINATORIAL PROOFS

$$(x+y)^p = x^p + y^p$$

for any x and y in the finite field **GF**(p).

Postponing the proof of the lemma for now, we proceed with the induction.

Proof. Assume $k^p \equiv k \pmod{p}$, and consider $(k+1)^p$. By the lemma we have

$$(k+1)^p \equiv k^p + 1^p \pmod{p}.$$

Using the induction hypothesis, we have that $k^p \equiv k \pmod{p}$; and, trivially, $1^p = 1$. Thus

$$(k+1)^p \equiv k+1 \pmod{p},$$

which is the statement of the theorem for $a = k+1$. ∎

In order to prove the lemma, we must introduce the binomial theorem, which states that for any positive integer n,

$$(x+y)^n = \sum_{i=0}^{n} \binom{n}{i} x^{n-i} y^i,$$

where the coefficients are the binomial coefficients,

$$\binom{n}{i} = \frac{n!}{i!(n-i)!},$$

described in terms of the factorial function, $n! = 1 \times 2 \times 3 \times \cdots \times n$.

Proof of lemma. The binomial coefficients are all integers and when $0 < i < p$, neither of the terms in the denominator includes a factor of p, leaving the coefficient itself to possess a prime factor of p which must exist in the numerator, implying that

$$\binom{p}{i} \equiv 0 \pmod{p}, \qquad 0 < i < p.$$

Modulo p, this eliminates all but the first and last terms of the sum on the right-hand side of the binomial theorem for prime p.

The primality of p is essential to the lemma; otherwise, we have examples like

$$\binom{4}{2} = 6,$$

which is not divisible by 4.

Proof using the multinomial expansion

The proof is a very simple application of the multinomial theorem which is brought here for the sake of simplicity.

$$(x_1 + x_2 + \cdots + x_m)^n = \sum_{k_1,k_2,\ldots,k_m} \binom{n}{k_1, k_2, \ldots, k_m} x_1^{k_1} x_2^{k_2} \cdots x_m^{k_m}.$$

The summation is taken over all sequences of nonnegative integer indices k_1 through k_m such the sum of all k_i is n.

Thus if we express a as a sum of 1s (ones), we obtain

$$a^p = \sum_{k_1,k_2,\ldots,k_a} \binom{p}{k_1, k_2, \ldots, k_a}$$

Clearly, if p is prime, and if k_j not equal to p for any j, we have

$$\binom{p}{k_1, k_2, \ldots, k_a} \equiv 0 \pmod{p}$$

and

$$\binom{p}{k_1, k_2, \ldots, k_a} \equiv 1 \pmod{p}$$

if k_j equal to p for some j

Since there are exactly a elements such that $k_j = p$ the theorem follows.

(This proof is essentially a coarser-grained variant of the necklace-counting proof given earlier; the multinomial coefficients count the number of ways a string can be permuted into arbitrary anagrams, while the necklace argument counts the number of ways a string can be rotated into cyclic anagrams. That is to say, that the nontrivial multinomial coefficients here are divisible by p can be seen as a consequence of the fact that each nontrivial necklace of length p can be unwrapped into a string in p many ways.

This multinomial expansion is also, of course, what essentially underlies the binomial theorem-based proof above)

7.2.4 Proof using power product expansions

An additive-combinatorial proof based on formal power product expansions was given by Giedrius Alkauskas.[2] This proof uses neither euclidean algorithm nor binomial theorem, but rather employs the interplay between an additive and multiplicative structures of integers.

7.3 Proofs using modular arithmetic

These proofs require some background in modular arithmetic.

Let us assume that a is positive and not divisible by p. The idea is that if we write down the sequence of numbers

$$a, 2a, 3a, \ldots, (p-1)a \quad (A)$$

and reduce each one modulo p, the resulting sequence turns out to be a rearrangement of

$$1, 2, 3, \ldots, p-1. \quad (B)$$

7.3. PROOFS USING MODULAR ARITHMETIC

Therefore, if we multiply together the numbers in each sequence, the results must be identical modulo p:

$$a \times 2a \times 3a \times \cdots \times (p-1)a \equiv 1 \times 2 \times 3 \times \cdots \times (p-1) \pmod{p}.$$

Collecting together the a terms yields

$$a^{p-1}(p-1)! \equiv (p-1)! \pmod{p}.$$

Finally, we may "cancel out" the numbers $1, 2, ..., p-1$ from both sides of this equation, obtaining

$$a^{p-1} \equiv 1 \pmod{p}.$$

There are two steps in the above proof that we need to justify:

- Why (A) is a rearrangement of (B), and
- Why it is valid to "cancel" in the setting of modular arithmetic.

We will prove these things below; let us first see an example of this proof in action.

7.3.1 An example

If $a = 3$ and $p = 7$, then the sequence in question is

$$3, 6, 9, 12, 15, 18;$$

reducing modulo 7 gives

$$3, 6, 2, 5, 1, 4,$$

which is just a rearrangement of

$$1, 2, 3, 4, 5, 6.$$

Multiplying them together gives

$$3 \times 6 \times 9 \times 12 \times 15 \times 18 \equiv 3 \times 6 \times 2 \times 5 \times 1 \times 4 \equiv 1 \times 2 \times 3 \times 4 \times 5 \times 6 \pmod{7};$$

that is,

$$3^6(1 \times 2 \times 3 \times 4 \times 5 \times 6) \equiv (1 \times 2 \times 3 \times 4 \times 5 \times 6) \pmod{7}.$$

Canceling out $1 \times 2 \times 3 \times 4 \times 5 \times 6$ yields

$$3^6 \equiv 1 \pmod{7},$$

which is Fermat's little theorem for the case $a = 3$ and $p = 7$.

7.3.2 The cancellation law

Let us first explain why it is valid, in certain situations, to "cancel". The exact statement is as follows. If u, x, and y are integers, and u is not divisible by a prime number p, and if

$$ux \equiv uy \pmod{p},$$

then we may "cancel" u to obtain

$$x \equiv y \pmod{p}.$$

Our use of this **cancellation law** in the above proof of Fermat's little theorem was valid, because the numbers $1, 2, \ldots, p-1$ are certainly not divisible by p (indeed they are *smaller* than p).

We can prove the cancellation law easily using Euclid's lemma, which generally states that if an integer b divides a product rs (where r and s are integers), and b is relatively prime to r, then b must divide s. Indeed, the equation

$$ux \equiv uy \pmod{p},$$

simply means that p divides $ux - uy = u(x - y)$. If p does not divide u and is a prime, Euclid's lemma tells us that it must divide $x - y$ instead; that is,

$$x \equiv y \pmod{p}.$$

(Note that the conditions under which the cancellation law holds are quite strict, and this explains why Fermat's little theorem demands that p be a prime in order to make a general case for all n. For example, $2 \times 2 \equiv 2 \times 5 \pmod{6}$, but we cannot conclude that $2 \equiv 5 \pmod{6}$, since 6 is not prime. See below for a more extensive proof for all p)

7.3.3 The rearrangement property

Finally, we must explain why the sequence

$$a, 2a, 3a, \ldots, (p-1)a,$$

when reduced modulo p, becomes a rearrangement of the sequence

$$1, 2, 3, \ldots, p-1.$$

To start with, none of the terms $a, 2a, \ldots, (p-1)a$ can be congruent to zero modulo p, since if k is one of the numbers $1, 2, \ldots, p-1$, then k is relatively prime with p, and so is a, so Euclid's lemma tells us that ka shares no factor with p. Therefore, at least we know that the numbers $a, 2a, \ldots, (p-1)a$, when reduced modulo p, must be found among the numbers $1, 2, 3, \ldots, p-1$.

Furthermore, the numbers $a, 2a, \ldots, (p-1)a$ must all be *distinct* after reducing them modulo p, because if

$$ka \equiv ma \pmod{p},$$

where k and m are one of $1, 2, \ldots, p-1$, then the cancellation law tells us that

$k \equiv m \pmod{p}$.

Since both k and m are between 1 and $p\text{-}1$, they must be equal. Therefore the terms $a, 2a, ..., (p-1)a$ when reduced modulo p must be distinct. To summarise: when we reduce the $p-1$ numbers $a, 2a, ..., (p-1)a$ modulo p, we obtain distinct members of the sequence $1, 2, ..., p-1$. Since there are exactly $p-1$ of these, the only possibility is that the former are a rearrangement of the latter.

7.3.4 Applications to Euler's theorem

This method can also be used to prove Euler's theorem, with a slight alteration in that 1 to (p-1) is substituted for the numbers less than and relatively prime to some base m (instead of p). Both the rearrangement property and the cancellation law are still satisfied and can be utilized.

For example:

$a \times 3a \times 7a \times 9a \equiv 1 \times 3 \times 7 \times 9 \pmod{10}$.

Therefore,

$a^{\varphi(10)} \equiv 1 \pmod{10}$.

7.4 Proof using group theory

This proof requires the most basic elements of group theory.

The idea is to recognise that the set $G = \{1, 2, ..., p-1\}$, with the operation of multiplication (taken modulo p), forms a group. The only group axiom that requires some effort to verify is that each element of G is invertible. Taking this on faith for the moment, let us assume that a is in the range $1 \leq a \leq p-1$, that is, a is an element of G. Let k be the order of a, so that k is the smallest positive integer such that

$a^k \equiv 1 \pmod{p}$.

By Lagrange's theorem, k divides the order of G, which is $p-1$, so $p-1 = km$ for some positive integer m. Then

$a^{p-1} \equiv a^{km} \equiv (a^k)^m \equiv 1^m \equiv 1 \pmod{p}$.

7.4.1 The invertibility property

To prove that every element b of G is invertible, we may proceed as follows. First, b is relatively prime to p. Then Bézout's identity assures us that there are integers x and y such that

$bx + py = 1$.

Reading this equation modulo p, we see that x is an inverse for b, since

$bx \equiv 1 \pmod{p}$.

Therefore every element of G is invertible, so as remarked earlier, G is a group.

For example, when $p = 11$, the inverses of each element are given as follows:

7.5 Notes

[1] Golomb, Solomon W. Combinatorial proof of Fermat's "Little" Theorem. The American Mathematical Monthly, Vol. 63, No. 10 (Dec., 1956), p. 718

[2] Alkauskas, Giedrius. A Curious Proof of Fermat's Little Theorem. The American Mathematical Monthly, Vol. 116, No. 4 (Apr., 2009), p. 362-364

Chapter 8

Gödel's completeness theorem

Gödel's completeness theorem is a fundamental theorem in mathematical logic that establishes a correspondence between semantic truth and syntactic provability in first-order logic. It makes a close link between model theory that deals with what is true in different models, and proof theory that studies what can be formally proven in particular formal systems.

It was first proved by Kurt Gödel in 1929. It was then simplified in 1947, when Leon Henkin observed in his Ph.D. thesis that the hard part of the proof can be presented as the Model Existence Theorem (published in 1949). Henkin's proof was simplified by Gisbert Hasenjaeger in 1953.

8.1 Statement of the theorem

8.1.1 Preliminaries

There are numerous deductive systems for first-order logic, including systems of natural deduction and Hilbert-style systems. Common to all deductive systems is the notion of a **formal deduction**. This is a sequence (or, in some cases, a finite tree) of formulas with a specially-designated **conclusion**. The definition of a deduction is such that it is finite and that it is possible to verify algorithmically (by a computer, for example, or by hand) that a given sequence (or tree) of formulas is indeed a deduction.

A first-order formula is called **logically valid** if it is true in every structure for the language of the formula (i.e. for any assignment of values to the variables of the formula). To formally state, and then prove, the completeness theorem, it is necessary to also define a deductive system. A deductive system is called **complete** if every logically valid formula is the conclusion of some formal deduction, and the completeness theorem for a particular deductive system is the theorem that it is complete in this sense. Thus, in a sense, there is a different completeness theorem for each deductive system. A converse to completeness is **soundness,** the fact that only logically valid formulas are provable in the deductive system.

If some specific deductive system of first-order logic is sound and complete, then it is "perfect" (a formula is provable if and only if it is a semantic consequence of the axioms), thus equivalent to any other deductive system with the same quality (any proof in one system can be converted into the other).

8.1.2 Gödel's original formulation

The completeness theorem says that if a formula is logically valid then there is a finite deduction (a formal proof) of the formula.

Gödel's completeness theorem says that a deductive system of first-order predicate calculus is "complete" in the sense that no additional inference rules are required to prove all the logically valid formulas. A converse to completeness is **soundness,** the fact that only logically valid formulas are provable in the deductive system. Together with soundness (whose verification is easy), this theorem implies that a formula is logically valid if and only if it is the conclusion of a formal deduction.

8.1.3 Model existence theorem

The simplest version of this theorem that suffices in practice for most needs, and has connections with the Löwenheim–Skolem theorem, says:

Every consistent, countable first-order theory has a finite or countable model

A more general version can be expressed as :

Every consistent first-order theory with a well-orderable language has a model.

Here, a consistent theory is defined as one in which, for no formula F, both F and ¬F can be proven. See Consistency, the syntactic definition; the semantic definition would be tautological in this context.

This theorem by Henkin is the most directly obtained version of the completeness theorem in its simplest proof.

Given Henkin's theorem, the proof of the completeness theorem is as follows: If $\models A$ is valid, then $\neg A$ does not have models. By the contrapositive of Henkin's, then $\neg A$ is an inconsistent formula. But, by the definition of consistency, if $\neg A$ is inconsistent then it's possible to build a proof of $\vdash A$

8.1.4 More general form

It says that for any first-order theory T with a well-orderable language, and any sentence s in the language of the theory, there is a formal proof of s in T if and only if s is satisfied by every model of T (s is a semantic consequence of T).

This more general theorem is used implicitly, for example, when a sentence is shown to be provable from the axioms of group theory by considering an arbitrary group and showing that the sentence is satisfied by that group. It is deduced from the model existence theorem as follows: if there is no formal proof of a formula then adding its negation to the axioms gives a consistent theory, which has thus a model, so that the formula is not a semantic consequence of the initial theory.

Gödel's original formulation is deduced by taking the particular case of a theory without any axiom.

8.1.5 As a theorem of arithmetic

The Model Existence Theorem and its proof can be formalized in the framework of Peano arithmetic. Precisely, we can systematically define a model of any consistent effective first-order theory T in Peano arithmetic by interpreting each symbol of T by an arithmetical formula whose free variables are the arguments of the symbol. However, the definition expressed by this formula is not recursive.

8.2 Consequences

An important consequence of the completeness theorem is that it is possible to recursively enumerate the semantic consequences of any effective first-order theory, by enumerating all the possible formal deductions from the axioms of the theory, and use this to produce an enumeration of their conclusions.

This comes in contrast with the direct meaning of the notion of semantic consequence, that quantifies over all structures in a particular language, which is clearly not a recursive definition.

Also, it makes the concept of "provability," and thus of "theorem," a clear concept that only depends on the chosen system of axioms of the theory, and not on the choice of a proof system.

8.3 Relationship to the incompleteness theorem

Gödel's incompleteness theorem, another celebrated result, shows that there are inherent limitations in what can be achieved with formal proofs in mathematics. The name for the incompleteness theorem refers to another meaning of *complete* (see model theory – Using the compactness and completeness theorems).

It shows that in any consistent effective theory T containing Peano arithmetic (PA), the formula CT expressing the consistency of T cannot be proven within T.

Applying the completeness theorem to this result, gives the existence of a model of T where the formula CT is false. Such a model (precisely, the set of "natural numbers" it contains) is necessarily non-standard, as it contains the code number of a proof of a contradiction of T. But T is consistent when viewed from the outside. Thus this code number of a proof of contradiction of T must be a non-standard number.

In fact, the model of *any* theory containing PA obtained by the systematic construction of the arithmetical model existence theorem, is *always* non-standard with a non-equivalent provability predicate and a non-equivalent way to interpret its own construction, so that this construction is non-recursive (as recursive definitions would be unambiguous).

Also, there is no recursive non-standard model of PA.

8.4 Relationship to the compactness theorem

The completeness theorem and the compactness theorem are two cornerstones of first-order logic. While neither of these theorems can be proven in a completely effective manner, each one can be effectively obtained from the other.

The compactness theorem says that if a formula φ is a logical consequence of a (possibly infinite) set of formulas Γ then it is a logical consequence of a finite subset of Γ. This is an immediate consequence of the completeness theorem, because only a finite number of axioms from Γ can be mentioned in a formal deduction of φ, and the soundness of the deduction system then implies φ is a logical consequence of this finite set. This proof of the compactness theorem is originally due to Gödel.

Conversely, for many deductive systems, it is possible to prove the completeness theorem as an effective consequence of the compactness theorem.

The ineffectiveness of the completeness theorem can be measured along the lines of reverse mathematics. When considered over a countable language, the completeness and compactness theorems are equivalent to each other and equivalent to a weak form of choice known as weak König's lemma, with the equivalence provable in RCA_0 (a second-order variant of Peano arithmetic restricted to induction over Σ^0_1 formulas). Weak König's lemma is provable in ZF, the system of Zermelo–Fraenkel set theory without axiom of choice, and thus the completeness and compactness theorems for countable languages are provable in ZF. However the situation is different when the language is of arbitrary large cardinality since then, though the completeness and compactness theorems remain provably equivalent to each other in ZF, they are also provably equivalent to a weak form of the axiom of choice known as the ultrafilter lemma. In particular, no theory extending ZF can prove either the completeness or compactness theorems over arbitrary (possibly uncountable) languages without also proving the ultrafilter lemma on a set of same cardinality, knowing that on countable sets, the ultrafilter lemma becomes equivalent to weak König's lemma.

8.5 Completeness in other logics

The completeness theorem is a central property of first-order logic that does not hold for all logics. Second-order logic, for example, does not have a completeness theorem for its standard semantics (but does have the completeness property for Henkin semantics), and the same is true of all higher-order logics. It is possible to produce sound deductive systems for higher-order logics, but no such system can be complete. The set of logically-valid formulas in second-order logic is not enumerable.

Lindström's theorem states that first-order logic is the strongest (subject to certain constraints) logic satisfying both compactness and completeness.

A completeness theorem can be proved for modal logic or intuitionistic logic with respect to Kripke semantics.

8.6 Proofs

Gödel's original proof of the theorem proceeded by reducing the problem to a special case for formulas in a certain syntactic form, and then handling this form with an *ad hoc* argument.

In modern logic texts, Gödel's completeness theorem is usually proved with Henkin's proof, rather than with Gödel's original proof. Henkin's proof directly constructs a term model for any consistent first-order theory. James Margetson (2004) developed a computerized formal proof using the Isabelle theorem prover. Other proofs are also known.

8.7 See also

- Gödel's incompleteness theorems
- Original proof of Gödel's completeness theorem

8.8 Further reading

- Gödel, K (1929). "Über die Vollständigkeit des Logikkalküls". Doctoral dissertation. University Of Vienna. The first proof of the completeness theorem.
- Gödel, K (1930). "Die Vollständigkeit der Axiome des logischen Funktionenkalküls". *Monatshefte für Mathematik* (in German) **37** (1): 349–360. doi:10.1007/BF01696781. JFM 56.0046.04. The same material as the dissertation, except with briefer proofs, more succinct explanations, and omitting the lengthy introduction.

8.9 External links

- Stanford Encyclopedia of Philosophy: "Kurt Gödel"—by Juliette Kennedy.
- MacTutor biography: Kurt Gödel.
- Detlovs, Vilnis, and Podnieks, Karlis, "Introduction to mathematical logic."

Chapter 9

Original proof of Gödel's completeness theorem

The proof of Gödel's completeness theorem given by Kurt Gödel in his doctoral dissertation of 1929 (and a rewritten version of the dissertation, published as an article in 1930) is not easy to read today; it uses concepts and formalism that are no longer used and terminology that is often obscure. The version given below attempts to represent all the steps in the proof and all the important ideas faithfully, while restating the proof in the modern language of mathematical logic. This outline should not be considered a rigorous proof of the theorem.

9.1 Definitions and assumptions

We work with first-order predicate calculus. Our languages allow constant, function and relation symbols. Structures consist of (non-empty) domains and interpretations of the relevant symbols as constant members, functions or relations over that domain.

We fix some axiomatization of the predicate calculus: logical axioms and rules of inference. Any of the several well-known axiomatisations will do; we assume without proof all the basic well-known results about our formalism (such as the normal form theorem or the soundness theorem) that we need.

We axiomatize predicate calculus *without equality*, i.e. there are no special axioms expressing the properties of equality as a special relation symbol. After the basic form of the theorem is proved, it will be easy to extend it to the case of predicate calculus *with equality*.

9.2 Statement of the theorem and its proof

In the following, we state two equivalent forms of the theorem, and show their equivalence.

Later, we prove the theorem. This is done in the following steps:

1. Reducing the theorem to sentences (formulas with no free variables) in prenex form, i.e. with all quantifiers ($\forall$ and $\exists$) at the beginning. Furthermore, we reduce it to formulas whose first quantifier is $\forall$. This is possible because for every sentence, there is an equivalent one in prenex form whose first quantifier is $\forall$.

2. Reducing the theorem to sentences of the form $\forall x_1 \forall x_2 ... \forall x_k \exists y_1 \exists y_2 ... \exists y_m \varphi(x_1...x_k, y_1...y_m)$. While we cannot do this by simply rearranging the quantifiers, we show that it is yet enough to prove the theorem for sentences of that form.

3. Finally we prove the theorem for sentences of that form.

- This is done by first noting that a sentence such as B = $\exists x_1 \exists x_2...\exists x_k \exists y_1 \exists y_2...\exists y_m \varphi(x_1...x_k, y_1...y_m)$ is either refutable or has some model in which it holds; this model is simply assigning truth values to the subpropositions from which B is built. The reason for that is the completeness of propositional logic, with the existential quantifiers playing no role.
- We extend this result to more and more complex and lengthy sentences, D_n (n=1,2...), built out from B, so that either any of them is refutable and therefore so is φ, or all of them are not refutable and therefore each holds in some model.
- We finally use the models in which the D_n hold (in case all are not refutable) in order to build a model in which φ holds.

9.2.1 Theorem 1. Every valid formula (true in all structures) is provable.

This is the most basic form of the completeness theorem. We immediately restate it in a form more convenient for our purposes:

9.2.2 Theorem 2. Every formula φ is either refutable or satisfiable in some structure.

"φ is refutable" means *by definition* "$\neg\varphi$ is provable".

9.2.3 Equivalence of both theorems

To see the equivalence, note first that if **Theorem 1** holds, and φ is not satisfiable in any structure, then $\neg\varphi$ is valid in all structures and therefore provable, thus φ is refutable and **Theorem 2** holds. If on the other hand **Theorem 2** holds and φ is valid in all structures, then $\neg\varphi$ is not satisfiable in any structure and therefore refutable; then $\neg\neg\varphi$ is provable and then so is φ, thus **Theorem 1** holds.

9.2.4 Proof of theorem 2: first step

We approach the proof of **Theorem 2** by successively restricting the class of all formulas φ for which we need to prove "φ is either refutable or satisfiable". At the beginning we need to prove this for all possible formulas φ in our language. However, suppose that for every formula φ there is some formula ψ taken from a more restricted class of formulas **C**, such that "ψ is either refutable or satisfiable" → "φ is either refutable or satisfiable". Then, once this claim (expressed in the previous sentence) is proved, it will suffice to prove "φ is either refutable or satisfiable" only for φ's belonging to the class **C**. Note also that if φ is provably equivalent to ψ (*i.e.*, ($\varphi\equiv\psi$) is provable), then it is indeed the case that "ψ is either refutable or satisfiable" → "φ is either refutable or satisfiable" (the soundness theorem is needed to show this).

There are standard techniques for rewriting an arbitrary formula into one that does not use function or constant symbols, at the cost of introducing additional quantifiers; we will therefore assume that all formulas are free of such symbols. Gödel's paper uses a version of first-order predicate calculus that has no function or constant symbols to begin with.

Next we consider a generic formula φ (which no longer uses function or constant symbols) and apply the prenex form theorem to find a formula ψ in *normal form* such that $\varphi\equiv\psi$ (ψ being in *normal form* means that all the quantifiers in ψ, if there are any, are found at the very beginning of ψ). It follows now that we need only prove **Theorem 2** for formulas φ in normal form.

Next, we eliminate all free variables from φ by quantifying them existentially: if, say, $x_1...x_n$ are free in φ, we form $\psi = \exists x_1...\exists x_n \phi$. If ψ is satisfiable in a structure M, then certainly so is φ and if ψ is refutable, then $\neg\psi = \forall x_1...\forall x_n \neg\phi$ is provable, and then so is $\neg\varphi$, thus φ is refutable. We see that we can restrict φ to be a *sentence*, that is, a formula with no free variables.

Finally, we would like, for reasons of technical convenience, that the *prefix* of φ (that is, the string of quantifiers at the beginning of φ, which is in normal form) begin with a universal quantifier and end with an existential quantifier. To achieve this for a generic φ (subject to restrictions we have already proved), we take some one-place relation symbol **F** unused in φ, and two new variables **y** and **z**. If φ = (P)Φ, where (P) stands for the prefix of φ and Φ for the *matrix* (the remaining, quantifier-free part of φ) we form $\psi = \forall y (P) \exists z (\Phi \wedge [F(y) \vee \neg F(z)])$. Since $\forall y \exists z (F(y) \vee \neg F(z))$ is clearly provable, it is easy to see that $\phi = \psi$ is provable.

9.2.5 Reducing the theorem to formulas of degree 1

Our generic formula φ now is a sentence, in normal form, and its prefix starts with a universal quantifier and ends with an existential quantifier. Let us call the class of all such formulas **R**. We are faced with proving that every formula in **R** is either refutable or satisfiable. Given our formula φ, we group strings of quantifiers of one kind together in blocks:

$$\phi = (\forall x_1 ... \forall x_{k_1})(\exists x_{k_1+1} ... \exists x_{k_2}) .... (\forall x_{k_{n-2}+1} ... \forall x_{k_{n-1}})(\exists x_{k_{n-1}+1} ... \exists x_{k_n})(\Phi)$$

We define the **degree** of ϕ to be the number of universal quantifier blocks, separated by existential quantifier blocks as shown above, in the prefix of ϕ. The following lemma, which Gödel adapted from Skolem's proof of the Löwenheim-Skolem theorem, lets us sharply reduce the complexity of the generic formula ϕ we need to prove the theorem for:

Lemma. Let **k**>=1. If every formula in **R** of degree **k** is either refutable or satisfiable, then so is every formula in **R** of degree **k+1**.

> **Comment**: Take a formula φ of degree k+1 of the form $\phi = (\forall x)(\exists y)(\forall u)(\exists v)(P)\psi$, where $(P)\psi$ is the remainder of ϕ (it is thus of degree **k-1**). φ states that for every x there is a y such that... (something). It would have been nice to have a predicate Q' so that for every x, $Q'(x,y)$ would be true if and only if y is the required one to make (something) true. Then we could have written a formula of degree k, which is equivalent to φ, namely $(\forall x')(\forall x)(\forall y)(\forall u)(\exists v)(\exists y')(P)Q'(x',y') \wedge (Q'(x,y) \rightarrow \psi)$. This formula is indeed equivalent to φ because it states that for every x, if there is a y that satisfies Q'(x,y), then (something) holds, and furthermore, we know that there is such a y, because for every x', there is a y' that satisfies Q'(x',y'). Therefore φ follows from this formula. It is also easy to show that if the formula is false, then so is φ. **Unfortunately**, in general there is no such predicate Q'. However, this idea can be understood as a basis for the following proof of the Lemma.

Proof. Let φ be a formula of degree **k+1**; then we can write it as

$$\phi = (\forall x)(\exists y)(\forall u)(\exists v)(P)\psi$$

where **(P)** is the remainder of the prefix of ϕ (it is thus of degree **k-1**) and ψ is the quantifier-free matrix of ϕ. **x, y, u** and **v** denote here *tuples* of variables rather than single variables; *e.g.* $(\forall x)$ really stands for $\forall x_1 \forall x_2 ... \forall x_n$ where $x_1...x_n$ are some distinct variables.

Let now **x'** and **y'** be tuples of previously unused variables of the same length as **x** and **y** respectively, and let **Q** be a previously unused relation symbol that takes as many arguments as the sum of lengths of **x** and **y**; we consider the formula

$$\Phi = (\forall x')(\exists y')Q(x',y') \wedge (\forall x)(\forall y)(Q(x,y) \rightarrow (\forall u)(\exists v)(P)\psi)$$

Clearly, $\Phi \rightarrow \phi$ is provable.

Now since the string of quantifiers $(\forall u)(\exists v)(P)$ does not contain variables from **x** or **y**, the following equivalence is easily provable with the help of whatever formalism we're using:

$$(Q(x,y) \rightarrow (\forall u)(\exists v)(P)\psi) \equiv (\forall u)(\exists v)(P)(Q(x,y) \rightarrow \psi)$$

And since these two formulas are equivalent, if we replace the first with the second inside Φ, we obtain the formula Φ' such that Φ≡Φ':

$$\Phi' = (\forall x')(\exists y')Q(x',y') \land (\forall x)(\forall y)(\forall u)(\exists v)(P)(Q(x,y) \to \psi)$$

Now Φ' has the form $(S)\rho \land (S')\rho'$, where **(S)** and **(S')** are some quantifier strings, ρ and ρ' are quantifier-free, and, **furthermore**, no variable of **(S)** occurs in ρ' and no variable of **(S')** occurs in ρ. Under such conditions every formula of the form $(T)(\rho \land \rho')$, where **(T)** is a string of quantifiers containing all quantifiers in (S) and (S') interleaved among themselves in any fashion, but maintaining the relative order inside (S) and (S'), will be equivalent to the original formula Φ'(this is yet another basic result in first-order predicate calculus that we rely on). To wit, we form Ψ as follows:

$$\Psi = (\forall x')(\forall x)(\forall y)(\forall u)(\exists y')(\exists v)(P)Q(x',y') \land (Q(x,y) \to \psi)$$

and we have $\Phi' \equiv \Psi$.

Now Ψ is a formula of degree **k** and therefore by assumption either refutable or satisfiable. If Ψ is satisfiable in a structure **M**, then, considering $\Psi \equiv \Phi' \equiv \Phi \land \Phi \to \phi$, we see that φ is satisfiable as well. If Ψ is refutable, then so is Φ, which is equivalent to it; thus ¬Φ is provable. Now we can replace all occurrences of Q inside the provable formula ¬Φ by some other formula dependent on the same variables, and we will still get a provable formula. (*This is yet another basic result of first-order predicate calculus. Depending on the particular formalism adopted for the calculus, it may be seen as a simple application of a "functional substitution" rule of inference, as in Gödel's paper, or it may be proved by considering the formal proof of* ¬Φ *, replacing in it all occurrences of Q by some other formula with the same free variables, and noting that all logical axioms in the formal proof remain logical axioms after the substitution, and all rules of inference still apply in the same way.*)

In this particular case, we replace Q(x',y') in ¬Φ with the formula $(\forall u)(\exists v)(P)\psi(x,y|x',y')$. Here (x,y|x',y') means that instead of ψ we are writing a different formula, in which x and y are replaced with x' and y'. Note that Q(x,y) is simply replaced by $(\forall u)(\exists v)(P)\psi$.

¬Φ then becomes

$$\neg((\forall x')(\exists y')(\forall u)(\exists v)(P)\psi(x,y|x',y') \land (\forall x)(\forall y)((\forall u)(\exists v)(P)\psi \to (\forall u)(\exists v)(P)\psi))$$

and this formula is provable; since the part under negation and after the ∧ sign is obviously provable, and the part under negation and before the ∧ sign is obviously φ, just with **x** and **y** replaced by **x'** and **y'**, we see that ¬φ is provable, and φ is refutable. We have proved that φ is either satisfiable or refutable, and this concludes the proof of the **Lemma**.

Notice that we could not have used $(\forall u)(\exists v)(P)\psi(x,y|x',y')$ instead of Q(x',y') from the beginning, because Ψ would not have been a well-formed formula in that case. This is why we cannot naively use the argument appearing at the comment that precedes the proof.

9.2.6 Proving the theorem for formulas of degree 1

As shown by the **Lemma** above, we only need to prove our theorem for formulas φ in **R** of degree 1. φ cannot be of degree 0, since formulas in R have no free variables and don't use constant symbols. So the formula φ has the general form:

$$(\forall x_1...x_k)(\exists y_1...y_m)\phi(x_1...x_k, y_1...y_m).$$

Now we define an ordering of the k-tuples of natural numbers as follows: $(x_1...x_k) < (y_1...y_k)$ should hold if either $\Sigma_k(x_1...x_k) < \Sigma_k(y_1...y_k)$, or $\Sigma_k(x_1...x_k) = \Sigma_k(y_1...y_k)$, and $(x_1...x_k)$ precedes $(y_1...y_k)$ in lexicographic order. [Here $\Sigma_k(x_1...x_k)$ denotes the sum of the terms of the tuple.] Denote the nth tuple in this order by $(a_1^n...a_k^n)$.

Set the formula B_n as $\phi(z_{a_1^n}...z_{a_k^n}, z_{(n-1)m+2}, z_{(n-1)m+3}...z_{nm+1})$. Then put D_n as

$(\exists z_1...z_{nm+1})(B_1 \wedge B_2... \wedge B_n)$.

Lemma: For every n, $\varphi \to D_n$.

Proof: By induction on n; we have

$$D_k \Leftarrow D_{k-1} \wedge (\forall z_1...z_{(n-1)m+1})(\exists z_{(n-1)m+2}...z_{nm+1})B_n \Leftarrow D_{k-1} \wedge (\forall z_{a_1^n}...z_{a_k^n})(\exists y_1...y_m)\phi(z_{a_1^n}...z_{a_k^n}, y_1...y_m)$$

, where the latter implication holds by variable substitution, since the ordering of the tuples is such that $(\forall k)(a_1^n...a_k^n) < (n-1)m+2$. But the last formula is equivalent to $D_{k-1} \wedge \varphi$.

For the base case, $D_1 \equiv (\exists z_1...z_{m+1})\phi(z_{a_1^1}...z_{a_k^1}, z_2, z_3...z_{m+1}) \equiv (\exists z_1...z_{m+1})\phi(z_1...z_1, z_2, z_3...z_{m+1})$ is obviously a corollary of φ as well. So the **Lemma** is proven.

Now if D_n is refutable for some n, it follows that φ is refutable. On the other hand, suppose that D_n is not refutable for any n. Then for each n there is some way of assigning truth values to the distinct subpropositions E_h (ordered by their first appearance in D_n; "distinct" here means either distinct predicates, or distinct bound variables) in B_k, such that D_n will be true when each proposition is evaluated in this fashion. This follows from the completeness of the underlying propositional logic.

We will now show that there is such an assignment of truth values to E_h, so that all D_n will be true: The E_h appear in the same order in every D_n; we will inductively define a general assignment to them by a sort of "majority vote": Since there are infinitely many assignments (one for each D_n) affecting E_1, either infinitely many make E_1 true, or infinitely many make it false and only finitely many make it true. In the former case, we choose E_1 to be true in general; in the latter we take it to be false in general. Then from the infinitely many n for which E_1 through E_{h-1} are assigned the same truth value as in the general assignment, we pick a general assignment to E_h in the same fashion.

This general assignment must lead to every one of the B_k and D_k being true, since if one of the B_k were false under the general assignment, D_n would also be false for every $n > k$. But this contradicts the fact that for the finite collection of general E_h assignments appearing in D_k, there are infinitely many n where the assignment making D_n true matches the general assignment.

From this general assignment, which makes all of the D_k true, we construct an interpretation of the language's predicates that makes φ true. The universe of the model will be the natural numbers. Each i-ary predicate Ψ should be true of the naturals $(u_1...u_i)$ precisely when the proposition $\Psi(z_{u_1}...z_{u_i})$ is either true in the general assignment, or not assigned by it (because it never appears in any of the D_k).

In this model, each of the formulas $(\exists y_1...y_m)\phi(a_1^n...a_k^n, y_1...y_m)$ is true by construction. But this implies that φ itself is true in the model, since the a^n range over all possible k-tuples of natural numbers. So φ is satisfiable, and we are done.

Intuitive explanation

We may write each B_i as $\Phi(x_1...x_k, y_1...y_m)$ for some x-s, which we may call "first arguments" and y-s that we may call "last arguments".

Take B_1 for example. Its "last arguments" are $z_2, z_3...z_{m+1}$, and for every possible combination of k of these variables there is some j so that they appear as "first arguments" in B_j. Thus for large enough n_1, D_{n1} has the property that the "last arguments" of B_1 appear, in every possible combinations of k of them, as "first arguments" in other B_j-s within D_n. For every B_i there is a D_{ni} with the corresponding property.

Therefore in a model that satisfies all the D_n-s, there are objects corresponding to $z_1, z_2...$ and each combination of k of these appear as "first arguments" in some B_j, meaning that for every k of these objects $z_{p1}...z_{pk}$ there are $z_{q1}...z_{qm}$, which makes $\Phi(z_{p1}...z_{pk}, z_{q1}...z_{qm})$ satisfied. By taking a submodel with only these $z_1, z_2...$ objects, we have a model satisfying φ.

9.3 Extensions

9.3.1 Extension to first-order predicate calculus with equality

Gödel reduced a formula containing instances of the equality predicate to ones without it in an extended language. His method involves replacing a formula φ containing some instances of equality with the formula

$$(\forall x) Eq(x,x) \wedge (\forall x,y,z)[Eq(x,y) \to (Eq(x,z) \to Eq(y,z))] \wedge (\forall x,y,z)[Eq(x,y) \to (Eq(z,x) \to Eq(z,y))] \wedge (\forall x_1...x_k, y_1...y_k)[(Eq(x_1,y_1) \wedge ... \wedge Eq(x_k,y_k)) \to (A(x_1...x_k) \equiv A(y_1...y_k))] \wedge ... \wedge (\forall x_1...x_m, y_1...x_m)[(Eq(x_1,y_1) \wedge ... \wedge Eq(x_m,y_m)) \to (Z(x_1...x_m) \equiv Z(y_1...y_m))] \wedge \varphi'.$$

Here $A...Z$ denote the predicates appearing in φ (with $k...m$ their respective arities), and φ' is the formula φ with all occurrences of equality replaced with the new predicate Eq. If this new formula is refutable, the original φ was as well; the same is true of satisfiability, since we may take a quotient of satisfying model of the new formula by the equivalence relation representing Eq. This quotient is well-defined with respect to the other predicates, and therefore will satisfy the original formula φ.

9.3.2 Extension to countable sets of formulas

Gödel also considered the case where there are a countably infinite collection of formulas. Using the same reductions as above, he was able to consider only those cases where each formula is of degree 1 and contains no uses of equality. For a countable collection of formulas ϕ^i of degree 1, we may define B_k^i as above; then define D_k to be the closure of $B_1^1...B_k^1, ..., B_1^k...B_k^k$. The remainder of the proof then went through as before.

9.3.3 Extension to arbitrary sets of formulas

When there is an uncountably infinite collection of formulas, the Axiom of Choice (or at least some weak form of it) is needed. Using the full AC, one can well-order the formulas, and prove the uncountable case with the same argument as the countable one, except with transfinite induction. Other approaches can be used to prove that the completeness theorem in this case is equivalent to the Boolean prime ideal theorem, a weak form of AC.

9.4 References

- Gödel, K (1929). "Über die Vollständigkeit des Logikkalküls". Doctoral dissertation. University Of Vienna. The first proof of the completeness theorem.

- Gödel, K (1930). "Die Vollständigkeit der Axiome des logischen Funktionenkalküls". *Monatshefte für Mathematik* (in German) **37** (1): 349–360. doi:10.1007/BF01696781. JFM 56.0046.04. The same material as the dissertation, except with briefer proofs, more succinct explanations, and omitting the lengthy introduction.

9.5 External links

- Stanford Encyclopedia of Philosophy: "Kurt Gödel"—by Juliette Kennedy.
- MacTutor biography: Kurt Gödel.

Chapter 10

Mathematical induction

Mathematical induction can be informally illustrated by reference to the sequential effect of falling dominoes.

Mathematical induction is a method of mathematical proof typically used to establish a given statement for all natural numbers. It is a form of direct proof, and it is done in two steps. The first step, known as the **base case**, is to prove the given statement for the first natural number. The second step, known as the **inductive step**, is to prove that the given statement for any one natural number implies the given statement for the next natural number. From these two steps, mathematical induction is the rule from which we infer that the given statement is established for all natural numbers.

The method can be extended to prove statements about more general well-founded structures, such as trees; this generalization, known as structural induction, is used in mathematical logic and computer science. Mathematical induction in this extended sense is closely related to recursion. Mathematical induction, in some form, is the foundation of all correctness proofs for computer programs.[1]

Although its name may suggest otherwise, mathematical induction should not be misconstrued as a form of inductive reasoning (also see Problem of induction). Mathematical induction is an inference rule used in proofs. In mathematics, proofs including those using mathematical induction are examples of deductive reasoning, and inductive reasoning is excluded from proofs.[2]

10.1 History

In 370 BC, Plato's Parmenides may have contained an early example of an implicit inductive proof.[3] The earliest implicit traces of mathematical induction can be found in Euclid's[4][5][6] proof that the number of primes is infinite and in Bhaskara's "cyclic method".[7] An opposite iterated technique, counting *down* rather than up, is found in the Sorites paradox, where one argued that if 1,000,000 grains of sand formed a heap, and removing one grain from a heap left it a heap, then a single grain of sand (or even no grains) forms a heap.

An implicit proof by mathematical induction for arithmetic sequences was introduced in the *al-Fakhri* written by al-Karaji around 1000 AD, who used it to prove the binomial theorem and properties of Pascal's triangle.[8]

None of these ancient mathematicians, however, explicitly stated the inductive hypothesis. Another similar case (contrary to what Vacca has written, as Freudenthal carefully showed) was that of Francesco Maurolico in his *Arithmeticorum libri duo* (1575), who used the technique to prove that the sum of the first n odd integers is n^2. The first explicit formulation of the principle of induction was given by Pascal in his *Traité du triangle arithmétique* (1665). Another Frenchman, Fermat, made ample use of a related principle, indirect proof by infinite descent. The inductive hypothesis was also employed by the Swiss Jakob Bernoulli, and from then on it became more or less well known. The modern rigorous and systematic treatment of the principle came only in the 19th century, with George Boole,[9] Augustus de Morgan, Charles Sanders Peirce,[10][11] Giuseppe Peano, and Richard Dedekind.[7]

10.2 Description

The simplest and most common form of mathematical induction infers that a statement involving a natural number n holds for all values of n. The proof consists of two steps:

1. The **basis** (**base case**): prove that the statement holds for the first natural number n. Usually, $n = 0$ or $n = 1$, rarely, $n = -1$ (although not a natural number, the extension of the natural numbers to -1 is still a well-ordered set).

2. The **inductive step**: prove that, if the statement holds for some natural number n, then the statement holds for $n + 1$.

The hypothesis in the inductive step that the statement holds for some n is called the **induction hypothesis** (or **inductive hypothesis**). To perform the inductive step, one assumes the induction hypothesis and then uses this assumption to prove the statement for $n + 1$.

Whether $n = 0$ or $n = 1$ depends on the definition of the natural numbers. If 0 is considered a natural number, as is common in the fields of combinatorics and mathematical logic, the base case is given by $n = 0$. If, on the other hand, 1 is taken as the first natural number, then the base case is given by $n = 1$.

10.3 Example

Mathematical induction can be used to prove that the following statement, which we will call $P(n)$, holds for all natural numbers n.

$$0 + 1 + 2 + \cdots + n = \frac{n(n+1)}{2}.$$

$P(n)$ gives a formula for the sum of the natural numbers less than or equal to number n. The proof that $P(n)$ is true for each natural number n proceeds as follows.

Basis: Show that the statement holds for $n = 0$.
$P(0)$ amounts to the statement:

$$0 = \frac{0 \cdot (0+1)}{2}.$$

In the left-hand side of the equation, the only term is 0, and so the left-hand side is simply equal to 0.
In the right-hand side of the equation, $0 \cdot (0+1)/2 = 0$.
The two sides are equal, so the statement is true for $n = 0$. Thus it has been shown that $P(0)$ holds.

Inductive step: Show that *if* $P(k)$ holds, then also $P(k+1)$ holds. This can be done as follows.

Assume $P(k)$ holds (for some unspecified value of k). It must then be shown that $P(k+1)$ holds, that is:

$$(0 + 1 + 2 + \cdots + k) + (k+1) = \frac{(k+1)((k+1)+1)}{2}.$$

Using the induction hypothesis that $P(k)$ holds, the left-hand side can be rewritten to:

$$\frac{k(k+1)}{2} + (k+1).$$

Algebraically:

$$\begin{aligned}
\frac{k(k+1)}{2} + (k+1) &= \frac{k(k+1) + 2(k+1)}{2} \\
&= \frac{(k+1)(k+2)}{2} \\
&= \frac{(k+1)((k+1)+1)}{2}
\end{aligned}$$

thereby showing that indeed $P(k+1)$ holds.

Since both the basis and the inductive step have been performed, by mathematical induction, the statement $P(n)$ holds for all natural numbers n. Q.E.D.

10.4 Axiom of induction

Mathematical induction as an inference rule can be formalized as a second-order axiom. The *axiom of induction* is, in logical symbols,

$$\forall P. \, [[P(0) \land \forall (k \in \mathbb{N}). \, [P(k) \Rightarrow P(k+1)]] \Rightarrow \forall (n \in \mathbb{N}). \, P(n)]$$

where P is any predicate and k and n are both natural numbers.

In words, the basis $P(0)$ and the inductive step (namely, that the inductive hypothesis $P(k)$ implies $P(k + 1)$) together imply that $P(n)$ for any natural number n. The axiom of induction asserts that the validity of inferring that $P(n)$ holds for any natural number n from the basis and the inductive step.

Note that the first quantifier in the axiom ranges over *predicates* rather than over individual numbers. This is a second-order quantifier, which means that this axiom is stated in second-order logic. Axiomatizing arithmetic induction in first-order logic requires an axiom schema containing a separate axiom for each possible predicate. The article Peano axioms contains further discussion of this issue.

10.4.1 Characterizing the structure of N by the induction axiom

Having proven the base case and the inductive step, then the structure of $\mathbb{N}$ is such that any value can be obtained by performing the inductive step repeatedly. It may be helpful to think of the domino effect. Consider a half line of dominoes each standing on end, and extending infinitely to the right (see picture). Suppose that:

1. The first domino falls right.

2. If a (fixed but arbitrary) domino falls right, then its next neighbor also falls right.

With these assumptions one can conclude (using mathematical induction) that all of the dominoes will fall right.

If the dominoes are arranged in another way, this conclusion needn't hold (see Peano axioms#Formulation for a counter example). Similarly, the induction axiom describes an essential property of $\mathbb{N}$, viz. that each of its members can be reached from 0 by sufficiently often adding 1. While there is only one structure that satisfies all Peano axioms (including induction),[12] there is no set of only first-order axioms that fulfils the same task.[13]

10.5 Variants

In practice, proofs by induction are often structured differently, depending on the exact nature of the property to be proved.

10.5.1 Induction basis other than 0 or 1

If we want to prove a statement not for all natural numbers but only for all numbers greater than or equal to a certain number b then the proof by induction consists of:

1. Showing that the statement holds when $n = b$.

2. Showing that if the statement holds for $n = m \geq b$ then the same statement also holds for $n = m + 1$.

This can be used, for example, to show that $n^2 \geq 3n$ for $n \geq 3$. A more substantial example is a proof that

$$\frac{n^n}{3^n} < n! < \frac{n^n}{2^n} \text{ for } n \geq 6.$$

In this way we can prove that $P(n)$ holds for all $n \geq 1$, or even $n \geq -5$. This form of mathematical induction is actually a special case of the previous form because if the statement that we intend to prove is $P(n)$ then proving it with these two rules is equivalent with proving $P(n + b)$ for all natural numbers n with the first two steps.

10.5. VARIANTS

10.5.2 Induction basis equal to 2

In mathematics, many standard functions, including operations such as "+" and relations such as "=", are binary, meaning that they take two arguments. Often these functions possess properties that implicitly extend them to more than two arguments. For example, once addition $a + b$ is defined and is known to satisfy the associativity property $(a + b) + c = a + (b + c)$, then the ternary addition $a + b + c$ makes sense, either as $(a + b) + c$ or as $a + (b + c)$. Similarly, many axioms and theorems in mathematics are stated only for the binary versions of mathematical operations and relations, and implicitly extend to higher-arity versions.

Suppose that we wish to prove a statement about an n-ary operation implicitly defined from a binary operation, using mathematical induction on n. In this case it is natural to take 2 for the induction basis.

Example: product rule for the derivative

In this example, the binary operation in question is multiplication (of functions). The usual product rule for the derivative taught in calculus states:

$$(fg)' = f'g + g'f.$$

or in logarithmic derivative form

$$(fg)'/(fg) = f'/f + g'/g.$$

This can be generalized to a product of n functions. One has

$$(f_1 f_2 f_3 \cdots f_n)'$$

$$= (f_1' f_2 f_3 \cdots f_n) + (f_1 f_2' f_3 \cdots f_n) + (f_1 f_2 f_3' \cdots f_n) + \cdots + (f_1 f_2 \cdots f_{n-1} f_n').$$

or in logarithmic derivative form

$$(f_1 f_2 f_3 \cdots f_n)'/(f_1 f_2 f_3 \cdots f_n)$$

$$= (f_1'/f_1) + (f_2'/f_2) + (f_3'/f_3) + \cdots + (f_n'/f_n).$$

In each of the n terms of the usual form, just one of the factors is a derivative; the others are not.

When this general fact is proved by mathematical induction, the $n = 0$ case is trivial, $(1)' = 0$ (since the empty product is 1, and the empty sum is 0). The $n = 1$ case is also trivial, $f_1' = f_1'$. And for each $n \geq 3$, the case is easy to prove from the preceding $n - 1$ case. The real difficulty lies in the $n = 2$ case, which is why that is the one stated in the standard product rule.

10.5.3 Induction on more than one counter

It is sometimes desirable to prove a statement involving two natural numbers, n and m, by iterating the induction process. That is, one performs a basis step and an inductive step for n, and in each of those performs a basis step and an inductive step for m. See, for example, the proof of commutativity accompanying *addition of natural numbers*. More complicated arguments involving three or more counters are also possible.

10.5.4 Infinite descent

Main article: Infinite descent

The method of infinite descent was one of Pierre de Fermat's favorites. This method of proof can assume several slightly different forms. For example, it might begin by showing that if a statement is true for a natural number n it must also be true for some smaller natural number m ($m < n$). Using mathematical induction (implicitly) with the inductive hypothesis being that the statement is false for all natural numbers less than or equal to m, we can conclude that the statement cannot be true for any natural number n.

Although this particular form of infinite-descent proof is clearly a mathematical induction, whether one holds all proofs "by infinite descent" to be mathematical inductions depends on how one defines the term "proof by infinite descent." One might, for example, use the term to apply to proofs in which the well-ordering of the natural numbers is assumed, but not the principle of induction. Such, for example, is the usual proof that 2 has no rational square root (see Infinite descent).

10.5.5 Prefix induction

The most common form of induction requires proving that

$$\forall k (P(k) \to P(k+1))$$

or equivalently

$$\forall k (P(k-1) \to P(k))$$

whereupon the induction principle "automates" n applications of this inference in getting from $P(0)$ to $P(n)$. This could be called "predecessor induction" because each step proves something about a number from something about that number's predecessor.

A variant of interest in computational complexity is "prefix induction", in which one needs to prove

$$\forall k (P(k) \to P(2k) \land P(2k+1))$$

or equivalently

$$\forall k \left(P\left(\left\lfloor \frac{k}{2} \right\rfloor\right) \to P(k) \right)$$

The induction principle then "automates" $\log n$ applications of this inference in getting from $P(0)$ to $P(n)$. (It is called "prefix induction" because each step proves something about a number from something about the "prefix" of that number formed by truncating the low bit of its binary representation.)

If traditional predecessor induction is interpreted computationally as an n-step loop, prefix induction corresponds to a $\log n$-step loop, and thus proofs using prefix induction are "more feasibly constructive" than proofs using predecessor induction.

Predecessor induction can trivially simulate prefix induction on the same statement. Prefix induction can simulate predecessor induction, but only at the cost of making the statement more syntactically complex (adding a bounded universal quantifier), so the interesting results relating prefix induction to polynomial-time computation depend on excluding unbounded quantifiers entirely, and limiting the alternation of bounded universal and existential quantifiers allowed in the statement. See [14]

10.5. VARIANTS

One could take it a step farther to "prefix of prefix induction": one must prove

$$\forall k \left(P\left(\lfloor \sqrt{k} \rfloor \right) \to P(k) \right)$$

whereupon the induction principle "automates" log log n applications of this inference in getting from $P(0)$ to $P(n)$. This form of induction has been used, analogously, to study log-time parallel computation.

10.5.6 Complete induction

Another variant, called **complete induction**, **course of values induction** or **strong induction** (in contrast to which the basic form of induction is sometimes known as **weak induction**) makes the inductive step easier to prove by using a stronger hypothesis: one proves the statement $P(m+1)$ under the assumption that $P(n)$ holds for **all** natural n less than $m+1$; by contrast, the basic form only assumes $P(m)$. The name "strong induction" does not mean that this method can prove more than "weak induction", but merely refers to the stronger hypothesis used in the inductive step; in fact the two methods are equivalent, as explained below. In this form of complete induction one still has to prove the base case, $P(0)$, and it may even be necessary to prove extra base cases such as $P(1)$ before the general argument applies, as in the example below of the Fibonacci number Fn.

Although the form just described requires one to prove the base case, this is unnecessary if one can prove $P(m)$ (assuming $P(n)$ for all lower n) for all $m \geq 0$. This is a special case of transfinite induction as described below. In this form the base case is subsumed by the case $m = 0$, where $P(0)$ is proved with no other $P(n)$ assumed; this case may need to be handled separately, but sometimes the same argument applies for $m = 0$ and $m > 0$, making the proof simpler and more elegant. In this method it is, however, vital to ensure that the proof of $P(m)$ does not implicitly assume that $m > 0$, e.g. by saying "choose an arbitrary $n < m$" or assuming that a set of m elements has an element.

Complete induction is equivalent to ordinary mathematical induction as described above, in the sense that a proof by one method can be transformed into a proof by the other. Suppose we have a proof of $P(n)$ by complete induction. Let $Q(n)$ mean "$P(m)$ holds for all m such that $0 \leq m \leq n$". Then $Q(n)$ holds for all n if and only if $P(n)$ holds for all n, and our proof of $P(n)$ is easily transformed into a proof of $Q(n)$ by (ordinary) induction. If, on the other hand, we have proved $P(n)$ by ordinary induction, we already effectively have a proof by complete induction: $P(0)$ is proved in the base case, using no assumptions, and $P(n+1)$ is proved in the inductive step, where we may assume all earlier cases but need only use the case $P(n)$.

Examples of complete induction

Complete induction is most useful when several instances of the inductive hypothesis are required for each inductive step. For example, complete induction can be used to show that

$$F_n = \frac{\varphi^n - \psi^n}{\varphi - \psi}$$

where Fn is the nth Fibonacci number, $\varphi = (1 + \sqrt{5})/2$ (the golden ratio) and $\psi = (1 - \sqrt{5})/2$ are the roots of the polynomial $x^2 - x - 1$. By using the fact that $F_{n+2} = F_{n+1} + F_n$ for each $n \in \mathbf{N}$, the identity above can be verified by direct calculation for F_{n+2} if we assume that it already holds for both F_{n+1} and Fn. To complete the proof, the identity must be verified in the two base cases $n = 0$ and $n = 1$.

Another proof by complete induction uses the hypothesis that the statement holds for *all* smaller n more thoroughly. Consider the statement that "every natural number greater than 1 is a product of (one or more) prime numbers", and assume that for a given $m > 1$ it holds for all smaller $n > 1$. If m is prime then it is certainly a product of primes, and if not, then by definition it is a product: $m = n_1 n_2$, where neither of the factors is equal to 1; hence neither is equal to m, and so both are smaller than m. The induction hypothesis now applies to n_1 and n_2, so each one is a product of primes. Thus m is a product of products of primes; i.e. a product of primes.

Transfinite induction

Main article: Transfinite induction

The last two steps can be reformulated as one step:

1. Showing that if the statement holds for all $n < m$ then the same statement also holds for $n = m$.

This form of mathematical induction is not only valid for statements about natural numbers, but for statements about elements of any well-founded set, that is, a set with an irreflexive relation < that contains no infinite descending chains.

This form of induction, when applied to ordinals (which form a well-ordered and hence well-founded class), is called *transfinite induction*. It is an important proof technique in set theory, topology and other fields.

Proofs by transfinite induction typically distinguish three cases:

1. when m is a minimal element, i.e. there is no element smaller than m
2. when m has a direct predecessor, i.e. the set of elements which are smaller than m has a largest element
3. when m has no direct predecessor, i.e. m is a so-called limit-ordinal

Strictly speaking, it is not necessary in transfinite induction to prove the basis, because it is a vacuous special case of the proposition that if P is true of all $n < m$, then P is true of m. It is vacuously true precisely because there are no values of $n < m$ that could serve as counterexamples.

10.6 Equivalence with the well-ordering principle

The principle of mathematical induction is usually stated as an axiom of the natural numbers; see Peano axioms. However, it can be proved from the well-ordering principle. Indeed, suppose the following:

- The set of natural numbers is well-ordered.
- Every natural number is either zero, or $n + 1$ for some natural number n.
- For any natural number n, $n + 1$ is greater than n.

To derive simple induction from these axioms, we must show that if $P(n)$ is some proposition predicated of n, and if:

- $P(0)$ holds and
- whenever $P(k)$ is true then $P(k + 1)$ is also true

then $P(n)$ holds for all n.

Proof. Let S be the set of all natural numbers for which $P(n)$ is false. Let us see what happens if we assert that S is nonempty. Well-ordering tells us that S has a least element, say t. Moreover, since $P(0)$ is true, t is not 0. Since every natural number is either zero or some $n + 1$, there is some natural number n such that $n + 1 = t$. Now n is less than t, and t is the least element of S. It follows that n is not in S, and so $P(n)$ is true. This means that $P(n + 1)$ is true, and so $P(t)$ is true. This is a contradiction, since t was in S. Therefore, S is empty.

It can also be proved that induction, given the other axioms, implies the well-ordering principle.

10.7 Example of error in the inductive step

Main article: All horses are the same color

This example demonstrated a subtle error in the proof of the inductive step.

Joel E. Cohen proposed the following argument, which purports to prove by mathematical induction that all horses are of the same color:[15]

- Basis: If there is only *one* horse, there is only one color.

- Induction step: Assume as induction hypothesis that within any set of n horses, there is only one color. Now look at any set of $n + 1$ horses. Number them: $1, 2, 3, ..., n, n + 1$. Consider the sets $\{1, 2, 3, ..., n\}$ and $\{2, 3, 4, ..., n + 1\}$. Each is a set of only n horses, therefore within each there is only one color. But the two sets overlap, so there must be only one color among all $n + 1$ horses.

The basis case $n = 1$ is trivial (as any horse is the same color as itself), and the inductive step is correct in all cases $n > 1$. However, the logic of the inductive step is incorrect for $n = 1$, because the statement that "the two sets overlap" is false (there are only $n + 1 = 2$ horses prior to either removal, and after removal the sets of one horse each do not overlap).

10.8 See also

- Combinatorial proof

- Recursion

- Recursion (computer science)

- Structural induction

10.9 Notes

[1] Anderson, Robert B. (1979). *Proving Programs Correct*. New York: John Wiley & Sons. p. 1. ISBN 0471033952.

[2] Suber, Peter. "Mathematical Induction". Earlham College. Retrieved 26 March 2011.

[3] Mathematical Induction: The Basis Step of Verification and Validation in a Modeling and Simulation Course

[4] Chris K. Caldwell. "Euclid's Proof of the Infinitude of Primes (c. 300 BC)". *utm.edu*.

[5] "Euclid's Primes". *mathsisgoodforyou.com*.

[6] "Proofs of the Infinity of the Prime Numbers". *hermetic.ch*.

[7] Cajori (1918), p. 197: 'The process of reasoning called "Mathematical Induction" has had several independent origins. It has been traced back to the Swiss Jakob (James) Bernoulli, the Frenchman B. Pascal and P. Fermat, and the Italian F. Maurolycus. [...] By reading a little between the lines one can find traces of mathematical induction still earlier, in the writings of the Hindus and the Greeks, as, for instance, in the "cyclic method" of Bhaskara, and in Euclid's proof that the number of primes is infinite.'

[8] Rashed, R. (1994), "Mathematical induction: al-Karajī and al-Samaw'al", *The Development of Arabic Mathematics: Between Arithmetic and Algebra*, Boston Studies in the Philosophy of Science **156**, Kluwer Academic Publishers, pp. 62–84, ISBN 9780792325659

[9] "It is sometimes required to prove a theorem which shall be true whenever a certain quantity *n* which it involves shall be an integer or whole number and the method of proof is usually of the following kind. *1st*. The theorem is proved to be true when *n* = 1. *2ndly*. It is proved that if the theorem is true when *n* is a given whole number, it will be true if *n* is the next greater integer. Hence the theorem is true universally. . . .. This species of argument may be termed a continued *sorites*" (Boole circa 1849 *Elementary Treatise on Logic not mathematical* pages 40–41 reprinted in Grattan-Guinness, Ivor and Bornet, Gérard (1997), *George Boole: Selected Manuscripts on Logic and its Philosophy*, Birkhäuser Verlag, Berlin, ISBN 3-7643-5456-9)

[10] Peirce, C. S. (1881). "On the Logic of Number". *American Journal of Mathematics* **4** (1–4). pp. 85–95. doi:10.2307/2369151. JSTOR 2369151. MR 1507856. Reprinted (CP 3.252-88), (W 4:299-309).

[11] Shields (1997)

[12] Hermes (1973), VI.3.1

[13] Hermes (1973), VI.4.3, presenting a theorem of Thoralf Skolem

[14] Buss, Samuel (1986). *Bounded Arithmetic*. Naples: Bibliopolis.

[15] Cohen, Joel E. (1961), "On the nature of mathematical proof", *Opus*. Reprinted in *A Random Walk in Science* (R. L. Weber, ed.), Crane, Russak & Co., 1973.

10.10 References

10.10.1 Introduction

- Franklin, J.; A. Daoud (2011). *Proof in Mathematics: An Introduction*. Sydney: Kew Books. ISBN 0-646-54509-4. (Ch. 8.)

- Hazewinkel, Michiel, ed. (2001), "Mathematical induction", *Encyclopedia of Mathematics*, Springer, ISBN 978-1-55608-010-4

- Hermes, Hans (1973). *Introduction to Mathematical Logic*. Hochschultext. London: Springer. ISBN 3540058192. ISSN 1431-4657.

- Knuth, Donald E. (1997). *The Art of Computer Programming, Volume 1: Fundamental Algorithms* (3rd ed.). Addison-Wesley. ISBN 0-201-89683-4. (Section 1.2.1: Mathematical Induction, pp. 11–21.)

- Kolmogorov, Andrey N.; Sergei V. Fomin (1975). *Introductory Real Analysis*. Silverman, R. A. (trans., ed.). New York: Dover. ISBN 0-486-61226-0. (Section 3.8: Transfinite induction, pp. 28–29.)

10.10.2 History

- Acerbi, F. (2000). "Plato: *Parmenides* 149a7-c3. A Proof by Complete Induction?". *Archive for History of Exact Sciences* **55**: 57–76. doi:10.1007/s004070000020.

- Bussey, W. H. (1917). "The Origin of Mathematical Induction". *The American Mathematical Monthly* **24** (5): 199–207. doi:10.2307/2974308. JSTOR 2974308.

- Cajori, Florian (1918). "Origin of the Name "Mathematical Induction"". *The American Mathematical Monthly* **25** (5): 197–201. doi:10.2307/2972638. JSTOR 2972638.

- Fowler D. (1994). "Could the Greeks Have Used Mathematical Induction? Did They Use It?". *Physis* **XXXI**: 253–265.

- Freudenthal, Hans (1953). "Zur Geschichte der vollständigen Induction". *Archives Internationales d'Histiore des Sciences* **6**: 17–37.

- Katz, Victor J. (1998). *History of Mathematics: An Introduction*. Addison-Wesley. ISBN 0-321-01618-1.

10.10. REFERENCES

- Peirce, C. S. (1881). "On the Logic of Number". *American Journal of Mathematics* **4** (1–4). pp. 85–95. doi:10.2307/2369151. JSTOR 2369151. MR 1507856. Reprinted (CP 3.252-88), (W 4:299-309).

- Rabinovitch, Nachum L. (1970). "Rabbi Levi Ben Gershon and the origins of mathematical induction". *Archive for History of Exact Sciences* **6** (3): 237–248. doi:10.1007/BF00327237.

- Rashed, Roshdi (1972). "L'induction mathématique: al-Karajī, as-Samaw'al". *Archive for History of Exact Sciences* (in French) **9** (1): 1–21. doi:10.1007/BF00348537.

- Shields, Paul (1997). "Peirce's Axiomatization of Arithmetic". In Houser; et al. *Studies in the Logic of Charles S. Peirce*.

- Ungure, S. (1991). "Greek Mathematics and Mathematical Induction". *Physis*. XXVIII: 273–289.

- Ungure, S. (1994). "Fowling after Induction". *Physis* **XXXI**: 267–272.

- Vacca, G. (1909). "Maurolycus, the First Discoverer of the Principle of Mathematical Induction". *Bulletin of the American Mathematical Society* **16** (2): 70–73. doi:10.1090/S0002-9904-1909-01860-9.

- Yadegari, Mohammad (1978). "The Use of Mathematical Induction by Abū Kāmil Shujā' Ibn Aslam (850-930)". *Isis* **69** (2): 259–262. doi:10.1086/352009. JSTOR 230435.

Chapter 11

0.999...

The repeating decimal continues with an infinite number of nines.

In mathematics, the repeating decimal **0.999...** (sometimes written with more or fewer 9s before the final ellipsis, for example as **0.9...**, or in a variety of other variants such as **0.9**, **0.(9)**, or $0.\dot{9}$) denotes a real number that can be shown to be the number *one*. In other words, the symbols "0.999..." and "1" represent the same number. Proofs of this equality have been formulated with varying degrees of mathematical rigor, taking into account preferred development of the real numbers, background assumptions, historical context, and target audience.

Every nonzero, terminating decimal (with infinitely many trailing 0s) has an equal twin representation with infinitely many trailing 9s (for example, 8.32 and 8.31999...). The terminating decimal representation is usually preferred, contributing to the misconception that it is the only representation. The same phenomenon occurs in all other bases (with a given base's largest digit) or in any similar representation of the real numbers.

The equality of 0.999... and 1 is closely related to the absence of nonzero infinitesimals in the real number system, the most commonly used system in mathematical analysis. Some alternative number systems, such as the hyperreals, do contain nonzero infinitesimals. In most such number systems, the standard interpretation of the expression 0.999... makes it equal to 1, but in some of these number systems, the symbol "0.999..." admits other interpretations that contain infinitely many 9s while falling infinitesimally short of 1.

The equality 0.999... = 1 has long been accepted by mathematicians and is part of general mathematical education. Nonetheless, some students find it sufficiently counterintuitive that they question or reject it. Such skepticism is common enough that the difficulty of convincing them of the validity of this identity has been the subject of several studies in mathematics education.

11.1 Algebraic proofs

Algebraic proofs showing that 0.999... represents the number 1 use concepts such as fractions, long division, and digit manipulation to build transformations preserving equality from 0.999... to 1. However, these proofs are not rigorous as they do not include a careful analytic definition of 0.999...

11.1. ALGEBRAIC PROOFS

11.1.1 Fractions and long division

One reason that infinite decimals are a necessary extension of finite decimals is to represent fractions. Using long division, a simple division of integers like $1/9$ becomes a recurring decimal, 0.111..., in which the digits repeat without end. This decimal yields a quick proof for 0.999... = 1. Multiplication of 9 times 1 produces 9 in each digit, so $9 \times 0.111...$ equals 0.999... and $9 \times 1/9$ equals 1, so 0.999... = 1:

$$\frac{1}{9} = 0.111\ldots$$
$$9 \times \frac{1}{9} = 9 \times 0.111\ldots$$
$$1 = 0.999\ldots$$

This result is consistent with other ninth fractions, all of which have repeating decimals, such as 3/9 and 8/9. If 0.999... is to be consistent, it must equal 9/9 = 1.

$$0.333\ldots = \frac{3}{9}$$
$$0.888\ldots = \frac{8}{9}$$
$$0.999\ldots = \frac{9}{9} = 1$$

11.1.2 Digit manipulation

When a number in decimal notation is multiplied by 10, the digits do not change but each digit moves one place to the left. Thus $10 \times 0.999...$ equals 9.999..., which is 9 greater than the original number. To see this, consider that in subtracting 0.999... from 9.999..., each of the digits after the decimal separator cancels, i.e. the result is $9 - 9 = 0$ for each such digit:

$$\begin{aligned}
x &= 0.999\ldots \\
10x &= 9.999\ldots & &\text{by multiply10} \\
10x &= 9 + 0.999\ldots \\
10x &= 9 + x & &\text{of definition}x \\
9x &= 9 & &\text{subtract}x \\
x &= 1 & &\text{by divide9}
\end{aligned}$$

11.1.3 Discussion

Although these proofs demonstrate that 0.999... = 1, the extent to which they *explain* the equation depends on the audience. In introductory arithmetic, such proofs help explain why 0.999... = 1 but 0.333... < 0.34. In introductory algebra, the proofs help explain why the general method of converting between fractions and repeating decimals works. But the proofs shed little light on the fundamental relationship between decimals and the numbers they represent, which underlies the question of how two different decimals can be said to be equal at all.[1]

Once a representation scheme is defined, it can be used to justify the rules of decimal arithmetic used in the above proofs. Moreover, one can directly demonstrate that the decimals 0.999... and 1.000... both represent the same real number; it is built into the definition. This is done below.

11.2 Analytic proofs

Since the question of 0.999... does not affect the formal development of mathematics, it can be postponed until one proves the standard theorems of real analysis. One requirement is to characterize real numbers that can be written in decimal notation, consisting of an optional sign, a finite sequence of one or more digits forming an integer part, a decimal separator, and a sequence of digits forming a fractional part. For the purpose of discussing 0.999..., the integer part can be summarized as b_0 and one can neglect negatives, so a decimal expansion has the form

$$b_0.b_1b_2b_3b_4b_5\ldots.$$

It should be noted that the fraction part, unlike the integer part, is not limited to a finite number of digits. This is a positional notation, so for example the digit 5 in 500 contributes ten times as much as the 5 in 50, and the 5 in 0.05 contributes one tenth as much as the 5 in 0.5.

11.2.1 Infinite series and sequences

Further information: Decimal representation

Perhaps the most common development of decimal expansions is to define them as sums of infinite series. In general:

$$b_0.b_1b_2b_3b_4\ldots = b_0 + b_1\left(\tfrac{1}{10}\right) + b_2\left(\tfrac{1}{10}\right)^2 + b_3\left(\tfrac{1}{10}\right)^3 + b_4\left(\tfrac{1}{10}\right)^4 + \cdots.$$

For 0.999... one can apply the convergence theorem concerning geometric series:[2]

If $|r| < 1$ then $ar + ar^2 + ar^3 + \cdots = \tfrac{ar}{1-r}$.

Since 0.999... is such a sum with a common ratio $r = \tfrac{1}{10}$, the theorem makes short work of the question:

$$0.999\ldots = 9\left(\tfrac{1}{10}\right) + 9\left(\tfrac{1}{10}\right)^2 + 9\left(\tfrac{1}{10}\right)^3 + \cdots = \frac{9\left(\tfrac{1}{10}\right)}{1 - \tfrac{1}{10}} = 1.$$

This proof (actually, that 10 equals 9.999...) appears as early as 1770 in Leonhard Euler's *Elements of Algebra*.[3]

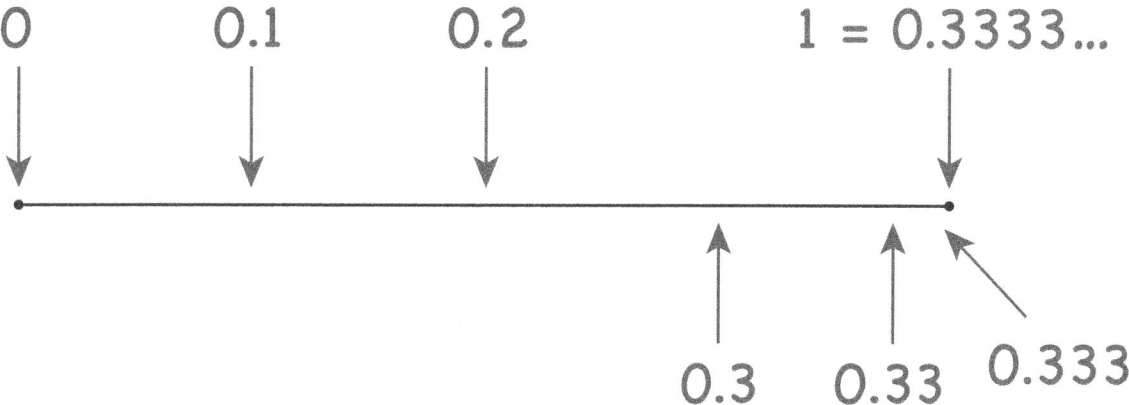

Limits: The unit interval, including the base-4 fraction sequence (.3, .33, .333, ...) converging to 1.

11.2. ANALYTIC PROOFS

The sum of a geometric series is itself a result even older than Euler. A typical 18th-century derivation used a term-by-term manipulation similar to the algebraic proof given above, and as late as 1811, Bonnycastle's textbook *An Introduction to Algebra* uses such an argument for geometric series to justify the same maneuver on 0.999...[4] A 19th-century reaction against such liberal summation methods resulted in the definition that still dominates today: the sum of a series is *defined* to be the limit of the sequence of its partial sums. A corresponding proof of the theorem explicitly computes that sequence; it can be found in any proof-based introduction to calculus or analysis.[5]

A sequence (x_0, x_1, x_2, ...) has a limit x if the distance |$x - x_n$| becomes arbitrarily small as n increases. The statement that 0.999... = 1 can itself be interpreted and proven as a limit:[6]

$$0.999\ldots = \lim_{n\to\infty} 0.\underbrace{99\ldots 9}_{n} = \lim_{n\to\infty} \sum_{k=1}^{n} \frac{9}{10^k} = \lim_{n\to\infty}\left(1 - \frac{1}{10^n}\right) = 1 - \lim_{n\to\infty} \frac{1}{10^n} = 1.$$

The last step, that $1/10^n \to 0$ as $n \to \infty$, is often justified by the Archimedean property of the real numbers. This limit-based attitude towards 0.999... is often put in more evocative but less precise terms. For example, the 1846 textbook *The University Arithmetic* explains, ".999 +, continued to infinity = 1, because every annexation of a 9 brings the value closer to 1"; the 1895 *Arithmetic for Schools* says, "...when a large number of 9s is taken, the difference between 1 and .99999... becomes inconceivably small".[7] Such heuristics are often interpreted by students as implying that 0.999... itself is less than 1.

11.2.2 Nested intervals and least upper bounds

Further information: Nested intervals

The series definition above is a simple way to define the real number named by a decimal expansion. A complementary approach is tailored to the opposite process: for a given real number, define the decimal expansion(s) to name it.

If a real number x is known to lie in the closed interval [0, 10] (i.e., it is greater than or equal to 0 and less than or equal to 10), one can imagine dividing that interval into ten pieces that overlap only at their endpoints: [0, 1], [1, 2], [2, 3], and so on up to [9, 10]. The number x must belong to one of these; if it belongs to [2, 3] then one records the digit "2" and subdivides that interval into [2, 2.1], [2.1, 2.2], ..., [2.8, 2.9], [2.9, 3]. Continuing this process yields an infinite sequence of nested intervals, labeled by an infinite sequence of digits b_0, b_1, b_2, b_3, ..., and one writes

$$x = b_0.b_1 b_2 b_3 \ldots.$$

In this formalism, the identities 1 = 0.999... and 1 = 1.000... reflect, respectively, the fact that 1 lies in both [0, 1] and [1, 2], so one can choose either subinterval when finding its digits. To ensure that this notation does not abuse the "=" sign, one needs a way to reconstruct a unique real number for each decimal. This can be done with limits, but other constructions continue with the ordering theme.[8]

One straightforward choice is the nested intervals theorem, which guarantees that given a sequence of nested, closed intervals whose lengths become arbitrarily small, the intervals contain exactly one real number in their intersection. So $b_0.b_1 b_2 b_3\ldots$ is defined to be the unique number contained within all the intervals [b_0, $b_0 + 1$], [$b_0.b_1$, $b_0.b_1 + 0.1$], and so on. 0.999... is then the unique real number that lies in all of the intervals [0, 1], [0.9, 1], [0.99, 1], and [0.99...9, 1] for every finite string of 9s. Since 1 is an element of each of these intervals, 0.999... = 1.[9]

The Nested Intervals Theorem is usually founded upon a more fundamental characteristic of the real numbers: the existence of least upper bounds or *suprema*. To directly exploit these objects, one may define $b_0.b_1 b_2 b_3\ldots$ to be the least upper bound of the set of approximants {b_0, $b_0.b_1$, $b_0.b_1 b_2$, ...}.[10] One can then show that this definition (or the nested intervals definition) is consistent with the subdivision procedure, implying 0.999... = 1 again. Tom Apostol concludes,

> The fact that a real number might have two different decimal representations is merely a reflection of the fact that two different sets of real numbers can have the same supremum.[11]

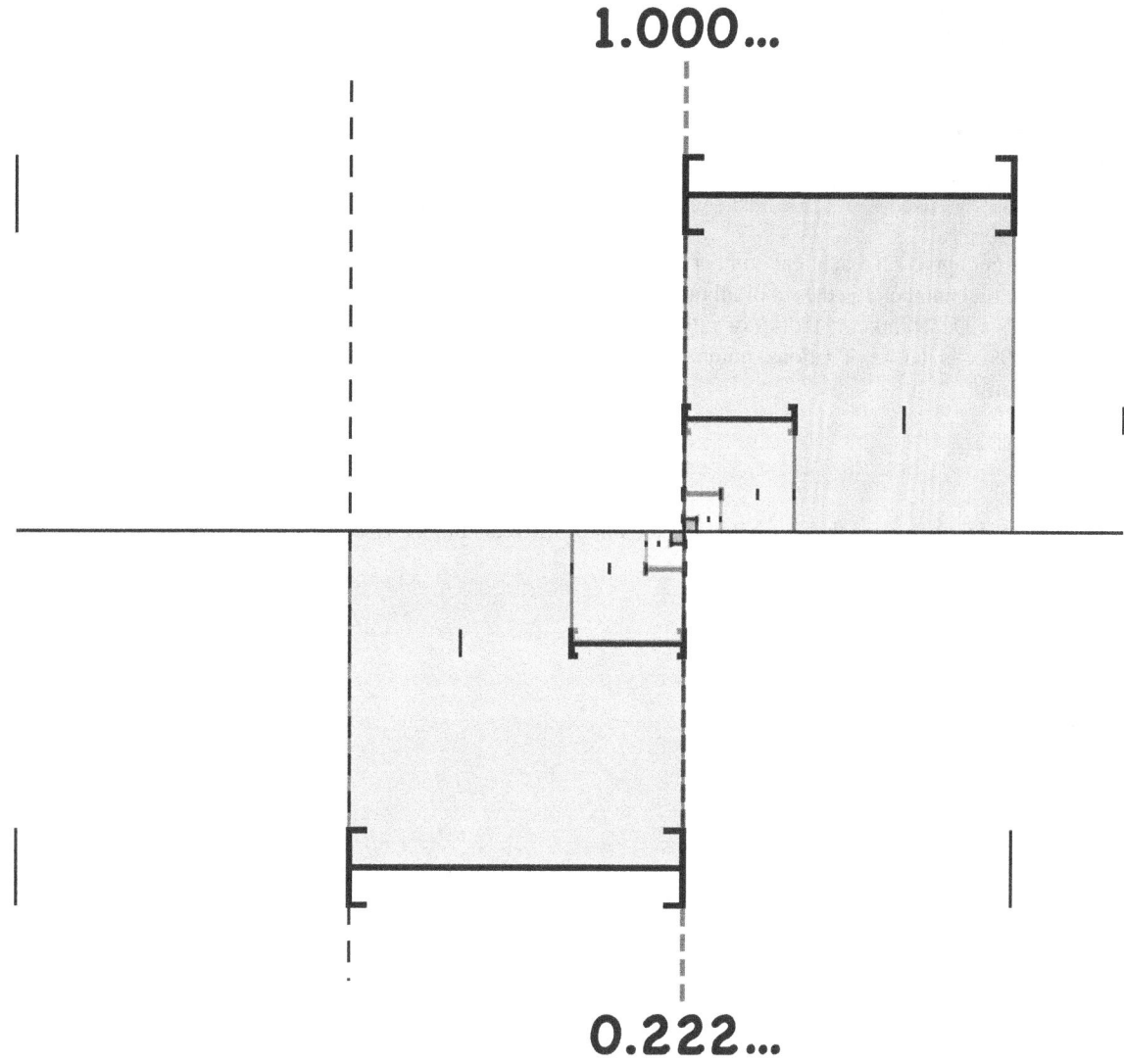

Nested intervals: in base 3, 1 = 1.000... = 0.222...

11.3 Proofs from the construction of the real numbers

Further information: Construction of the real numbers

Some approaches explicitly define real numbers to be certain structures built upon the rational numbers, using axiomatic set theory. The natural numbers – 0, 1, 2, 3, and so on – begin with 0 and continue upwards, so that every number has a successor. One can extend the natural numbers with their negatives to give all the integers, and to further extend to ratios, giving the rational numbers. These number systems are accompanied by the arithmetic of addition, subtraction, multiplication, and division. More subtly, they include ordering, so that one number can be compared to another and found to be less than, greater than, or equal to another number.

The step from rationals to reals is a major extension. There are at least two popular ways to achieve this step, both published in 1872: Dedekind cuts and Cauchy sequences. Proofs that 0.999... = 1 which directly use these constructions are not found in textbooks on real analysis, where the modern trend for the last few decades has been to use an axiomatic analysis. Even when a construction is offered, it is usually applied towards proving the axioms of the real numbers, which

11.3. PROOFS FROM THE CONSTRUCTION OF THE REAL NUMBERS

then support the above proofs. However, several authors express the idea that starting with a construction is more logically appropriate, and the resulting proofs are more self-contained.[12]

11.3.1 Dedekind cuts

Further information: Dedekind cut

In the Dedekind cut approach, each real number x is defined as the **infinite set of all rational numbers less than x**.[13] In particular, the real number 1 is the set of all rational numbers that are less than 1.[14] Every positive decimal expansion easily determines a Dedekind cut: the set of rational numbers which are less than some stage of the expansion. So the real number 0.999... is the set of rational numbers r such that $r < 0$, or $r < 0.9$, or $r < 0.99$, or r is less than some other number of the form

$$1 - \left(\tfrac{1}{10}\right)^n.$$ [15]

Every element of 0.999... is less than 1, so it is an element of the real number 1. Conversely, an element of 1 is a rational number

$$\tfrac{a}{b} < 1,$$

which implies

$$\tfrac{a}{b} < 1 - \left(\tfrac{1}{10}\right)^b.$$

Since 0.999... and 1 contain the same rational numbers, they are the same set: 0.999... = 1.

The definition of real numbers as Dedekind cuts was first published by Richard Dedekind in 1872.[16] The above approach to assigning a real number to each decimal expansion is due to an expository paper titled "Is 0.999 ... = 1?" by Fred Richman in *Mathematics Magazine*,[17] which is targeted at teachers of collegiate mathematics, especially at the junior/senior level, and their students.[18] Richman notes that taking Dedekind cuts in any dense subset of the rational numbers yields the same results; in particular, he uses decimal fractions, for which the proof is more immediate. He also notes that typically the definitions allow { x : x < 1 } to be a cut but not { x : x ≤ 1 } (or vice versa) "Why do that? Precisely to rule out the existence of distinct numbers 0.9* and 1. [...] So we see that in the traditional definition of the real numbers, the equation 0.9* = 1 is built in at the beginning."[19] A further modification of the procedure leads to a different structure where the two are not equal. Although it is consistent, many of the common rules of decimal arithmetic no longer hold, for example the fraction 1/3 has no representation; see "Alternative number systems" below.

11.3.2 Cauchy sequences

Further information: Cauchy sequence

Another approach is to define a real number as the **limit of a Cauchy sequence of rational numbers**. This construction of the real numbers uses the ordering of rationals less directly. First, the distance between x and y is defined as the absolute value $|x - y|$, where the absolute value $|z|$ is defined as the maximum of z and $-z$, thus never negative. Then the reals are defined to be the sequences of rationals that have the Cauchy sequence property using this distance. That is, in the sequence $(x_0, x_1, x_2, ...)$, a mapping from natural numbers to rationals, for any positive rational δ there is an N such that $|x_m - x_n| \leq \delta$ for all $m, n > N$. (The distance between terms becomes smaller than any positive rational.)[20]

If (x_n) and (y_n) are two Cauchy sequences, then they are defined to be equal as real numbers if the sequence $(x_n - y_n)$ has the limit 0. Truncations of the decimal number $b_0.b_1b_2b_3...$ generate a sequence of rationals which is Cauchy; this is taken to define the real value of the number.[21] Thus in this formalism the task is to show that the sequence of rational numbers

$$\left(1-0, 1-\frac{9}{10}, 1-\frac{99}{100}, \ldots\right) = \left(1, \frac{1}{10}, \frac{1}{100}, \ldots\right)$$

has the limit 0. Considering the nth term of the sequence, for $n \in \mathbb{N}$, it must therefore be shown that

$$\lim_{n \to \infty} \frac{1}{10^n} = 0.$$

This limit is plain[22] if one understands the definition of limit. So again 0.999... = 1.

The definition of real numbers as Cauchy sequences was first published separately by Eduard Heine and Georg Cantor, also in 1872.[16] The above approach to decimal expansions, including the proof that 0.999... = 1, closely follows Griffiths & Hilton's 1970 work *A comprehensive textbook of classical mathematics: A contemporary interpretation*. The book is written specifically to offer a second look at familiar concepts in a contemporary light.[23]

11.3.3 Infinite decimal representation

Further information: Stevin's construction

Commonly in secondary schools' mathematics education, the real numbers are constructed by defining a number using an integer followed by a radix point and an infinite sequence written out as a string to represent the fractional part of any given real number. In this construction, the set of any combination of integers and digits after the decimal point (or radix point in non-base 10 systems) are the set of real numbers. This construction can too be rigorously shown to satisfy all of the real axioms after defining an equivalence relation over the set that **defines** $1 =_{eq} 0.999...$ as well as for any other nonzero decimals with only finitely many nonzero terms in the decimal string with its trailing 9s version.[24] With this construction of the reals, all proofs of the statement 1 = .999... can be viewed as implicitly assuming the equality when any operations are performed on the real numbers.

11.4 Generalizations

The result that 0.999... = 1 generalizes readily in two ways. First, every nonzero number with a finite decimal notation (equivalently, endless trailing 0s) has a counterpart with trailing 9s. For example, 0.24999... equals 0.25, exactly as in the special case considered. These numbers are exactly the decimal fractions, and they are dense.[25]

Second, a comparable theorem applies in each radix or base. For example, in base 2 (the binary numeral system) 0.111... equals 1, and in base 3 (the ternary numeral system) 0.222... equals 1. In general, any terminating base b expression has a counterpart with repeated trailing digits equal to $b - 1$. Textbooks of real analysis are likely to skip the example of 0.999... and present one or both of these generalizations from the start.[26]

Alternative representations of 1 also occur in non-integer bases. For example, in the golden ratio base, the two standard representations are 1.000... and 0.101010..., and there are infinitely many more representations that include adjacent 1s. Generally, for almost all q between 1 and 2, there are uncountably many base-q expansions of 1. On the other hand, there are still uncountably many q (including all natural numbers greater than 1) for which there is only one base-q expansion of 1, other than the trivial 1.000.... This result was first obtained by Paul Erdős, Miklos Horváth, and István Joó around 1990. In 1998 Vilmos Komornik and Paola Loreti determined the smallest such base, the Komornik–Loreti constant $q = 1.787231650...$. In this base, 1 = 0.11010011001011010010110011010011...; the digits are given by the Thue–Morse sequence, which does not repeat.[27]

A more far-reaching generalization addresses the most general positional numeral systems. They too have multiple representations, and in some sense the difficulties are even worse. For example:[28]

- In the balanced ternary system, $1/2 = 0.111... = 1.111....$

- In the reverse factorial number system (using bases $2!,3!,4!,...$ for positions *after* the decimal point), $1 = 1.000... = 0.1234....$

11.4.1 Impossibility of unique representation

That all these different number systems suffer from multiple representations for some real numbers can be attributed to a fundamental difference between the real numbers as an ordered set and collections of infinite strings of symbols, ordered lexicographically. Indeed, the following two properties account for the difficulty:

- If an interval of the real numbers is partitioned into two non-empty parts L, R, such that every element of L is (strictly) less than every element of R, then either L contains a largest element or R contains a smallest element, but not both.

- The collection of infinite strings of symbols taken from any finite "alphabet", lexicographically ordered, can be partitioned into two non-empty parts L, R, such that every element of L is less than every element of R, while L contains a largest element *and* R contains a smallest element. Indeed, it suffices to take two finite prefixes (initial substrings) p_1, p_2 of elements from the collection such that they differ only in their final symbol, for which symbol they have successive values, and take for L the set of all strings in the collection whose corresponding prefix is at most p_1, and for R the remainder, the strings in the collection whose corresponding prefix is at least p_2. Then L has a largest element, starting with p_1 and choosing the largest available symbol in all following positions, while R has a smallest element obtained by following p_2 by the smallest symbol in all positions.

The first point follows from basic properties of the real numbers: L has a supremum and R has an infimum, which are easily seen to be equal; being a real number it either lies in R or in L, but not both since L and R are supposed to be disjoint. The second point generalizes the 0.999.../1.000... pair obtained for $p_1 =$ "0", $p_2 =$ "1". In fact one need not use the same alphabet for all positions (so that for instance mixed radix systems can be included) or consider the full collection of possible strings; the only important points are that at each position a finite set of symbols (which may even depend on the previous symbols) can be chosen from (this is needed to ensure maximal and minimal choices), and that making a valid choice for any position should result in a valid infinite string (so one should not allow "9" in each position while forbidding an infinite succession of "9"s). Under these assumptions, the above argument shows that an order preserving map from the collection of strings to an interval of the real numbers cannot be a bijection: either some numbers do not correspond to any string, or some of them correspond to more than one string.

Marko Petkovšek has proven that for any positional system that names all the real numbers, the set of reals with multiple representations is always dense. He calls the proof "an instructive exercise in elementary point-set topology"; it involves viewing sets of positional values as Stone spaces and noticing that their real representations are given by continuous functions.[29]

11.5 Applications

One application of 0.999... as a representation of 1 occurs in elementary number theory. In 1802, H. Goodwin published an observation on the appearance of 9s in the repeating-decimal representations of fractions whose denominators are certain prime numbers. Examples include:

- $1/7 = 0.142857142857...$ and $142 + 857 = 999$.

- $1/73 = 0.0136986301369863...$ and $0136 + 9863 = 9999$.

- $1/77 = 0.012987012987...$ and $0129 + 870 = 999$.

E. Midy proved a general result about such fractions, now called *Midy's theorem*, in 1836. The publication was obscure, and it is unclear if his proof directly involved 0.999..., but at least one modern proof by W. G. Leavitt does. If it can be proved that a decimal of the form $0.b_1b_2b_3...$ is a positive integer, then it must be 0.999..., which is then the source of the 9s in the theorem.[30] Investigations in this direction can motivate such concepts as greatest common divisors, modular arithmetic, Fermat primes, order of group elements, and quadratic reciprocity.[31]

Returning to real analysis, the base-3 analogue 0.222... = 1 plays a key role in a characterization of one of the simplest fractals, the middle-thirds Cantor set:

- A point in the unit interval lies in the Cantor set if and only if it can be represented in ternary using only the digits 0 and 2.

The *n*th digit of the representation reflects the position of the point in the *n*th stage of the construction. For example, the point $\frac{2}{3}$ is given the usual representation of 0.2 or 0.2000..., since it lies to the right of the first deletion and to the left of every deletion thereafter. The point $\frac{1}{3}$ is represented not as 0.1 but as 0.0222..., since it lies to the left of the first deletion and to the right of every deletion thereafter.[32]

Repeating nines also turn up in yet another of Georg Cantor's works. They must be taken into account to construct a valid proof, applying his 1891 diagonal argument to decimal expansions, of the uncountability of the unit interval. Such a proof needs to be able to declare certain pairs of real numbers to be different based on their decimal expansions, so one needs to avoid pairs like 0.2 and 0.1999... A simple method represents all numbers with nonterminating expansions; the opposite method rules out repeating nines.[33] A variant that may be closer to Cantor's original argument actually uses base 2, and by turning base-3 expansions into base-2 expansions, one can prove the uncountability of the Cantor set as well.[34]

11.6 Skepticism in education

Students of mathematics often reject the equality of 0.999... and 1, for reasons ranging from their disparate appearance to deep misgivings over the limit concept and disagreements over the nature of infinitesimals. There are many common contributing factors to the confusion:

- Students are often "mentally committed to the notion that a number can be represented in one and only one way by a decimal." Seeing two manifestly different decimals representing the same number appears to be a paradox, which is amplified by the appearance of the seemingly well-understood number 1.[35]

- Some students interpret "0.999..." (or similar notation) as a large but finite string of 9s, possibly with a variable, unspecified length. If they accept an infinite string of nines, they may still expect a last 9 "at infinity".[36]

- Intuition and ambiguous teaching lead students to think of the limit of a sequence as a kind of infinite process rather than a fixed value, since a sequence need not reach its limit. Where students accept the difference between a sequence of numbers and its limit, they might read "0.999..." as meaning the sequence rather than its limit.[37]

These ideas are mistaken in the context of the standard real numbers, although some may be valid in other number systems, either invented for their general mathematical utility or as instructive counterexamples to better understand 0.999...

Many of these explanations were found by David Tall, who has studied characteristics of teaching and cognition that lead to some of the misunderstandings he has encountered in his college students. Interviewing his students to determine why the vast majority initially rejected the equality, he found that "students continued to conceive of 0.999... as a sequence of numbers getting closer and closer to 1 and not a fixed value, because 'you haven't specified how many places there are' or 'it is the nearest possible decimal below 1'".[38]

Of the elementary proofs, multiplying 0.333... = $\frac{1}{3}$ by 3 is apparently a successful strategy for convincing reluctant students that 0.999... = 1. Still, when confronted with the conflict between their belief of the first equation and their disbelief of the second, some students either begin to disbelieve the first equation or simply become frustrated.[39] Nor are more sophisticated methods foolproof: students who are fully capable of applying rigorous definitions may still fall back on intuitive images when they are surprised by a result in advanced mathematics, including 0.999.... For example, one real analysis student was able to prove that 0.333... = $\frac{1}{3}$ using a supremum definition, but then insisted that 0.999... < 1 based on her earlier understanding of long division.[40] Others still are able to prove that $\frac{1}{3}$ = 0.333..., but, upon being confronted by the fractional proof, insist that "logic" supersedes the mathematical calculations.

Joseph Mazur tells the tale of an otherwise brilliant calculus student of his who "challenged almost everything I said in class but never questioned his calculator," and who had come to believe that nine digits are all one needs to do mathematics, including calculating the square root of 23. The student remained uncomfortable with a limiting argument that 9.99... = 10, calling it a "wildly imagined infinite growing process."[41]

As part of Ed Dubinsky's APOS theory of mathematical learning, he and his collaborators (2005) propose that students who conceive of 0.999... as a finite, indeterminate string with an infinitely small distance from 1 have "not yet constructed a complete process conception of the infinite decimal". Other students who have a complete process conception of 0.999... may not yet be able to "encapsulate" that process into an "object conception", like the object conception they have of 1, and so they view the process 0.999... and the object 1 as incompatible. Dubinsky *et al.* also link this mental ability of encapsulation to viewing $\frac{1}{3}$ as a number in its own right and to dealing with the set of natural numbers as a whole.[42]

11.7 In popular culture

With the rise of the Internet, debates about 0.999... have escaped the classroom and are commonplace on newsgroups and message boards, including many that nominally have little to do with mathematics. In the newsgroup sci.math, arguing over 0.999... is described as a "popular sport", and it is one of the questions answered in its FAQ.[43] The FAQ briefly covers $\frac{1}{3}$, multiplication by 10, and limits, and it alludes to Cauchy sequences as well.

A 2003 edition of the general-interest newspaper column *The Straight Dope* discusses 0.999... via $\frac{1}{3}$ and limits, saying of misconceptions,

> The lower primate in us still resists, saying: .999~ doesn't really represent a *number*, then, but a *process*. To find a number we have to halt the process, at which point the .999~ = 1 thing falls apart. Nonsense.[44]

The Straight Dope cites a discussion on its own message board that grew out of an unidentified "other message board ... mostly about video games". In the same vein, the question of 0.999... proved such a popular topic in the first seven years of Blizzard Entertainment's Battle.net forums that the company issued a "press release" on April Fools' Day 2004 that it is 1:

> We are very excited to close the book on this subject once and for all. We've witnessed the heartache and concern over whether .999~ does or does not equal 1, and we're proud that the following proof finally and conclusively addresses the issue for our customers.[45]

Two proofs are then offered, based on limits and multiplication by 10.

0.999... features also in mathematical folklore, specifically in the following joke:[46]

> Q: How many mathematicians does it take to screw in a lightbulb?
> A: 0.999999....

11.8 In alternative number systems

Although the real numbers form an extremely useful number system, the decision to interpret the notation "0.999..." as naming a real number is ultimately a convention, and Timothy Gowers argues in *Mathematics: A Very Short Introduction* that the resulting identity 0.999... = 1 is a convention as well:

> However, it is by no means an arbitrary convention, because not adopting it forces one either to invent strange new objects or to abandon some of the familiar rules of arithmetic.[47]

One can define other number systems using different rules or new objects; in some such number systems, the above proofs would need to be reinterpreted and one might find that, in a given number system, 0.999... and 1 might not be identical. However, many number systems are extensions of —rather than independent alternatives to— the real number system, so 0.999... = 1 continues to hold. Even in such number systems, though, it is worthwhile to examine alternative number systems, not only for how 0.999... behaves (if, indeed, a number expressed as "0.999..." is both meaningful and unambiguous), but also for the behavior of related phenomena. If such phenomena differ from those in the real number system, then at least one of the assumptions built into the system must break down.

11.8.1 Infinitesimals

Main article: Infinitesimal

Some proofs that 0.999... = 1 rely on the Archimedean property of the real numbers: that there are no nonzero infinitesimals. Specifically, the difference 1 − 0.999... must be smaller than any positive rational number, so it must be an infinitesimal; but since the reals do not contain nonzero infinitesimals, the difference is therefore zero, and therefore the two values are the same.

However, there are mathematically coherent ordered algebraic structures, including various alternatives to the real numbers, which are non-Archimedean. Non-standard analysis provides a number system with a full array of infinitesimals (and their inverses).[48] A. H. Lightstone developed a decimal expansion for hyperreal numbers in $(0, 1)^*$.[49] Lightstone shows how to associate to each number a sequence of digits,

$$0.d_1 d_2 d_3 \ldots; \ldots d_{\infty-1} d_\infty d_{\infty+1} \ldots ,$$

indexed by the hypernatural numbers. While he does not directly discuss 0.999..., he shows the real number 1/3 is represented by 0.333...;...333... which is a consequence of the transfer principle. As a consequence the number 0.999...;...999... = 1. With this type of decimal representation, not every expansion represents a number. In particular "0.333...;...000..." and "0.999...;...000..." do not correspond to any number.

The standard definition of the number 0.999... is the limit of the sequence 0.9, 0.99, 0.999, ... A different definition involves what Terry Tao refers to as *ultralimit*, i.e., the equivalence class [(0.9, 0.99, 0.999, ...)] of this sequence in the ultrapower construction, which is a number that falls short of 1 by an infinitesimal amount. More generally, the hyperreal number uH=0.999...;...999000..., with last digit 9 at infinite hypernatural rank H, satisfies a strict inequality $uH < 1$. Accordingly, an alternative interpretation for "zero followed by infinitely many 9s" could be

$$0.\underbrace{999\ldots}_{H} = 1 - \frac{1}{10^H}.{}^{[50]}$$

All such interpretations of "0.999..." are infinitely close to 1. Ian Stewart characterizes this interpretation as an "entirely reasonable" way to rigorously justify the intuition that "there's a little bit missing" from 1 in 0.999....[51] Along with Katz & Katz, Robert Ely also questions the assumption that students' ideas about 0.999... < 1 are erroneous intuitions about the real numbers, interpreting them rather as *nonstandard* intuitions that could be valuable in the learning of calculus.[52][53] Jose Benardete in his book *Infinity: An essay in metaphysics* argues that some natural pre-mathematical intuitions cannot be expressed if one is limited to an overly restrictive number system:

> The intelligibility of the continuum has been found—many times over—to require that the domain of real numbers be enlarged to include infinitesimals. This enlarged domain may be styled the domain of continuum numbers. It will now be evident that .9999... does not equal 1 but falls infinitesimally short of it. I think that .9999... should indeed be admitted as a *number* ... though not as a *real* number.[54]

11.8.2 Hackenbush

Combinatorial game theory provides alternative reals as well, with infinite Blue-Red Hackenbush as one particularly relevant example. In 1974, Elwyn Berlekamp described a correspondence between Hackenbush strings and binary expansions of real numbers, motivated by the idea of data compression. For example, the value of the Hackenbush string LRRLRLRL... is $0.010101_2... = 1/3$. However, the value of LRLLL... (corresponding to $0.111..._2$) is infinitesimally less than 1. The difference between the two is the surreal number $1/\omega$, where ω is the first infinite ordinal; the relevant game is LRRRR... or $0.000..._2$.[55]

This is in fact true of the binary expansions of many rational numbers, where the values of the numbers are equal but the corresponding binary tree paths are different. For example, $0.10111..._2 = 0.11000..._2$, which are both equal to $3/4$, but the first representation corresponds to the binary tree path LRLRLLL... while the second corresponds to the different path LRLLRRR....

11.8.3 Revisiting subtraction

Another manner in which the proofs might be undermined is if $1 - 0.999...$ simply does not exist, because subtraction is not always possible. Mathematical structures with an addition operation but not a subtraction operation include commutative semigroups, commutative monoids and semirings. Richman considers two such systems, designed so that $0.999... < 1$.

First, Richman defines a nonnegative *decimal number* to be a literal decimal expansion. He defines the lexicographical order and an addition operation, noting that $0.999... < 1$ simply because $0 < 1$ in the ones place, but for any nonterminating x, one has $0.999... + x = 1 + x$. So one peculiarity of the decimal numbers is that addition cannot always be cancelled; another is that no decimal number corresponds to $1/3$. After defining multiplication, the decimal numbers form a positive, totally ordered, commutative semiring.[56]

In the process of defining multiplication, Richman also defines another system he calls "cut D", which is the set of Dedekind cuts of decimal fractions. Ordinarily this definition leads to the real numbers, but for a decimal fraction d he allows both the cut $(-\infty, d)$ and the "principal cut" $(-\infty, d)$. The result is that the real numbers are "living uneasily together with" the decimal fractions. Again $0.999... < 1$. There are no positive infinitesimals in cut D, but there is "a sort of negative infinitesimal," 0^-, which has no decimal expansion. He concludes that $0.999... = 1 + 0^-$, while the equation "$0.999... + x = 1$" has no solution.[57]

11.8.4 *p*-adic numbers

Main article: p-adic number

When asked about 0.999..., novices often believe there should be a "final 9," believing $1 - 0.999...$ to be a positive number which they write as "0.000...1". Whether or not that makes sense, the intuitive goal is clear: adding a 1 to the final 9 in 0.999... would carry all the 9s into 0s and leave a 1 in the ones place. Among other reasons, this idea fails because there is no "final 9" in 0.999....[58] However, there is a system that contains an infinite string of 9s including a last 9.

The *p*-adic numbers are an alternative number system of interest in number theory. Like the real numbers, the *p*-adic numbers can be built from the rational numbers via Cauchy sequences; the construction uses a different metric in which 0 is closer to p, and much closer to p^n, than it is to 1. The *p*-adic numbers form a field for prime p and a ring for other p, including 10. So arithmetic can be performed in the *p*-adics, and there are no infinitesimals.

In the 10-adic numbers, the analogues of decimal expansions run to the left. The 10-adic expansion ...999 does have a last 9, and it does not have a first 9. One can add 1 to the ones place, and it leaves behind only 0s after carrying through: $1 + ...999 = ...000 = 0$, and so $...999 = -1$.[59] Another derivation uses a geometric series. The infinite series implied by "...999" does not converge in the real numbers, but it converges in the 10-adics, and so one can re-use the familiar formula:

$$...999 = 9 + 9(10) + 9(10)^2 + 9(10)^3 + \cdots = \frac{9}{1-10} = -1.$$ [60]

(Compare with the series above.) A third derivation was invented by a seventh-grader who was doubtful over her teacher's limiting argument that 0.999... = 1 but was inspired to take the multiply-by-10 proof above in the opposite direction: if $x = \ldots 999$ then $10x = \ldots 990$, so $10x = x - 9$, hence $x = -1$ again.[59]

As a final extension, since 0.999... = 1 (in the reals) and ...999 = −1 (in the 10-adics), then by "blind faith and unabashed juggling of symbols"[61] one may add the two equations and arrive at ...999.999... = 0. This equation does not make sense either as a 10-adic expansion or an ordinary decimal expansion, but it turns out to be meaningful and true if one develops a theory of "double-decimals" with eventually repeating left ends to represent a familiar system: the real numbers.[62]

11.9 Related questions

- Zeno's paradoxes, particularly the paradox of the runner, are reminiscent of the apparent paradox that 0.999... and 1 are equal. The runner paradox can be mathematically modelled and then, like 0.999..., resolved using a geometric series. However, it is not clear if this mathematical treatment addresses the underlying metaphysical issues Zeno was exploring.[63]

- Division by zero occurs in some popular discussions of 0.999..., and it also stirs up contention. While most authors choose to define 0.999..., almost all modern treatments leave division by zero undefined, as it can be given no meaning in the standard real numbers. However, division by zero is defined in some other systems, such as complex analysis, where the extended complex plane, i.e. the Riemann sphere, has a "point at infinity". Here, it makes sense to define $1/0$ to be infinity;[64] and, in fact, the results are profound and applicable to many problems in engineering and physics. Some prominent mathematicians argued for such a definition long before either number system was developed.[65]

- Negative zero is another redundant feature of many ways of writing numbers. In number systems, such as the real numbers, where "0" denotes the additive identity and is neither positive nor negative, the usual interpretation of "−0" is that it should denote the additive inverse of 0, which forces −0 = 0.[66] Nonetheless, some scientific applications use separate positive and negative zeroes, as do some computing binary number systems (for example integers stored in the sign and magnitude or ones' complement formats, or floating point numbers as specified by the IEEE floating-point standard).[67][68]

11.10 See also

- Limit (mathematics)
- Series (mathematics)
- Naive mathematics
- Finitism

11.11 Notes

[1] This argument is found in Peressini and Peressini p. 186. William Byers argues that a student who agrees that 0.999... = 1 because of the above proofs, but hasn't resolved the ambiguity, doesn't really understand the equation (Byers pp. 39–41). Fred Richman argues that the first argument "gets its force from the fact that most people have been indoctrinated to accept the first equation without thinking".(p. 396)

[2] Rudin p. 61, Theorem 3.26; J. Stewart p. 706

[3] Euler p. 170

[4] Grattan-Guinness p. 69; Bonnycastle p. 177

11.11. NOTES

[5] For example, J. Stewart p. 706, Rudin p. 61, Protter and Morrey p. 213, Pugh p. 180, J.B. Conway p. 31

[6] The limit follows, for example, from Rudin p. 57, Theorem 3.20e. For a more direct approach, see also Finney, Weir, Giordano (2001) *Thomas' Calculus: Early Transcendentals* 10ed, Addison-Wesley, New York. Section 8.1, example 2(a), example 6(b).

[7] Davies p. 175; Smith and Harrington p. 115

[8] Beals p. 22; I. Stewart p. 34

[9] Bartle and Sherbert pp. 60–62; Pedrick p. 29; Sohrab p. 46

[10] Apostol pp. 9, 11–12; Beals p. 22; Rosenlicht p. 27

[11] Apostol p. 12

[12] The historical synthesis is claimed by Griffiths and Hilton (p.xiv) in 1970 and again by Pugh (p. 10) in 2001; both actually prefer Dedekind cuts to axioms. For the use of cuts in textbooks, see Pugh p. 17 or Rudin p. 17. For viewpoints on logic, Pugh p. 10, Rudin p.ix, or Munkres p. 30

[13] Enderton (p. 113) qualifies this description: "The idea behind Dedekind cuts is that a real number x can be named by giving an infinite set of rationals, namely all the rationals less than x. We will in effect define x to be the set of rationals smaller than x. To avoid circularity in the definition, we must be able to characterize the sets of rationals obtainable in this way..."

[14] Rudin pp. 17–20, Richman p. 399, or Enderton p. 119. To be precise, Rudin, Richman, and Enderton call this cut 1^*, 1^-, and $1R$, respectively; all three identify it with the traditional real number 1. Note that what Rudin and Enderton call a Dedekind cut, Richman calls a "nonprincipal Dedekind cut".

[15] Richman p. 399

[16] O'Connor, J. J.; Robertson, E. F. (October 2005). "History topic: The real numbers: Stevin to Hilbert". *MacTutor History of Mathematics*. Archived from the original on 2007-09-29. Retrieved 2006-08-30.

[17] "Is 0.999... = 1?".

[18] Richman

[19] Richman pp. 398–399

[20] Griffiths & Hilton §24.2 "Sequences" p. 386

[21] Griffiths & Hilton pp. 388, 393

[22] Griffiths & Hilton p. 395

[23] Griffiths & Hilton pp.viii, 395

[24] Liangpan Li (March 2011). "A new approach to the real numbers". arXiv:1101.1800 [math.CA].

[25] Petkovšek p. 408

[26] Protter and Morrey p. 503; Bartle and Sherbert p. 61

[27] Komornik and Loreti p. 636

[28] Kempner p. 611; Petkovšek p. 409

[29] Petkovšek pp. 410–411

[30] Leavitt 1984 p. 301

[31] Lewittes pp. 1–3; Leavitt 1967 pp. 669, 673; Shrader-Frechette pp. 96–98

[32] Pugh p. 97; Alligood, Sauer, and Yorke pp. 150–152. Protter and Morrey (p. 507) and Pedrick (p. 29) assign this description as an exercise.

[33] Maor (p. 60) and Mankiewicz (p. 151) review the former method; Mankiewicz attributes it to Cantor, but the primary source is unclear. Munkres (p. 50) mentions the latter method.

[34] Rudin p. 50, Pugh p. 98

[35] Bunch p. 119; Tall and Schwarzenberger p. 6. The last suggestion is due to Burrell (p. 28): "Perhaps the most reassuring of all numbers is 1 ... So it is particularly unsettling when someone tries to pass off 0.9~ as 1."

[36] Tall and Schwarzenberger pp. 6–7; Tall 2000 p. 221

[37] Tall and Schwarzenberger p. 6; Tall 2000 p. 221

[38] Tall 2000 p. 221

[39] Tall 1976 pp. 10–14

[40] Pinto and Tall p. 5, Edwards and Ward pp. 416–417

[41] Mazur pp. 137–141

[42] Dubinsky *et al.* pp. 261–262

[43] As observed by Richman (p. 396). de Vreught, Hans (1994). "sci.math FAQ: Why is 0.9999... = 1?". Archived from the original on 2007-09-29. Retrieved 2006-06-29.

[44] Adams, Cecil (2003-07-11). "An infinite question: Why doesn't .999~ = 1?". *The Straight Dope*. Chicago Reader. Archived from the original on 15 August 2006. Retrieved 2006-09-06.

[45] "Blizzard Entertainment Announces .999~ (Repeating) = 1" (Press release). Blizzard Entertainment. 2004-04-01. Archived from the original on 4 November 2009. Retrieved 2009-11-16.

[46] Renteln and Dundes, p. 27

[47] Gowers p. 60

[48] For a full treatment of non-standard numbers see for example Robinson's *Non-standard Analysis*.

[49] Lightstone pp. 245–247

[50] Katz & Katz 2010

[51] Stewart 2009, p. 175; the full discussion of 0.999... is spread through pp. 172–175.

[52] Katz & Katz (2010b)

[53] R. Ely (2010)

[54] Benardete, José Amado (1964). *Infinity: An essay in metaphysics*. Clarendon Press. p. 279. Retrieved 27 November 2011.

[55] Berlekamp, Conway, and Guy (pp. 79–80, 307–311) discuss 1 and $1/3$ and touch on $1/\omega$. The game for $0.111..._2$ follows directly from Berlekamp's Rule.

[56] Richman pp. 397–399

[57] Richman pp. 398–400. Rudin (p. 23) assigns this alternative construction (but over the rationals) as the last exercise of Chapter 1.

[58] Gardiner p. 98; Gowers p. 60

[59] Fjelstad p. 11

[60] Fjelstad pp. 14–15

[61] DeSua p. 901

[62] DeSua pp. 902–903

[63] Wallace p. 51, Maor p. 17

[64] See, for example, J.B. Conway's treatment of Möbius transformations, pp. 47–57

[65] Maor p. 54

[66] Munkres p. 34, Exercise 1(c)

[67] Kroemer, Herbert; Kittel, Charles (1980). *Thermal Physics* (2e ed.). W. H. Freeman. p. 462. ISBN 0-7167-1088-9.

[68] "Floating point types". *MSDN C# Language Specification*. Archived from the original on 24 August 2006. Retrieved 2006-08-29.

11.12 References

- Alligood, K. T.; Sauer, T. D.; Yorke, J. A. (1996). "4.1 Cantor Sets". *Chaos: An introduction to dynamical systems*. Springer. ISBN 0-387-94677-2.

  This introductory textbook on dynamical systems is aimed at undergraduate and beginning graduate students. (p. ix)

- Apostol, Tom M. (1974). *Mathematical analysis* (2e ed.). Addison-Wesley. ISBN 0-201-00288-4.

  A transition from calculus to advanced analysis, *Mathematical analysis* is intended to be "honest, rigorous, up to date, and, at the same time, not too pedantic." (pref.) Apostol's development of the real numbers uses the least upper bound axiom and introduces infinite decimals two pages later. (pp. 9–11)

- Bartle, R. G.; Sherbert, D. R. (1982). *Introduction to real analysis*. Wiley. ISBN 0-471-05944-7.

  This text aims to be "an accessible, reasonably paced textbook that deals with the fundamental concepts and techniques of real analysis." Its development of the real numbers relies on the supremum axiom. (pp. vii–viii)

- Beals, Richard (2004). *Analysis*. Cambridge UP. ISBN 0-521-60047-2.

- Berlekamp, E. R.; Conway, J. H.; Guy, R. K. (1982). *Winning Ways for your Mathematical Plays*. Academic Press. ISBN 0-12-091101-9.

- Berz, Martin (1992). *Automatic differentiation as nonarchimedean analysis*. Computer Arithmetic and Enclosure Methods. Elsevier. pp. 439–450. CiteSeerX: 10.1.1.31.3019.

- Beswick, Kim (2004). "Why Does 0.999... = 1?: A Perennial Question and Number Sense". *Australian Mathematics Teacher* **60** (4): 7–9.

- Bunch, Bryan H. (1982). *Mathematical fallacies and paradoxes*. Van Nostrand Reinhold. ISBN 0-442-24905-5.

  This book presents an analysis of paradoxes and fallacies as a tool for exploring its central topic, "the rather tenuous relationship between mathematical reality and physical reality". It assumes first-year high-school algebra; further mathematics is developed in the book, including geometric series in Chapter 2. Although 0.999... is not one of the paradoxes to be fully treated, it is briefly mentioned during a development of Cantor's diagonal method. (pp. ix-xi, 119)

- Burrell, Brian (1998). *Merriam-Webster's Guide to Everyday Math: A Home and Business Reference*. Merriam-Webster. ISBN 0-87779-621-1.

- Byers, William (2007). *How Mathematicians Think: Using Ambiguity, Contradiction, and Paradox to Create Mathematics*. Princeton UP. ISBN 0-691-12738-7.

- Conway, John B. (1978) [1973]. *Functions of one complex variable I* (2e ed.). Springer-Verlag. ISBN 0-387-90328-3.

  This text assumes "a stiff course in basic calculus" as a prerequisite; its stated principles are to present complex analysis as "An Introduction to Mathematics" and to state the material clearly and precisely. (p. vii)

- Davies, Charles (1846). *The University Arithmetic: Embracing the Science of Numbers, and Their Numerous Applications*. A.S. Barnes. Retrieved 4 July 2011.

- DeSua, Frank C. (November 1960). "A system isomorphic to the reals". *The American Mathematical Monthly* **67** (9): 900–903. doi:10.2307/2309468. JSTOR 2309468.

- Dubinsky, Ed; Weller, Kirk; McDonald, Michael; Brown, Anne (2005). "Some historical issues and paradoxes regarding the concept of infinity: an APOS analysis: part 2". *Educational Studies in Mathematics* **60** (2): 253–266. doi:10.1007/s10649-005-0473-0.

- Edwards, Barbara; Ward, Michael (May 2004). "Surprises from mathematics education research: Student (mis)use of mathematical definitions" (PDF). *The American Mathematical Monthly* **111** (5): 411–425. doi:10.2307/4145268. JSTOR 4145268. Archived from the original (PDF) on 22 July 2011. Retrieved 4 July 2011.

- Enderton, Herbert B. (1977). *Elements of set theory*. Elsevier. ISBN 0-12-238440-7.

    An introductory undergraduate textbook in set theory that "presupposes no specific background". It is written to accommodate a course focusing on axiomatic set theory or on the construction of number systems; the axiomatic material is marked such that it may be de-emphasized. (pp. xi–xii)

- Euler, Leonhard (1822) [1770]. John Hewlett and Francis Horner, English translators., ed. *Elements of Algebra* (3rd English ed.). Orme Longman. ISBN 0-387-96014-7. Retrieved 4 July 2011.

- Fjelstad, Paul (January 1995). "The repeating integer paradox". *The College Mathematics Journal* **26** (1): 11–15. doi:10.2307/2687285. JSTOR 2687285.

- Gardiner, Anthony (2003) [1982]. *Understanding Infinity: The Mathematics of Infinite Processes*. Dover. ISBN 0-486-42538-X.

- Gowers, Timothy (2002). *Mathematics: A Very Short Introduction*. Oxford UP. ISBN 0-19-285361-9.

- Grattan-Guinness, Ivor (1970). *The development of the foundations of mathematical analysis from Euler to Riemann*. MIT Press. ISBN 0-262-07034-0.

- Griffiths, H. B.; Hilton, P. J. (1970). *A Comprehensive Textbook of Classical Mathematics: A Contemporary Interpretation*. London: Van Nostrand Reinhold. ISBN 0-442-02863-6. LCC QA37.2 G75.

    This book grew out of a course for Birmingham-area grammar school mathematics teachers. The course was intended to convey a university-level perspective on school mathematics, and the book is aimed at students "who have reached roughly the level of completing one year of specialist mathematical study at a university". The real numbers are constructed in Chapter 24, "perhaps the most difficult chapter in the entire book", although the authors ascribe much of the difficulty to their use of ideal theory, which is not reproduced here. (pp. vii, xiv)

- Katz, K.; Katz, M. (2010a). "When is .999... less than 1?". *The Montana Mathematics Enthusiast* **7** (1): 3–30. arXiv:1007.3018. Bibcode:2010arXiv1007.3018U. Archived from the original on 20 July 2011. Retrieved 4 July 2011.

- Kempner, A. J. (December 1936). "Anormal Systems of Numeration". *The American Mathematical Monthly* **43** (10): 610–617. doi:10.2307/2300532. JSTOR 2300532.

- Komornik, Vilmos; Loreti, Paola (1998). "Unique Developments in Non-Integer Bases". *The American Mathematical Monthly* **105** (7): 636–639. doi:10.2307/2589246. JSTOR 2589246.

- Leavitt, W. G. (1967). "A Theorem on Repeating Decimals". *The American Mathematical Monthly* **74** (6): 669–673. doi:10.2307/2314251. JSTOR 2314251.

- Leavitt, W. G. (September 1984). "Repeating Decimals". *The College Mathematics Journal* **15** (4): 299–308. doi:10.2307/2686394. JSTOR 2686394.

11.12. REFERENCES

- Lightstone, A. H. (March 1972). "Infinitesimals". *The American Mathematical Monthly* 79(3): 242–251. JSTOR 2316619.

- Mankiewicz, Richard (2000). *The story of mathematics*. Cassell. ISBN 0-304-35473-2.

  Mankiewicz seeks to represent "the history of mathematics in an accessible style" by combining visual and qualitative aspects of mathematics, mathematicians' writings, and historical sketches. (p. 8)

- Maor, Eli (1987). *To infinity and beyond: a cultural history of the infinite*. Birkhäuser. ISBN 3-7643-3325-1.

  A topical rather than chronological review of infinity, this book is "intended for the general reader" but "told from the point of view of a mathematician". On the dilemma of rigor versus readable language, Maor comments, "I hope I have succeeded in properly addressing this problem." (pp. x-xiii)

- Mazur, Joseph (2005). *Euclid in the Rainforest: Discovering Universal Truths in Logic and Math*. Pearson: Pi Press. ISBN 0-13-147994-6.

- Munkres, James R. (2000) [1975]. *Topology* (2e ed.). Prentice-Hall. ISBN 0-13-181629-2.

  Intended as an introduction "at the senior or first-year graduate level" with no formal prerequisites: "I do not even assume the reader knows much set theory." (p. xi) Munkres' treatment of the reals is axiomatic; he claims of bare-hands constructions, "This way of approaching the subject takes a good deal of time and effort and is of greater logical than mathematical interest." (p. 30)

- Núñez, Rafael (2006). "Do Real Numbers Really Move? Language, Thought, and Gesture: The Embodied Cognitive Foundations of Mathematics". *18 Unconventional Essays on the Nature of Mathematics*. Springer. pp. 160–181. ISBN 978-0-387-25717-4. Archived from the original on 18 July 2011. Retrieved 4 July 2011.

- Pedrick, George (1994). *A First Course in Analysis*. Springer. ISBN 0-387-94108-8.

- Peressini, Anthony; Peressini, Dominic (2007). "Philosophy of Mathematics and Mathematics Education". In Bart van Kerkhove, Jean Paul van Bendegem. *Perspectives on Mathematical Practices*. Logic, Epistemology, and the Unity of Science **5**. Springer. ISBN 978-1-4020-5033-6.

- Petkovšek, Marko (May 1990). "Ambiguous Numbers are Dense". *American Mathematical Monthly* **97** (5): 408–411. doi:10.2307/2324393. JSTOR 2324393.

- Pinto, Márcia; Tall, David (2001). *PME25: Following students' development in a traditional university analysis course* (PDF). pp. v4: 57–64. Archived from the original (PDF) on 30 May 2009. Retrieved 2009-05-03.

- Protter, M. H.; Morrey, Jr., Charles B. (1991). *A first course in real analysis* (2e ed.). Springer. ISBN 0-387-97437-7.

  This book aims to "present a theoretical foundation of analysis that is suitable for students who have completed a standard course in calculus." (p. vii) At the end of Chapter 2, the authors assume as an axiom for the real numbers that bounded, nondecreasing sequences converge, later proving the nested intervals theorem and the least upper bound property. (pp. 56–64) Decimal expansions appear in Appendix 3, "Expansions of real numbers in any base". (pp. 503–507)

- Pugh, Charles Chapman (2001). *Real mathematical analysis*. Springer-Verlag. ISBN 0-387-95297-7.

  While assuming familiarity with the rational numbers, Pugh introduces Dedekind cuts as soon as possible, saying of the axiomatic treatment, "This is something of a fraud, considering that the entire structure of analysis is built on the real number system." (p. 10) After proving the least upper bound property and some allied facts, cuts are not used in the rest of the book.

- Renteln, Paul; Dundes, Allan (January 2005). "Foolproof: A Sampling of Mathematical Folk Humor" (PDF). *Notices of the AMS* **52** (1): 24–34. Archived from the original (PDF) on 25 February 2009. Retrieved 2009-05-03.

- Richman, Fred (December 1999). "Is 0.999... = 1?". *Mathematics Magazine volume=72* **72** (5): 396–400. doi:10.2307/2690798. JSTOR 2690798. Free HTML preprint: Richman, Fred (June 1999). "Is 0.999... = 1?". Archived from the original on 2 September 2006. Retrieved 2006-08-23. Note: the journal article contains material and wording not found in the preprint.

- Robinson, Abraham (1996). *Non-standard analysis* (Revised ed.). Princeton University Press. ISBN 0-691-04490-2.

- Rosenlicht, Maxwell (1985). *Introduction to Analysis*. Dover. ISBN 0-486-65038-3. This book gives a "careful rigorous" introduction to real analysis. It gives the axioms of the real numbers and then constructs them (pp. 27–31) as infinite decimals with 0.999... = 1 as part of the definition.

- Rudin, Walter (1976) [1953]. *Principles of mathematical analysis* (3e ed.). McGraw-Hill. ISBN 0-07-054235-X.

  A textbook for an advanced undergraduate course. "Experience has convinced me that it is pedagogically unsound (though logically correct) to start off with the construction of the real numbers from the rational ones. At the beginning, most students simply fail to appreciate the need for doing this. Accordingly, the real number system is introduced as an ordered field with the least-upper-bound property, and a few interesting applications of this property are quickly made. However, Dedekind's construction is not omitted. It is now in an Appendix to Chapter 1, where it may be studied and enjoyed whenever the time is ripe." (p. ix)

- Shrader-Frechette, Maurice (March 1978). "Complementary Rational Numbers". *Mathematics Magazine* **51** (2): 90–98. doi:10.2307/2690144. JSTOR 2690144.

- Smith, Charles; Harrington, Charles (1895). *Arithmetic for Schools*. Macmillan. ISBN 0-665-54808-7. Retrieved 4 July 2011.

- Sohrab, Houshang (2003). *Basic Real Analysis*. Birkhäuser. ISBN 0-8176-4211-0.

- Starbird, M.; Starbird, T. (March 1992). "Required Redundancy in the Representation of Reals". *Proceedings of the American Mathematical Society* (AMS) **114** (3): 769–774. doi:10.1090/S0002-9939-1992-1086343-5. JSTOR 2159403.

- Stewart, Ian (1977). *The Foundations of Mathematics*. Oxford UP. ISBN 0-19-853165-6.

- Stewart, Ian (2009). *Professor Stewart's Hoard of Mathematical Treasures*. Profile Books. ISBN 978-1-84668-292-6.

- Stewart, James (1999). *Calculus: Early transcendentals* (4e ed.). Brooks/Cole. ISBN 0-534-36298-2.

  This book aims to "assist students in discovering calculus" and "to foster conceptual understanding". (p. v) It omits proofs of the foundations of calculus.

- Tall, D. O.; Schwarzenberger, R. L. E. (1978). "Conflicts in the Learning of Real Numbers and Limits" (PDF). *Mathematics Teaching* **82**: 44–49. Archived from the original (PDF) on 30 May 2009. Retrieved 2009-05-03.

- Tall, David (1977). "Conflicts and Catastrophes in the Learning of Mathematics" (PDF). *Mathematical Education for Teaching* **2** (4): 2–18. Archived from the original (PDF) on 26 March 2009. Retrieved 2009-05-03.

- Tall, David (2000). "Cognitive Development In Advanced Mathematics Using Technology" (PDF). *Mathematics Education Research Journal* **12** (3): 210–230. Bibcode:2000MEdRJ..12..196T. doi:10.1007/BF03217085. Archived from the original (PDF) on 30 May 2009. Retrieved 2009-05-03.

- von Mangoldt, Dr. Hans (1911). "Reihenzahlen". *Einführung in die höhere Mathematik* (in German) (1st ed.). Leipzig: Verlag von S. Hirzel.

- Wallace, David Foster (2003). *Everything and more: a compact history of infinity*. Norton. ISBN 0-393-00338-8.

11.13 Further reading

- Burkov, S. E. (1987). "One-dimensional model of the quasicrystalline alloy". *Journal of Statistical Physics* **47** (3/4): 409. Bibcode:1987JSP....47..409B. doi:10.1007/BF01007518.

- Burn, Bob (March 1997). "81.15 A Case of Conflict".*The Mathematical Gazette***81**(490): 109–112. doi:10.2307/JSTOR 3618786.

- Calvert, J. B.; Tuttle, E. R.; Martin, Michael S.; Warren, Peter (February 1981). "The Age of Newton: An Intensive Interdisciplinary Course". *The History Teacher* **14** (2): 167–190. doi:10.2307/493261. JSTOR 493261.

- Choi, Younggi; Do, Jonghoon (November 2005). "Equality Involved in 0.999... and $(-8)^{1/3}$". *For the Learning of Mathematics* **25** (3): 13–15, 36. JSTOR 40248503.

- Choong, K. Y.; Daykin, D. E.; Rathbone, C. R. (April 1971). "Rational Approximations to π". *Mathematics of Computation* **25** (114): 387–392. doi:10.2307/2004936. JSTOR 2004936.

- Edwards, B. (1997). "An undergraduate student's understanding and use of mathematical definitions in real analysis". In Dossey, J., Swafford, J.O., Parmentier, M., Dossey, A.E. *Proceedings of the 19th Annual Meeting of the North American Chapter of the International Group for the Psychology of Mathematics Education* **1**. Columbus, OH: ERIC Clearinghouse for Science, Mathematics and Environmental Education. pp. 17–22.

- Eisenmann, Petr (2008). "Why is it not true that 0.999... < 1?" (PDF). *The Teaching of Mathematics* **11** (1): 35–40. Retrieved 4 July 2011.

- Ely, Robert (2010). "Nonstandard student conceptions about infinitesimals". *Journal for Research in Mathematics Education* **41** (2): 117–146.

  This article is a field study involving a student who developed a Leibnizian-style theory of infinitesimals to help her understand calculus, and in particular to account for 0.999... falling short of 1 by an infinitesimal 0.000...1.

- Ferrini-Mundy, J.; Graham, K. (1994). Kaput, J.; Dubinsky, E., eds. "Research in calculus learning: Understanding of limits, derivatives and integrals". *MAA Notes: Research issues in undergraduate mathematics learning* **33**: 31–45.

- Lewittes, Joseph (2006). "Midy's Theorem for Periodic Decimals". arXiv:math.NT/0605182.

- Katz, Karin Usadi; Katz, Mikhail G. (2010b). "Zooming in on infinitesimal 1 − .9.. in a post-triumvirate era". *Educational Studies in Mathematics* **74** (3): 259. arXiv:1003.1501. doi:10.1007/s10649-010-9239-4.

- Gardiner, Tony (June 1985). "Infinite processes in elementary mathematics: How much should we tell the children?". *The Mathematical Gazette* **69** (448): 77–87. doi:10.2307/3616921. JSTOR 3616921.

- Monaghan, John (December 1988). "Real Mathematics: One Aspect of the Future of A-Level". *The Mathematical Gazette* **72** (462): 276–281. doi:10.2307/3619940. JSTOR 3619940.

- Navarro, Maria Angeles; Carreras, Pedro Pérez (2010). "A Socratic methodological proposal for the study of the equality 0.999...=1" (PDF). *The Teaching of Mathematics* **13** (1): 17–34. Retrieved 4 July 2011.

- Przenioslo, Malgorzata (March 2004). "Images of the limit of function formed in the course of mathematical studies at the university".*Educational Studies in Mathematics***55**(1–3): 103–132. doi:10.1023/B:EDUC.0000017667.7

- Sandefur, James T. (February 1996). "Using Self-Similarity to Find Length, Area, and Dimension". *The American Mathematical Monthly* **103** (2): 107–120. doi:10.2307/2975103. JSTOR 2975103.

- Sierpińska, Anna (November 1987). "Humanities students and epistemological obstacles related to limits". *Educational Studies in Mathematics* **18** (4): 371–396. doi:10.1007/BF00240986. JSTOR 3482354.

- Szydlik, Jennifer Earles (May 2000). "Mathematical Beliefs and Conceptual Understanding of the Limit of a Function". *Journal for Research in Mathematics Education* **31** (3): 258–276. doi:10.2307/749807. JSTOR 749807.

- Tall, David O. (2009). "Dynamic mathematics and the blending of knowledge structures in the calculus". *ZDM Mathematics Education* **41** (4): 481–492. doi:10.1007/s11858-009-0192-6.

- Tall, David O. (May 1981). "Intuitions of infinity". *Mathematics in School* **10** (3): 30–33. JSTOR 30214290.

11.14 External links

- .999999... = 1? from cut-the-knot
- Why does 0.9999... = 1 ?
- Proof of the equality based on arithmetic
- David Tall's research on mathematics cognition
- What is so wrong with thinking of real numbers as infinite decimals?
- Theorem 0.999... on Metamath

11.14. EXTERNAL LINKS

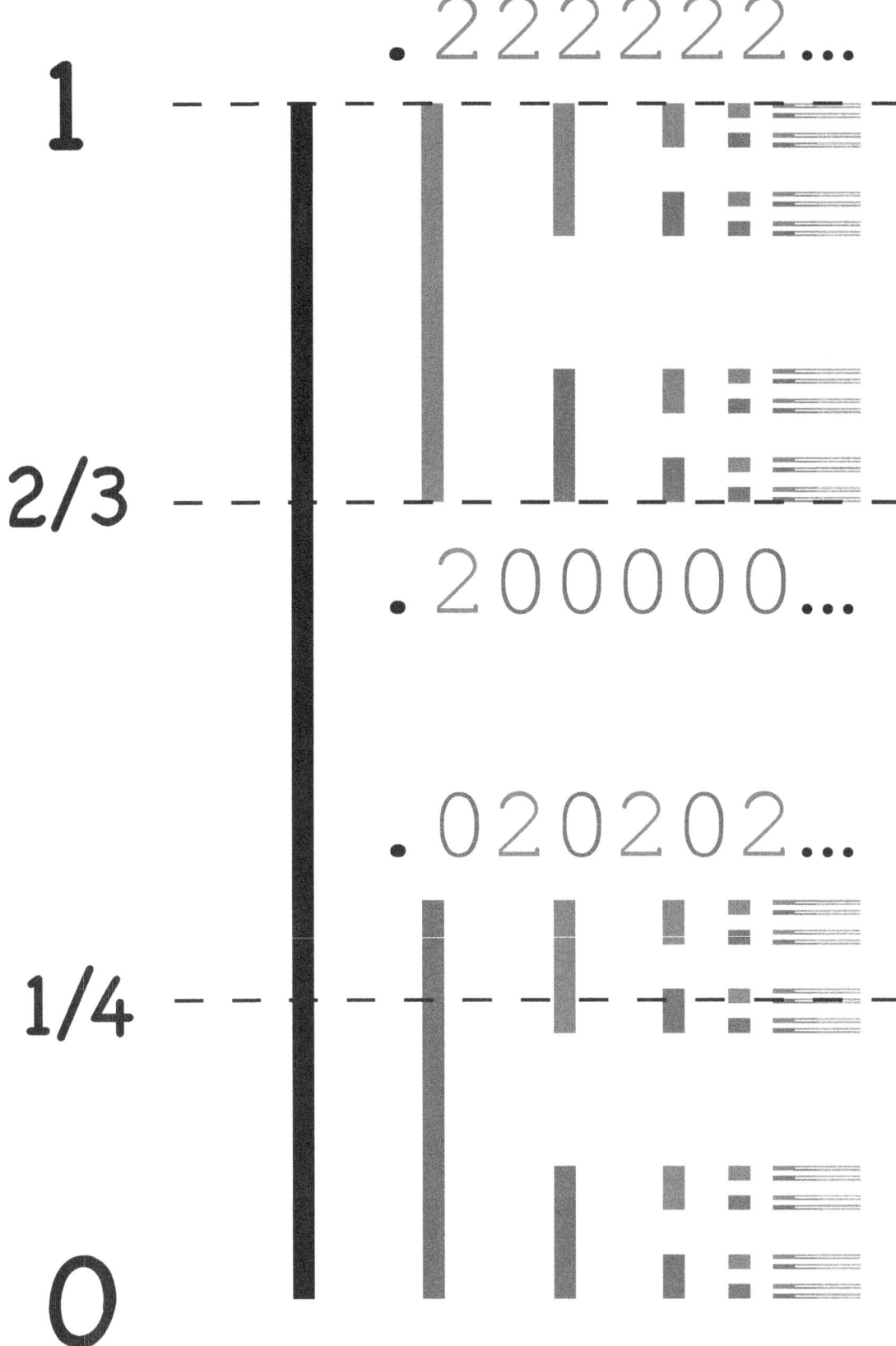

Positions of $\frac{1}{4}$, $\frac{2}{3}$, and 1 in the Cantor set

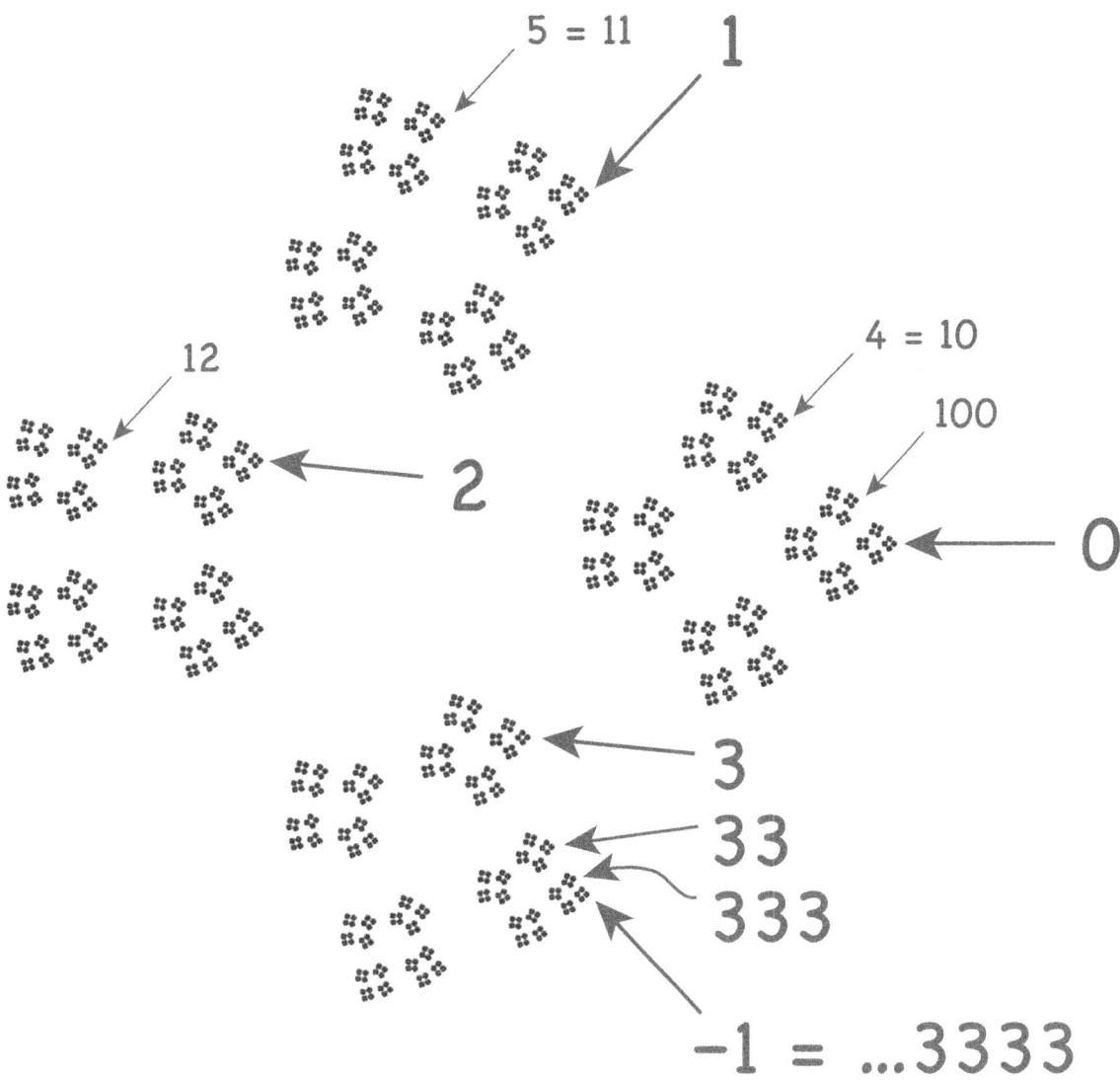

The 4-adic integers (black points), including the sequence (3, 33, 333, ...) converging to −1. The 10-adic analogue is ...999 = −1.

Chapter 12

Proof that 22/7 exceeds π

Proofs of the famous mathematical result that the rational number 22/7 is greater than π (pi) date back to antiquity. One of these proofs, more recently developed but requiring only elementary techniques from calculus, has attracted attention in modern mathematics due to its mathematical elegance and its connections to the theory of diophantine approximations. Stephen Lucas calls this proof, "One of the more beautiful results related to approximating π".[1] Julian Havil ends a discussion of continued fraction approximations of π with the result, describing it as "impossible to resist mentioning" in that context.[2]

The purpose of the proof is not primarily to convince its readers that 22/7 (or $3\frac{1}{7}$) is indeed bigger than π; systematic methods of computing the value of π exist. If one knows that π is approximately 3.14159, then it trivially follows that π < 22/7, which is approximately 3.142857. But it takes much less work to show that π < 22/7 by the method used in this proof than to show that π is approximately 3.14159.

12.1 Background

22/7 is a widely used Diophantine approximation of π. It is a convergent in the simple continued fraction expansion of π. It is greater than π, as can be readily seen in the decimal expansions of these values:

$$\frac{22}{7} = 3.\overline{142\,857},$$
$$\pi = 3.141\,592\,65\ldots$$

The approximation has been known since antiquity. Archimedes wrote the first known proof that 22/7 is an overestimate in the 3rd century BCE, although he may not have been the first to use that approximation. His proof proceeds by showing that 22/7 is greater than the ratio of the perimeter of a circumscribed regular polygon with 96 sides to the diameter of the circle. Another rational approximation of π that is far more accurate is 355/113.

12.2 The proof

The proof can be expressed very succinctly:

$$0 < \int_0^1 \frac{x^4(1-x)^4}{1+x^2}\,dx = \frac{22}{7} - \pi.$$

Therefore 22/7 > π.

The evaluation of this integral was the first problem in the 1968 Putnam Competition.[3] It is easier than most Putnam Competition problems, but the competition often features seemingly obscure problems that turn out to refer to something very familiar. This integral also has been used in the entrance examinations for the Indian Institutes of Technology.[4]

12.3 Details of evaluation of the integral

That the integral is positive follows from the fact that the integrand is a quotient whose numerator and denominator are both non-negative, being sums or products of powers of non-negative real numbers. Since the integrand is positive, the integral from 0 to 1 is positive because the lower limit of integration is less than the upper limit of integration.

It remains to show that the integral in fact evaluates to the desired quantity:

$$0 < \int_0^1 \frac{x^4(1-x)^4}{1+x^2} dx$$

$$= \int_0^1 \frac{x^4 - 4x^5 + 6x^6 - 4x^7 + x^8}{1+x^2} dx \quad \text{numerator) the in terms of (expansion}$$

$$= \int_0^1 \left(x^6 - 4x^5 + 5x^4 - 4x^2 + 4 - \frac{4}{1+x^2} \right) dx$$

division) long (polynomial

$$= \left(\frac{x^7}{7} - \frac{2x^6}{3} + x^5 - \frac{4x^3}{3} + 4x - 4\arctan x \right) \Big|_0^1 \quad \text{integration) (definite}$$

$$= \frac{1}{7} - \frac{2}{3} + 1 - \frac{4}{3} + 4 - \pi \quad (\text{ since } \arctan(1) = \pi/4 \text{ and } \arctan(0) = 0)$$

$$= \frac{22}{7} - \pi. \quad \text{(addition)}$$

(See polynomial long division.)

12.4 Quick upper and lower bounds

In Dalzell (1944), it is pointed out that if 1 is substituted for x in the denominator, one gets a lower bound on the integral, and if 0 is substituted for x in the denominator, one gets an upper bound:[5]

$$\frac{1}{1260} = \int_0^1 \frac{x^4(1-x)^4}{2} dx < \int_0^1 \frac{x^4(1-x)^4}{1+x^2} dx < \int_0^1 \frac{x^4(1-x)^4}{1} dx = \frac{1}{630}.$$

Thus we have

$$\frac{22}{7} - \frac{1}{630} < \pi < \frac{22}{7} - \frac{1}{1260},$$

hence $3.1412 < \pi < 3.1421$ in decimal expansion. The bounds deviate by less than 0.015% from π. See also Dalzell (1971).[6]

12.5 Proof that 355/113 exceeds π

As discussed in Lucas (2005), the well-known Diophantine approximation and far better upper estimate 355/113 for π follows from the relation

$$0 < \int_0^1 \frac{x^8(1-x)^8(25+816x^2)}{3164(1+x^2)}\, dx = \frac{355}{113} - \pi.$$

Note that

$$\frac{355}{113} = 3.141\,592\,92\ldots,$$

where the first six digits after the period agree with those of π. Substituting 1 for x in the denominator, we get the lower bound

$$\int_0^1 \frac{x^8(1-x)^8(25+816x^2)}{6328}\, dx = \frac{911}{5\,261\,111\,856} = 0.000\,000\,173\ldots,$$

substituting 0 for x in the denominator, we get twice this value as an upper bound, hence

$$\frac{355}{113} - \frac{911}{2\,630\,555\,928} < \pi < \frac{355}{113} - \frac{911}{5\,261\,111\,856}.$$

In decimal expansion, this means $\mathbf{3.141\,592}\,57 < \pi < \mathbf{3.141\,592}\,74$, where the bold digits of the lower and upper bound are those of π.

12.6 Extensions

The above ideas can be generalized to get better approximations of π; see also Backhouse (1995)[7] and Lucas (2005) (in both references, however, no calculations are given). For explicit calculations, consider, for every integer $n \geq 1$,

$$\frac{1}{2^{2n-1}}\int_0^1 x^{4n}(1-x)^{4n}\, dx < \frac{1}{2^{2n-2}}\int_0^1 \frac{x^{4n}(1-x)^{4n}}{1+x^2}\, dx < \frac{1}{2^{2n-2}}\int_0^1 x^{4n}(1-x)^{4n}\, dx,$$

where the middle integral evaluates to

$$\frac{1}{2^{2n-2}}\int_0^1 \frac{x^{4n}(1-x)^{4n}}{1+x^2}\, dx$$
$$= \sum_{j=0}^{2n-1} \frac{(-1)^j}{2^{2n-j-2}(8n-j-1)\binom{8n-j-2}{4n+j}} + (-1)^n\left(\pi - 4\sum_{j=0}^{3n-1}\frac{(-1)^j}{2j+1}\right)$$

involving π. The last sum also appears in Leibniz' formula for π. The correction term and error bound is given by

$$\frac{1}{2^{2n-1}}\int_0^1 x^{4n}(1-x)^{4n}\, dx = \frac{1}{2^{2n-1}(8n+1)\binom{8n}{4n}}$$
$$\sim \frac{\sqrt{\pi n}}{2^{10n-2}(8n+1)},$$

where the approximation (the tilde means that the quotient of both sides tends to one for large n) of the central binomial coefficient follows from Stirling's formula and shows the fast convergence of the integrals to π.

Calculation of these integrals

The results for $n = 1$ are given above. For $n = 2$ we get

$$\frac{1}{4}\int_0^1 \frac{x^8(1-x)^8}{1+x^2}\,dx = \pi - \frac{47\,171}{15\,015}$$

and

$$\frac{1}{8}\int_0^1 x^8(1-x)^8\,dx = \frac{1}{1\,750\,320},$$

hence **3.141 592** 31 $< \pi <$ **3.141 592** 89, where the bold digits of the lower and upper bound are those of π. Similarly for $n = 3$,

$$\frac{1}{16}\int_0^1 \frac{x^{12}(1-x)^{12}}{1+x^2}\,dx = \frac{431\,302\,721}{137\,287\,920} - \pi$$

with correction term and error bound

$$\frac{1}{32}\int_0^1 x^{12}(1-x)^{12}\,dx = \frac{1}{2\,163\,324\,800},$$

hence **3.141 592 653** 40 $< \pi <$ **3.141 592 653** 87. The next step for $n = 4$ is

$$\frac{1}{64}\int_0^1 \frac{x^{16}(1-x)^{16}}{1+x^2}\,dx = \pi - \frac{741\,269\,838\,109}{235\,953\,517\,800}$$

with

$$\frac{1}{128}\int_0^1 x^{16}(1-x)^{16}\,dx = \frac{1}{2\,538\,963\,567\,360},$$

which gives **3.141 592 653 589** 55 $< \pi <$ **3.141 592 653 589** 96.

12.7 See also

- Approximations of π
- Chronology of computation of π
- Proof that π is irrational
- Lindemann–Weierstrass theorem (proof that π is transcendental)
- List of topics related to π

12.8 References

[1] Lucas, Stephen (2005), "Integral proofs that 355/113 > π" (PDF), *Australian Mathematical Society Gazette* **32** (4): 263–266, MR 2176249, Zbl 1181.11077

[2] Havil, Julian (2003), *Gamma. Exploring Euler's Constant*, Princeton, NJ: Princeton University Press, p. 96, ISBN 0-691-09983-9, MR 1968276, Zbl 1023.11001

[3] Alexanderson, Gerald L.; Klosinski, Leonard F.; Larson, Loren C. (editors) (1985), *The William Lowell Putnam Mathematical Competition: Problems and Solutions: 1965–1984*, Washington, D.C.: The Mathematical Association of America, ISBN 0-88385-463-5, Zbl 0584.00003

[4] 2010 IIT Joint Entrance Exam, question 38 on page 15 of the mathematics section.

[5] Dalzell, D. P. (1944), "On 22/7", *Journal of the London Mathematical Society* **19**: 133–134, doi:10.1112/jlms/19.75_part_3.133, MR 0013425, Zbl 0060.15306.

[6] Dalzell, D. P. (1971), "On 22/7 and 355/113", *Eureka; the Archimedeans' Journal* **34**: 10–13, ISSN 0071-2248.

[7] Backhouse, Nigel (July 1995), "Note 79.36, Pancake functions and approximations to π", *The Mathematical Gazette* **79** (485): 371–374, JSTOR 3618318

12.9 External links

- The problems of the 1968 Putnam competition, with this proof listed as question A1.
- The Life of Pi by Jonathan Borwein—see page 5 for this integral.

Chapter 13

Proof that e is irrational

The number e was introduced by Jacob Bernoulli in 1683. More than half a century later, Euler, who had been a student of Jacob's younger brother Johann, proved that e is irrational, that is, that it can not be expressed as the quotient of two integers.

13.1 Euler's proof

Euler wrote the first proof of the fact that e is irrational in 1737 (but the text was only published seven years later).[1][2][3] He computed the representation of e as a simple continued fraction, which is

$$e = [2; 1, 2, 1, 1, 4, 1, 1, 6, 1, 1, 8, 1, 1, \ldots, 2n, 1, 1, \ldots].$$

Since this continued fraction is infinite and rational numbers can't be written as infinite continued fractions, e is irrational. A short proof of the previous equality is known.[4] Since the simple continued fraction of e is not periodic, this also proves that e is not a root of second degree polynomial with rational coefficients; in particular, e^2 is irrational.

13.2 Fourier's proof

The most well-known proof is Joseph Fourier's proof by contradiction,[5] which is based upon the equality

$$e = \sum_{n=0}^{\infty} \frac{1}{n!}.$$

Initially e is assumed to be a rational number of the form a/b. Note that b couldn't be equal to one as e is not an integer. It can be shown using the above equality that e is strictly between 2 and 3.

$$\frac{1}{1} + \frac{1}{1} < e = \frac{1}{1} + \frac{1}{1} + \frac{1}{1 \cdot 2} + \frac{1}{1 \cdot 2 \cdot 3} + \ldots < \frac{1}{1} + \frac{1}{1} + \frac{1}{1 \cdot 2} + \frac{1}{1 \cdot 2 \cdot 2} + \ldots = 3$$

We then analyze a blown-up difference x of the series representing e and its strictly smaller b^{th} partial sum, which approximates the limiting value e. By choosing the magnifying factor to be the factorial of b, the fraction a/b and the bth partial sum are turned into integers, hence x must be a positive integer. However, the fast convergence of the series representation implies that the magnified approximation error x is still strictly smaller than 1. From this contradiction we deduce that e is irrational.

Suppose that e is a rational number. Then there exist positive integers a and b such that $e = a/b$. Define the number

$$x = b!\left(e - \sum_{n=0}^{b}\frac{1}{n!}\right)$$

To see that if e is rational, then x is an integer, substitute $e = a/b$ into this definition to obtain

$$x = b!\left(\frac{a}{b} - \sum_{n=0}^{b}\frac{1}{n!}\right) = a(b-1)! - \sum_{n=0}^{b}\frac{b!}{n!}.$$

The first term is an integer, and every fraction in the sum is actually an integer because $n \leq b$ for each term. Therefore x is an integer.

We now prove that $0 < x < 1$. First, to prove that x is strictly positive, we insert the above series representation of e into the definition of x and obtain

$$x = b!\left(\sum_{n=0}^{\infty}\frac{1}{n!} - \sum_{n=0}^{b}\frac{1}{n!}\right) = \sum_{n=b+1}^{\infty}\frac{b!}{n!} > 0,$$

because all the terms are strictly positive.

We now prove that $x < 1$. For all terms with $n \geq b+1$ we have the upper estimate

$$\frac{b!}{n!} = \frac{1}{(b+1)(b+2)\cdots(b+(n-b))} < \frac{1}{(b+1)^{n-b}}.$$

This inequality is strict for every $n \geq b+2$. Changing the index of summation to $k = n - b$ and using the formula for the infinite geometric series, we obtain

$$x = \sum_{n=b+1}^{\infty}\frac{b!}{n!} < \sum_{n=b+1}^{\infty}\frac{1}{(b+1)^{n-b}} = \sum_{k=1}^{\infty}\frac{1}{(b+1)^k} = \frac{1}{b+1}\left(\frac{1}{1-\frac{1}{b+1}}\right) = \frac{1}{b} < 1.$$

Since there is no integer strictly between 0 and 1, we have reached a contradiction, and so e must be irrational. Q.E.D.

13.3 Alternate proofs

Another proof[6] can be obtained from the previous one by noting that

$$(b+1)x = 1 + \frac{1}{b+2} + \frac{1}{(b+2)(b+3)} + \cdots < 1 + \frac{1}{b+1} + \frac{1}{(b+1)(b+2)} + \cdots = 1 + x,$$

and this inequality is equivalent to the assertion that $bx < 1$. This is impossible, of course, since b and x are natural numbers.

Still another proof[7] can be obtained from the fact that

$$\frac{1}{e} = e^{-1} = \sum_{n=0}^{\infty}\frac{(-1)^n}{n!}.$$

13.4 Generalizations

In 1840, Liouville published a proof of the fact that e^2 is irrational[8] followed by a proof that e^2 is not a root of a second degree polynomial with rational coefficients.[9] This last fact implies that e^4 is irrational. His proofs are similar to Fourier's proof of the irrationality of e. In 1891, Hurwitz explained how it is possible to prove along the same line of ideas that e is not a root of a third degree polynomial with rational coefficients.[10] In particular, e^3 is irrational.

More generally, e^q is irrational for any non-zero rational q.[11]

13.5 See also

- Characterizations of the exponential function
- Transcendental number, including a proof that e is transcendental
- Lindemann–Weierstrass theorem

13.6 References

[1] Euler, Leonhard (1744). "De fractionibus continuis dissertatio" [A dissertation on continued fractions] (PDF). *Commentarii academiae scientiarum Petropolitanae* **9**: 98–137.

[2] Euler, Leonhard (1985). "An essay on continued fractions". *Mathematical Systems Theory* **18**: 295–398. doi:10.1007/bf01699475.

[3] Sandifer, C. Edward (2007). "Chapter 32: Who proved e is irrational?". *How Euler did it*. Mathematical Association of America. pp. 185–190. ISBN 978-0-88385-563-8. LCCN 2007927658.

[4] Cohn, Henry (2006). "A short proof of the simple continued fraction expansion of e". *American Mathematical Monthly* (Mathematical Association of America) **113** (1): 57–62. JSTOR 27641837.

[5] de Stainville, Janot (1815). *Mélanges d'Analyse Algébrique et de Géométrie* [*A mixture of Algebraic Analysis and Geometry*]. Veuve Courcier. pp. 340–341.

[6] MacDivitt, A. R. G.; Yanagisawa, Yukio (1987), "An elementary proof that e is irrational", *The Mathematical Gazette* (London: Mathematical Association) **71** (457): 217, JSTOR 3616765

[7] Penesi, L. L. (1953). "Elementary proof that e is irrational". *American Mathematical Monthly* (Mathematical Association of America) **60** (7): 474. JSTOR 2308411.

[8] Liouville, Joseph (1840). "Sur l'irrationalité du nombre $e = 2,718...$". *Journal de Mathématiques Pures et Appliquées*. 1 (in French) **5**: 192.

[9] Liouville, Joseph (1840). "Addition à la note sur l'irrationnalité du nombre e". *Journal de Mathématiques Pures et Appliquées*. 1 (in French) **5**: 193–194.

[10] Hurwitz, Adolf (1933) [1891]. "Über die Kettenbruchentwicklung der Zahl e". *Mathematische Werke* (in German) **2**. Basel: Birkhäuser. pp. 129–133.

[11] Aigner, Martin; Ziegler, Günter M. (1998), *Proofs from THE BOOK* (4th ed.), Berlin, New York: Springer-Verlag, pp. 27–36, doi:10.1007/978-3-642-00856-6, ISBN 978-3-642-00855-9.

Chapter 14

Proof that π is irrational

In the 18th century, Johann Heinrich Lambert proved that the number π (pi) is irrational. That is, it cannot be expressed as a fraction *a/b*, where *a* is an integer and *b* is a non-zero integer. In the 19th century, Charles Hermite found a proof that requires no prerequisite knowledge beyond basic calculus. Three simplifications of Hermite's proof are due to Mary Cartwright, Ivan Niven and Bourbaki. Another proof, which is a simplification of Lambert's proof, is due to Miklós Laczkovich.

In 1882, Ferdinand von Lindemann proved that π is not just irrational, but transcendental as well.[1]

14.1 Lambert's proof

$$\operatorname{tang}\left(\frac{\varphi}{\omega}\right) = \cfrac{\varphi}{\omega - \cfrac{\varphi\varphi}{3\omega - \cfrac{\varphi\varphi}{5\omega - \cfrac{\varphi\varphi}{7\omega - \cfrac{\varphi\varphi}{9\omega - \&c.}}}}}$$

Scan of formula on page 288 of Lambert's "Mémoires sur quelques propriétés remarquables des quantités transcendantes, circulaires et logarithmiques", Mémoires de l'Académie royale des sciences de Berlin (1768), 265–322.

In 1761, Lambert proved that π is irrational by first showing that this continued fraction expansion holds:

$$\tan(x) = \cfrac{x}{1 - \cfrac{x^2}{3 - \cfrac{x^2}{5 - \cfrac{x^2}{7 - \ddots}}}}.$$

Then Lambert proved that if x is non-zero and rational then this expression must be irrational. Since $\tan(\pi/4) = 1$, it follows that $\pi/4$ is irrational and therefore that π is irrational.[2] A simplification of Lambert's proof is given below.

14.2 Hermite's proof

This proof uses the characterization of π as the smallest positive number whose half is a zero of the cosine function and it actually proves that π^2 is irrational.[3][4] As in many proofs of irrationality, the argument proceeds by reductio ad absurdum.

Consider the sequences $(An)n \geq 0$ and $(Un)n \geq 0$ of functions from **R** into **R** thus defined:

1. $A_0(x) = \sin(x)$;
2. $(\forall n \in \mathbb{Z}_+) : A_{n+1}(x) = \int_0^x y A_n(y)\, dy$;
3. $U_0(x) = \frac{\sin(x)}{x}$;
4. $(\forall n \in \mathbb{Z}_+) : U_{n+1}(x) = -\frac{U_n'(x)}{x}$.

It can be proved by induction that

$$(\forall n \in \mathbb{Z}_+) : A_n(x) = \frac{x^{2n+1}}{(2n+1)!!} - \frac{x^{2n+3}}{2 \times (2n+3)!!} + \frac{x^{2n+5}}{2 \times 4 \times (2n+5)!!} \mp \cdots$$

and that

$$(\forall n \in \mathbb{Z}_+) : U_n(x) = \frac{1}{(2n+1)!!} - \frac{x^2}{2 \times (2n+3)!!} + \frac{x^4}{2 \times 4 \times (2n+5)!!} \mp \cdots$$

and therefore that

$$U_n(x) = \frac{A_n(x)}{x^{2n+1}}.$$

So

$$\frac{A_{n+1}(x)}{x^{2n+3}} = U_{n+1}(x) = -\frac{U_n'(x)}{x} = -\frac{1}{x}\frac{d}{dx}\left(\frac{A_n(x)}{x^{2n+1}}\right),$$

which is equivalent to

$$A_{n+1}(x) = (2n+1)A_n(x) - xA_n'(x) = (2n+1)A_n(x) - x^2 A_{n-1}(x).$$

It follows by induction from this, together with the fact that $A_0(x) = \sin(x)$ and that $A_1(x) = -x\cos(x) + \sin(x)$, that $An(x)$ can be written as $Pn(x^2)\sin(x) + xQn(x^2)\cos(x)$, where Pn and Qn are polynomial functions with integer coefficients and where the degree of Pn is smaller than or equal to $\lfloor n/2 \rfloor$. In particular, $An(\pi/2) = Pn(\pi^2/4)$.

Hermite also gave a closed expression for the function An, namely

$$A_n(x) = \frac{x^{2n+1}}{2^n n!} \int_0^1 (1-z^2)^n \cos(xz)\, dz.$$

14.2. HERMITE'S PROOF

He did not justify this assertion, but it can be proved easily. First of all, this assertion is equivalent to

$$\frac{1}{2^n n!}\int_0^1 (1-z^2)^n \cos(xz)\,dz = \frac{A_n(x)}{x^{2n+1}} = U_n(x).$$

Proceeding by induction, take $n = 0$.

$$\int_0^1 \cos(xz)\,dz = \frac{\sin(x)}{x} = U_0(x)$$

and, for the inductive step, consider any $n \in \mathbf{Z}_+$. If

$$\frac{1}{2^n n!}\int_0^1 (1-z^2)^n \cos(xz)\,dz = U_n(x),$$

then, using integration by parts and Leibniz's rule, one gets

$$\frac{1}{2^{n+1}(n+1)!}\int_0^1 (1-z^2)^{n+1} \cos(xz)\,dz$$

$$= \frac{1}{2^{n+1}(n+1)!}\left(\overbrace{(1-z^2)^{n+1}\frac{\sin(xz)}{x}\Big|_{z=0}^{z=1}}^{=0} + \int_0^1 2(n+1)(1-z^2)^n z\frac{\sin(xz)}{x}\,dz \right)$$

$$= \frac{1}{x}\cdot\frac{1}{2^n n!}\int_0^1 (1-z^2)^n z \sin(xz)\,dz$$

$$= -\frac{1}{x}\cdot\frac{d}{dx}\left(\frac{1}{2^n n!}\int_0^1 (1-z^2)^n \cos(xz)\,dz\right)$$

$$= -\frac{U_n'(x)}{x} = U_{n+1}(x).$$

If $\pi^2/4 = p/q$, with p and q in $\mathbf{N}$, then, since the coefficients of P_n are integers and its degree is smaller than or equal to $\lfloor n/2 \rfloor$, $q^{\lfloor n/2 \rfloor}P_n(\pi^2/4)$ is some integer N. In other words,

$$N = q^{\lfloor \frac{n}{2} \rfloor}A_n\left(\frac{\pi}{2}\right)$$

$$= q^{\lfloor \frac{n}{2} \rfloor}\frac{\left(\frac{p}{q}\right)^{n+\frac{1}{2}}}{2^n n!}\int_0^1 (1-z^2)\cos\left(\frac{\pi}{2}z\right)\,dz.$$

But this number is clearly greater than 0; therefore, $N \in \mathbf{N}$. On the other hand, the integral that appears here is not greater than 1 and

$$\lim_{n \in \mathbf{N}} q^{\lfloor \frac{n}{2} \rfloor}\frac{\left(\frac{p}{q}\right)^{n+\frac{1}{2}}}{2^n n!} = 0.$$

So, if n is large enough, $N < 1$. Thereby, a contradiction is reached.

Hermite did not present his proof as an end in itself but as an afterthought within his search for a proof of the transcendence of π. He discussed the recurrence relations to motivate and to obtain a convenient integral representation. Once this integral representation is obtained, there are various ways to present a succinct and self-contained proof starting from the integral (as in Cartwright's, Bourbaki's or Niven's presentations), which Hermite could easily see (as he did in his proof of the transcendence of e[5]).

Moreover, Hermite's proof is closer to Lambert's proof than it seems. In fact, $An(x)$ is the "residue" (or "remainder") of Lambert's continued fraction for $\tan(x)$.[6]

14.3 Cartwright's proof

Harold Jeffreys wrote that this proof was set as an example in an exam at Cambridge University in 1945 by Mary Cartwright, but that she had not traced its origin.[7]

Consider the integrals

$$I_n(x) = \int_{-1}^{1} (1-z^2)^n \cos(xz)\,dz,$$

where n is a non-negative integer.

Two integrations by parts give the recurrence relation

$$(\forall n \in \mathbb{N} \setminus \{1\}) : x^2 I_n(x) = 2n(2n-1)I_{n-1}(x) - 4n(n-1)I_{n-2}(x).$$

If

$$J_n(x) = x^{2n+1} I_n(x),$$

then this becomes

$$J_n(x) = 2n(2n-1)J_{n-1}(x) - 4n(n-1)x^2 J_{n-2}(x).$$

Furthermore, $J_0(x) = 2\sin(x)$ and $J_1(x) = -4x\cos(x) + 4\sin(x)$. Hence for all $n \in \mathbb{Z}_+$,

$$J_n(x) = x^{2n+1} I_n(x) = n!\big(P_n(x)\sin(x) + Q_n(x)\cos(x)\big),$$

where $Pn(x)$ and $Qn(x)$ are polynomials of degree $\leq 2n$, and with integer coefficients (depending on n).

Take $x = \pi/2$, and suppose if possible that $\pi/2 = b/a$, where a and b are natural numbers (i.e., assume that π is rational). Then

$$\frac{b^{2n+1}}{n!} I_n\left(\frac{\pi}{2}\right) = P_n\left(\frac{\pi}{2}\right) a^{2n+1}.$$

The right side is an integer. But $0 < In(\pi/2) < 2$ since the interval $[-1, 1]$ has length 2 and the function that is being integrated takes only values between 0 and 1. On the other hand,

$$\frac{b^{2n+1}}{n!} \to 0 \text{ as } n \to \infty.$$

Hence, for sufficiently large n

$$0 < \frac{b^{2n+1} I_n\left(\frac{\pi}{2}\right)}{n!} < 1,$$

that is, we could find an integer between 0 and 1. That is the contradiction that follows from the assumption that π is rational.

This proof is similar to Hermite's proof. Indeed,

$$\begin{aligned} J_n(x) &= x^{2n+1} \int_{-1}^{1} (1-z^2)^n \cos(xz)\, dz \\ &= 2x^{2n+1} \int_{0}^{1} (1-z^2)^n \cos(xz)\, dz \\ &= 2^{n+1} n! A_n(x). \end{aligned}$$

However, it is clearly simpler. This is achieved by passing the inductive definition of the functions A_n and taking as a starting point their expression as an integral.

14.4 Niven's proof

This proof uses the characterization of π as the smallest positive zero of the sine function.[8]

Suppose that π is rational, i.e. $\pi = a/b$ for some integers a and $b \neq 0$, which may be taken without loss of generality to be positive. Given any positive integer n, we define the polynomial function f from **R** into **R** defined by

$$f(x) = \frac{x^n(a-bx)^n}{n!}$$

and, for each $x \in$ **R** denote by

$$F(x) = f(x) - f''(x) + f^{(4)}(x) + \cdots + (-1)^j f^{(2j)}(x) + \cdots + (-1)^n f^{(2n)}(x)$$

the alternating sum of f and its first n derivatives of even order.

Claim 1: $F(0) + F(\pi)$ is an integer.

Proof: Expanding f as a sum of monomials, the coefficient of x^k is a number of the form $c_k/n!$ where c_k is an integer, which is 0 if $k < n$. Therefore, $f^{(k)}(0)$ is 0 when $k < n$ and it is equal to $(k!/n!)\, c_k$ if $n \leq k \leq 2n$; in each case, $f^{(k)}(0)$ is an integer and therefore $F(0)$ is an integer.

On the other hand, $f(\pi - x) = f(x)$ and so $(-1)^k f^{(k)}(\pi - x) = f^{(k)}(x)$ for each non-negative integer k. In particular, $(-1)^k f^{(k)}(\pi) = f^{(k)}(0)$. Therefore, $f^{(k)}(\pi)$ is also an integer and so $F(\pi)$ is an integer (in fact, it is easy to see that $F(\pi) = F(0)$, but that is not relevant to the proof). Since $F(0)$ and $F(\pi)$ are integers, so is their sum.

Claim 2:

$$\int_0^\pi f(x) \sin(x)\, dx = F(0) + F(\pi)$$

Proof: Since $f^{(2n+2)}$ is the zero polynomial, we have

$$F'' + F = f.$$

The derivatives of the sine and cosine function are given by sin' = cos and cos' = −sin. Hence the product rule implies

$$(F' \cdot \sin - F \cdot \cos)' = f \cdot \sin$$

By the fundamental theorem of calculus

$$\int_0^\pi f(x) \sin(x)\, dx = \left(F'(x) \sin x - F(x) \cos x\right)\Big|_0^\pi.$$

Since sin 0 = sin π = 0 and cos 0 = − cos π = 1 (here we use the above-mentioned characterization of π as a zero of the sine function), Claim 2 follows.

Conclusion: Since $f(x) > 0$ and sin $x > 0$ for $0 < x < \pi$ (because π is the *smallest* positive zero of the sine function), Claims 1 and 2 show that $F(0) + F(\pi)$ is a *positive* integer. Since $0 \leq x(a - bx) \leq \pi a$ and $0 \leq \sin x \leq 1$ for $0 \leq x \leq \pi$, we have, by the original definition of f,

$$\int_0^\pi f(x) \sin(x)\, dx \leq \pi \frac{(\pi a)^n}{n!}$$

which is smaller than 1 for large n, hence $F(0) + F(\pi) < 1$ for these n, by Claim 2. This is impossible for the positive integer $F(0) + F(\pi)$.

The above proof is a polished version, which is kept as simple as possible concerning the prerequisites, of an analysis of the formula

$$\int_0^\pi f(x) \sin(x)\, dx = \sum_{j=0}^n (-1)^j \left(f^{(2j)}(\pi) + f^{(2j)}(0)\right)$$
$$+ (-1)^{n+1} \int_0^\pi f^{(2n+2)}(x) \sin(x)\, dx,$$

which is obtained by $2n + 2$ integrations by parts. Claim 2 essentially establishes this formula, where the use of F hides the iterated integration by parts. The last integral vanishes because $f^{(2n+2)}$ is the zero polynomial. Claim 1 shows that the remaining sum is an integer.

Niven's proof is closer to Cartwright's (and therefore Hermite's) proof than it appears at first sight.[6] In fact,

$$J_n(x) = x^{2n+1} \int_{-1}^1 (1 - z^2)^n \cos(xz)\, dz$$
$$= \int_{-1}^1 \left(x^2 - (xz)^2\right)^n x \cos(xz)\, dz.$$

Therefore, the substitution $xz = y$ turns this integral into

$$\int_{-x}^x (x^2 - y^2)^n \cos(y)\, dy.$$

In particular,

$$J_n\left(\frac{\pi}{2}\right) = \int_{-\pi/2}^{\pi/2} \left(\frac{\pi^2}{4} - y^2\right)^n \cos(y)\,dy$$
$$= \int_0^{\pi} \left(\frac{\pi^2}{4} - \left(y - \frac{\pi}{2}\right)^2\right)^n \cos\left(y - \frac{\pi}{2}\right) dy$$
$$= \int_0^{\pi} y^n(\pi - y)^n \sin(y)\,dy$$
$$= \frac{n!}{b^n} \int_0^{\pi} f(x) \sin(x)\,dx.$$

Another connection between the proofs lies in the fact that Hermite already mentions[3] that if f is a polynomial function and

$$F = f - f^{(2)} + f^{(4)} \mp \cdots,$$

then

$$\int f(x) \sin(x)\,dx = F'(x) \sin(x) - F(x) \cos(x),$$

from which it follows that

$$\int_0^{\pi} f(x) \sin(x)\,dx = F(\pi) + F(0).$$

14.5 Bourbaki's proof

Bourbaki's proof is outlined as an exercise in his Calculus treatise.[9] For each natural number b and each non-negative integer n, define

$$A_n(b) = b^n \int_0^{\pi} \frac{x^n(\pi - x)^n}{n!} \sin(x)\,dx.$$

Since $An(b)$ is the integral of a function which defined on $[0,\pi]$ that takes the value 0 on 0 and on π and which is greater than 0 otherwise, $An(b) > 0$. Besides, for each natural number b, $An(b) < 1$ if n is large enough, because

$$(\forall x \in [0, \pi]) : x(\pi - x) \le \left(\frac{\pi}{2}\right)^2$$

and therefore

$$A_n(b) \le \pi b^n \frac{1}{n!} \left(\frac{\pi}{2}\right)^{2n} = \pi \frac{(b\pi^2/4)^n}{n!}.$$

On the other hand, recursive integration by parts allows us to deduce that, if a and b are natural number such that $\pi = a/b$ and f is the polynomial function from $[0,\pi]$ into **R** defined by

$$f(x) = \frac{x^n(a-bx)^n}{n!},$$

then:

$$A_n(b) = \int_0^\pi f(x)\sin(x)\,dx$$
$$= [-f(x)\cos(x)]_{x=0}^{x=\pi} - [-f'(x)\sin(x)]_{x=0}^{x=\pi} + \cdots \pm \left[f^{(2n)}(x)\cos(x)\right]_{x=0}^{x=\pi} \pm \int_0^\pi f^{(2n+1)}(x)\cos(x)\,dx.$$

This last integral is 0, since $f^{(2n+1)}$ is the null function (because f is a polynomial function whose degree is $2n$). Since each function $f^{(k)}$ (with $0 \leq k \leq 2n$) takes integer values on 0 and on π (see Claim 1 from Niven's proof) and since the same thing happens with the sine and the cosine functions, this proves that $A_n(b)$ is an integer. Since it is also greater than 0, it must be a natural number. But it was also proved that $A_n(b) < 1$ if n is large enough, thereby reaching a contradiction.

This proof is quite close to Niven's proof, the main difference between them being the way of proving that the numbers $A_n(b)$ are integers.

14.6 Laczkovich's proof

Miklós Laczkovich's proof is a simplification of Lambert's original proof.[10] He considers the functions

$$f_k(x) = 1 - \frac{x^2}{k} + \frac{x^4}{2!k(k+1)} - \frac{x^6}{3!k(k+1)(k+2)} + \cdots$$
$$(k \notin \{0, -1, -2, \ldots\}).$$

These functions are clearly defined for all $x \in \mathbf{R}$. Besides

$$f_{1/2}(x) = \cos(2x) \text{ and } f_{3/2}(x) = \frac{\sin(2x)}{2x}.$$

Claim 1: The following recurrence relation holds:

$$(\forall x \in \mathbb{R}) : \frac{x^2}{k(k+1)} f_{k+2}(x) = f_{k+1}(x) - f_k(x).$$

Proof: This can be proved by comparing the coefficients of the powers of x.

Claim 2: For each $x \in \mathbf{R}$, $\lim_{k \to +\infty} f_k(x) = 1$.

Proof: In fact, the sequence $x^{2n}/n!$ is bounded (since it converges to 0) and if C is an upper bound and if $k > 1$, then

$$|f_k(x) - 1| \leq \sum_{n=1}^\infty \frac{C}{k^n} = C\frac{1/k}{1-1/k} = \frac{C}{k-1}.$$

Claim 3: If $x \neq 0$ and if x^2 is rational, then

$$(\forall k \in \mathbb{Q} \setminus \{0, -1, -2, \ldots\}) : f_k(x) \neq 0 \text{ and } \frac{f_{k+1}(x)}{f_k(x)} \notin \mathbb{Q}.$$

Proof: Otherwise, there would be a number $y \neq 0$ and integers a and b such that $f_k(x) = ay$ and $f_{k+1}(x) = by$. In order to see why, take $y = f_{k+1}(x)$, $a = 0$ and $b = 1$ if $f_k(x) = 0$; otherwise, choose integers a and b such that $f_{k+1}(x)/f_k(x) = b/a$ and define $y = f_k(x)/a = f_{k+1}(x)/b$. In each case, y cannot be 0, because otherwise it would follow from claim 1 that each $f_{k+n}(x)$ ($n \in \mathbf{N}$) would be 0, which would contradict claim 2. Now, take a natural number c such that all three numbers bc/k, ck/x^2 and c/x^2 are integers and consider the sequence

$$g_n = \begin{cases} f_k(x) & \text{if } n = 0 \\ \frac{c^n}{k(k+1)\cdots(k+n-1)} f_{k+n}(x) & \text{otherwise.} \end{cases}$$

Then

$$g_0 = f_k(x) = ay \in \mathbb{Z}y \text{ and } g_1 = \frac{c}{k} f_{k+1}(x) = \frac{bc}{k} y \in \mathbb{Z}y.$$

On the other hand, it follows from claim 1 that

$$\begin{aligned} g_{n+2} &= \frac{c^{n+2}}{x^2 k(k+1)\cdots(k+n-1)} \cdot \frac{x^2}{(k+n)(k+n+1)} f_{k+n+2}(x) \\ &= \frac{c^{n+2}}{x^2 k(k+1)\cdots(k+n-1)} f_{k+n+1}(x) - \frac{c^{n+2}}{x^2 k(k+1)\cdots(k+n-1)} f_{k+n}(x) \\ &= \frac{c(k+n)}{x^2} g_{n+1} - \frac{c^2}{x^2} g_n \\ &= \left(\frac{ck}{x^2} + \frac{c}{x^2} n \right) g_{n+1} - \frac{c^2}{x^2} g_n, \end{aligned}$$

which is a linear combination of g_{n+1} and g_n with integer coefficients. Therefore, each g_n is an integer multiple of y. Besides, it follows from claim 2 that each g_n is greater than 0 (and therefore that $g_n \geq |y|$) if n is large enough and that the sequence of all g_n's converges to 0. But a sequence of numbers greater than or equal to $|y|$ cannot converge to 0.

Since $f_{1/2}(\pi/4) = \cos(\pi/2) = 0$, it follows from claim 3 that $\pi^2/16$ is irrational and therefore that π is irrational.

On the other hand, since

$$\tan x = \frac{\sin x}{\cos x} = x \frac{f_{3/2}(x/2)}{f_{1/2}(x/2)},$$

another consequence of claim 3 is that, if $x \in \mathbf{Q} \setminus \{0\}$, then $\tan x$ is irrational.

Laczkovich's proof is really about the hypergeometric function. In fact, $f_k(x) = {}_0F_1(k; -x^2)$ and Gauss found a continued fraction expansion of the hypergeometric function using its functional equation.[11] This allowed Laczkovich to find a new and simpler proof of the fact that the tangent function has the continued fraction expansion that Lambert had discovered.

Laczkovich's result can also be expressed in Bessel functions of the first kind $J_\nu(x)$. In fact, $\Gamma(k)J_{k-1}(2x) = x^{k-1}f_k(x)$. So Laczkovich's result is equivalent to: If $x \neq 0$ and if x^2 is rational, then

$$(\forall k \in \mathbb{Q} \setminus \{0, -1, -2, \ldots\}) : \frac{x J_k(x)}{J_{k-1}(x)} \notin \mathbb{Q}.$$

14.7 See also

- Proof that e is irrational
- Proof that π is transcendental

14.8 References

[1] Lindemann, Ferdinand von (2004) [1882], "Ueber die Zahl π", in Berggren, Lennart; Borwein, Jonathan M.; Borwein, Peter B., *Pi, a source book* (3rd ed.), New York: Springer-Verlag, pp. 194–225, ISBN 0-387-20571-3

[2] Lambert, Johann Heinrich (2004) [1768], "Mémoire sur quelques propriétés remarquables des quantités transcendantes circulaires et logarithmiques", in Berggren, Lennart; Borwein, Jonathan M.; Borwein, Peter B., *Pi, a source book* (3rd ed.), New York: Springer-Verlag, pp. 129–140, ISBN 0-387-20571-3

[3] Hermite, Charles (1873), "Extrait d'une lettre de Monsieur Ch. Hermite à Monsieur Paul Gordan", *Journal für die reine und angewandte Mathematik* (in French) **76**: 303–311

[4] Hermite, Charles (1873), "Extrait d'une lettre de Mr. Ch. Hermite à Mr. Carl Borchardt", *Journal für die reine und angewandte Mathematik* (in French) **76**: 342–344

[5] Hermite, Charles (1912) [1873], "Sur la fonction exponentielle", in Picard, Émile, *Œuvres de Charles Hermite* (in French) **III**, Gauthier-Villars, pp. 150–181

[6] Zhou, Li (2011), "Irrationality proofs à la Hermite", *Math. Gazette* (November), arXiv:0911.1929

[7] Jeffreys, Harold (1973), *Scientific Inference* (3rd ed.), Cambridge University Press, p. 268, ISBN 0-521-08446-6

[8] Niven, Ivan (1947), "A simple proof that π is irrational" (PDF), *Bulletin of the American Mathematical Society* **53** (6): 509

[9] Bourbaki, Nicolas (1949), *Fonctions d'une variable réelle, chap. I–II–III*, Actualités Scientifiques et Industrielles (in French) **1074**, Hermann, pp. 137–138

[10] Laczkovich, Miklós (1997), "On Lambert's proof of the irrationality of π", *American Mathematical Monthly* **104** (5): 439–443, JSTOR 2974737

[11] Gauss, Carl Friedrich (1811–1813), "Disquisitiones generales circa seriem infinitam
$$1 + \frac{\alpha\beta}{1\cdot\gamma}x + \frac{\alpha(\alpha+1)\beta(\beta+1)}{1\cdot 2\cdot\gamma(\gamma+1)}xx + \frac{\alpha(\alpha+1)(\alpha+2)\beta(\beta+1)(\beta+2)}{1\cdot 2\cdot 3\cdot\gamma(\gamma+1)(\gamma+1)}x^3 +$$
etc", *Commentationes Societatis Regiae Scientiarum Gottingensis recentiores* (in Latin) **2**

Chapter 15

Divergence of the sum of the reciprocals of the primes

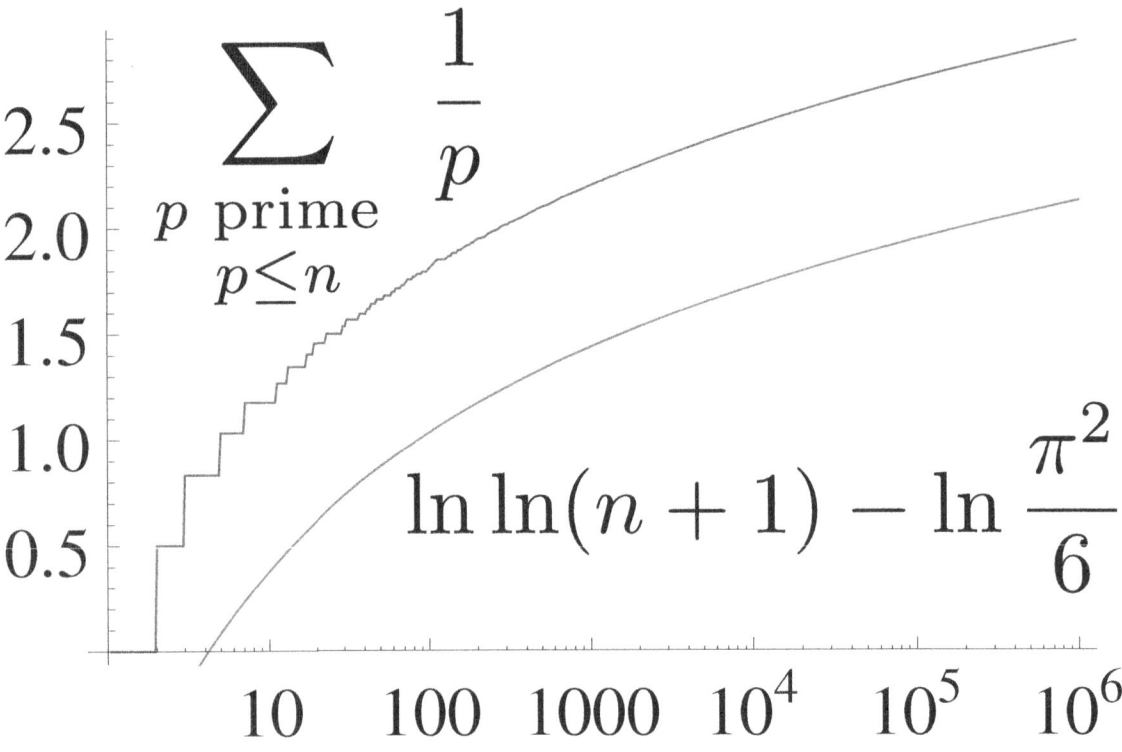

The sum of the reciprocal of the primes increasing without bound. The x axis is in log scale, showing that the divergence is very slow. The red function is a lower bound that also diverges.

The **sum of the reciprocals of all prime numbers diverges**; that is:

$$\sum_{p\text{ prime}} \frac{1}{p} = \frac{1}{2} + \frac{1}{3} + \frac{1}{5} + \frac{1}{7} + \frac{1}{11} + \frac{1}{13} + \frac{1}{17} + \cdots = \infty$$

This was proved by Leonhard Euler in 1737, and strengthens Euclid's 3rd-century-BC result that there are infinitely many prime numbers.

There are a variety of proofs of Euler's result, including a lower bound for the partial sums stating that

$$\sum_{\substack{p \text{ prime} \\ p \leq n}} \frac{1}{p} \geq \log \log(n+1) - \log \frac{\pi^2}{6}$$

for all natural numbers n. The double natural logarithm indicates that the divergence might be very slow, which is indeed the case, see Meissel–Mertens constant.

15.1 The harmonic series

First, we describe how Euler originally discovered the result. He was considering the harmonic series

$$\sum_{n=1}^{\infty} \frac{1}{n} = 1 + \frac{1}{2} + \frac{1}{3} + \frac{1}{4} + \cdots$$

He had already used the following "product formula" to show the existence of infinitely many primes.

$$\sum_{n=1}^{\infty} \frac{1}{n} = \prod_p \frac{1}{1-p^{-1}} = \prod_p \left(1 + \frac{1}{p} + \frac{1}{p^2} + \cdots \right)$$

(Here, the product is taken over all primes p; in the following, a sum or product taken over p always represents a sum or product taken over a specified set of primes, unless noted otherwise.)

Such infinite products are today called Euler products. The product above is a reflection of the fundamental theorem of arithmetic. Of course, the above "equation" is not necessary because the harmonic series is known (by other means) to diverge.

Euler noted that if there were only a finite number of primes, then the product on the right would clearly converge, contradicting the divergence of the harmonic series. (In modern language, we now say that the existence of infinitely many primes is reflected by the fact that the Riemann zeta function has a simple pole at $s = 1$.)

15.2 Proofs

15.2.1 First

Euler took the above product formula and proceeded to make a sequence of audacious leaps of logic. First, he took the natural logarithm of each side, then he used the Taylor series expansion for $\ln(x)$ as well as the sum of a geometric series:

$$\ln\left(\sum_{n=1}^{\infty}\frac{1}{n}\right) = \ln\left(\prod_p \frac{1}{1-p^{-1}}\right) = -\sum_p \ln\left(1-\frac{1}{p}\right)$$
$$= \sum_p \left(\frac{1}{p} + \frac{1}{2p^2} + \frac{1}{3p^3} + \cdots\right)$$
$$= \left(\sum_p \frac{1}{p}\right) + \sum_p \frac{1}{p^2}\left(\frac{1}{2} + \frac{1}{3p} + \frac{1}{4p^2} + \cdots\right)$$
$$< \left(\sum_p \frac{1}{p}\right) + \sum_p \frac{1}{p^2}\left(1 + \frac{1}{p} + \frac{1}{p^2} + \cdots\right)$$
$$= \left(\sum_p \frac{1}{p}\right) + \left(\sum_p \frac{1}{p(p-1)}\right)$$
$$= \left(\sum_p \frac{1}{p}\right) + C$$

for a fixed constant $C < 1$. Since the sum of the reciprocals of the first n positive integers is asymptotic to $\ln(n)$, (i.e. their ratio approaches one as n approaches infinity), Euler then concluded

$$\frac{1}{2} + \frac{1}{3} + \frac{1}{5} + \frac{1}{7} + \frac{1}{11} + \cdots = \ln\ln(+\infty)$$

It is almost certain that Euler meant that the sum of the reciprocals of the primes less than n is asymptotic to $\ln(\ln(n))$ as n approaches infinity. It turns out this is indeed the case; Euler had reached a correct result by questionable means.

A variation

$$\log\left(\sum_{n=1}^{\infty}\frac{1}{n}\right) = \log\left(\prod_p \frac{1}{1-p^{-1}}\right) = \sum_p \log\left(\frac{p}{p-1}\right) = \sum_p \log\left(1 + \frac{1}{p-1}\right).$$

Since the Maclaurin series for e^x is,

$e^x = 1 + x + \sum_{n=2}^{\infty}\frac{x^n}{n!}$ dropping the summation shows that $e^x \geq 1 + x$, and we see $\log(e^x) \geq \log(1 + x)$, and $x \geq \log(1 + x)$. So,

$$\sum_p \frac{1}{p-1} \geq \sum_p \log\left(1 + \frac{1}{p-1}\right) = \log\left(\sum_{n=1}^{\infty}\frac{1}{n}\right).$$

Hence $\sum_p \frac{1}{p-1}$ diverges. But, with $p_i \geq 3$ and p_i is the ith prime,

$$\frac{1}{p_{i-1}} \geq \frac{1}{p_i - 1}$$

Hence $\sum_p \frac{1}{p}$ diverges.

15.2.2 Second

The following proof by contradiction is due to Paul Erdős.

Let p_i denote the ith prime number. Assume that the sum of the reciprocals of the primes converges; i.e.,

$$\sum_{i=1}^{\infty} \frac{1}{p_i} < \infty$$

Then there exists a smallest positive integer k such that

$$\sum_{i=k+1}^{\infty} \frac{1}{p_i} < \frac{1}{2} \qquad (1)$$

For a positive integer x let M_x denote the set of those n in $\{1, 2, \ldots, x\}$ which are not divisible by any prime greater than p_k. We will now derive an upper and a lower estimate for the number $|M_x|$ of elements in M_x. For large x, these bounds will turn out to be contradictory.

Upper estimate

Every n in M_x can be written as $n = r\,m^2$ with positive integers m and r, where r is square-free. Since only the k primes $p_1, \ldots, p_k$ can show up (with exponent 1) in the prime factorization of r, there are at most 2^k different possibilities for r. Furthermore, there are at most $\sqrt{x}$ possible values for m. This gives us the upper estimate

$$|M_x| \leq 2^k \sqrt{x} \qquad (2)$$

Lower estimate

The remaining $x - |M_x|$ numbers in the set difference $\{1, 2, \ldots, x\} \setminus M_x$ are all divisible by a prime greater than p_k. Let $N_{i,x}$ denote the set of those n in $\{1, 2, \ldots, x\}$ which are divisible by the ith prime p_i. Then

$$\{1, 2, \ldots, x\} \setminus M_x = \bigcup_{i=k+1}^{\infty} N_{i,x}$$

Since the number of integers in $N_{i,x}$ is at most x/p_i (actually zero for $p_i > x$), we get

$$x - |M_x| \leq \sum_{i=k+1}^{\infty} |N_{i,x}| < \sum_{i=k+1}^{\infty} \frac{x}{p_i}$$

Using (1), this implies

$$\frac{x}{2} < |M_x| \qquad (3)$$

Contradiction

When $x \geq 2^{2k+2}$, the estimates (2) and (3) cannot both hold, because $\frac{x}{2} \geq 2^k \sqrt{x}$.

15.2. PROOFS

15.2.3 Third

Here is another proof that actually gives a lower estimate for the partial sums; in particular, it shows that these sums grow at least as fast as log(log(n)). The proof is an adaptation of the product expansion idea of Euler. In the following, a sum or product taken over p always represents a sum or product taken over a specified set of primes.

The proof rests upon the following four inequalities:

- Every positive integer i can be uniquely expressed as the product of a square-free integer and a square. This gives the inequality

$$\sum_{i=1}^{n} \frac{1}{i} \leq \prod_{p \leq n}\left(1 + \frac{1}{p}\right) \sum_{k=1}^{n} \frac{1}{k^2}$$

where for every i between 1 and n the (expanded) product corresponds to the square-free part of i and the sum corresponds to the square part of i (see fundamental theorem of arithmetic).

- The upper estimate for the natural logarithm

$$\log(n+1) = \int_{1}^{n+1} \frac{dx}{x} = \sum_{i=1}^{n} \underbrace{\int_{i}^{i+1} \frac{dx}{x}}_{< 1/i} < \sum_{i=1}^{n} \frac{1}{i}$$

- The lower estimate $1 + x < \exp(x)$ for the exponential function, which holds for all $x > 0$.

- Let n ≥ 2. The upper bound (using a telescoping sum) for the partial sums (convergence is all we really need)

$$\sum_{k=1}^{n} \frac{1}{k^2} < 1 + \sum_{k=2}^{n} \underbrace{\left(\frac{1}{k-\frac{1}{2}} - \frac{1}{k+\frac{1}{2}}\right)}_{= 1/(k^2 - 1/4) > 1/k^2} = 1 + \frac{2}{3} - \frac{1}{n+\frac{1}{2}} < \frac{5}{3}$$

Combining all these inequalities, we see that

$$\log(n+1) < \sum_{i=1}^{n} \frac{1}{i}$$

$$\leq \prod_{p \leq n}\left(1 + \frac{1}{p}\right) \sum_{k=1}^{n} \frac{1}{k^2}$$

$$< \frac{5}{3} \prod_{p \leq n} \exp\left(\frac{1}{p}\right)$$

$$= \frac{5}{3} \exp\left(\sum_{p \leq n} \frac{1}{p}\right)$$

Dividing through by 5/3 and taking the natural logarithm of both sides gives

$$\log\log(n+1) - \log\frac{5}{3} < \sum_{p \leq n} \frac{1}{p}$$

as desired. ∎

Using

$$\sum_{k=1}^{\infty} \frac{1}{k^2} = \frac{\pi^2}{6}$$

(see Basel problem), the above constant ln (5/3) = 0.51082... can be improved to ln(π^2/6) = 0.4977...; in fact it turns out that

$$\lim_{n \to \infty} \left(\sum_{p \leq n} \frac{1}{p} - \log\log(n) \right) = M$$

where $M = 0.261497...$ is the Meissel–Mertens constant (somewhat analogous to the much more famous Euler–Mascheroni constant).

15.2.4 Fourth

From Dusart's inequality, we get

$$p_n < n \log n + n \log \log n \quad \text{for } n \geq 6$$

Then

$$\begin{aligned}
\sum_{n=1}^{\infty} \frac{1}{p_n} &\geq \sum_{n=6}^{\infty} \frac{1}{p_n} \\
&\geq \sum_{n=6}^{\infty} \frac{1}{n \log n + n \log \log n} \\
&\geq \sum_{n=6}^{\infty} \frac{1}{2n \log n} \\
&= \infty
\end{aligned}$$

by the integral test for convergence. This shows that the series on the left diverges.

15.3 Partial sums

While the partial sums of the reciprocals of the primes eventually exceed any integer value, they never equal an integer. One proof[1] is by induction: The first partial sum is $\frac{1}{2}$, which has the form $\frac{\text{odd}}{\text{even}} = \frac{O_1}{E_1}$ where O and E refer to odd and even numbers respectively. If the nth partial sum has the form $\frac{O_n}{E_n}$, then the $(n+1)^{\text{st}}$ one equals $\frac{O_n}{E_n} + \frac{1}{p_{n+1}} = \frac{O_n p_{n+1} + E_n}{E_n p_{n+1}} =$

$\frac{O_{n+1}}{E_{n+1}}$ for the $(n+1)^{\text{st}}$ prime p_{n+1}; since this repeats the odd-over-even form, this partial sum cannot be an integer (because 2 divides the denominator but not the numerator), and the induction continues.

Another proof rewrites the expression for the sum of the first n reciprocals of primes (or indeed the sum of the reciprocals of *any* set of primes) in terms of the least common denominator, which is the product of all these primes. Then each of these primes divides all but one of the numerator terms and hence does not divide the numerator itself; but each prime *does* divide the denominator. Thus the expression is irreducible and is non-integer.

15.4 See also

- Euclid's theorem that there are infinitely many primes
- Small set (combinatorics)
- Brun's theorem, on the convergent sum of reciprocals of the twin primes
- List of sums of reciprocals

15.5 References

[1] Lord, Nick. "Quick proofs that certain sums of fractions are not integers", *The Mathematical Gazette* 99, March 2015, 128-130. doi: http://dx.doi.org/10.1017/mag.2014.16

15.6 References

- William Dunham (1999). *Euler The Master of Us All*. MAA. pp. 61–79. ISBN 0-88385-328-0.

15.7 External links

- Chris K. Caldwell: There are infinitely many primes, but, how big of an infinity?

Chapter 16

Gödel's ontological proof

Gödel's ontological proof is a formal argument for God's existence by the mathematician Kurt Gödel (1906–1978).

It is in a line of development that goes back to Anselm of Canterbury (1033–1109). St. Anselm's ontological argument, in its most succinct form, is as follows: "God, by definition, is that for which no greater can be conceived. God exists in the understanding. If God exists in the understanding, we could imagine Him to be greater by existing in reality. Therefore, God must exist." A more elaborate version was given by Gottfried Leibniz (1646–1716); this is the version that Gödel studied and attempted to clarify with his ontological argument.

Gödel left a fourteen-point outline of his philosophical beliefs in his papers. Points relevant to the ontological proof include

> 4. There are other worlds and rational beings of a different and higher kind.
>
> 5. The world in which we live is not the only one in which we shall live or have lived.
>
> 13. There is a scientific (exact) philosophy and theology, which deals with concepts of the highest abstractness; and this is also most highly fruitful for science.
>
> 14. Religions are, for the most part, bad—but religion is not.

16.1 History of Gödel's proof

The first version of the ontological proof in Gödel's papers is dated "around 1941". Gödel is not known to have told anyone about his work on the proof until 1970, when he thought he was dying. In February, he allowed Dana Scott to copy out a version of the proof, which circulated privately. In August 1970, Gödel told Oskar Morgenstern that he was "satisfied" with the proof, but Morgenstern recorded in his diary entry for 29 August 1970, that Gödel would not publish because he was afraid that others might think "that he actually believes in God, whereas he is only engaged in a logical investigation (that is, in showing that such a proof with classical assumptions (completeness, etc.) correspondingly axiomatized, is possible)."[1] Gödel died January 14, 1978. Another version, slightly different from Scott's, was found in his papers. It was finally published, together with Scott's version, in 1987.[2]

Morgenstern's diary is an important and usually reliable source for Gödel's later years, but the implication of the August 1970 diary entry—that Gödel did not believe in God—is not consistent with the other evidence. In letters to his mother, who was not a churchgoer and had raised Kurt and his brother as freethinkers,[3] Gödel argued at length for a belief in an afterlife.[4] He did the same in an interview with a skeptical Hao Wang, who said: "I expressed my doubts as G spoke [...] Gödel smiled as he replied to my questions, obviously aware that his answers were not convincing me."[5] Wang reports that Gödel's wife, Adele, two days after Gödel's death, told Wang that "Gödel, although he did not go to church, was religious and read the Bible in bed every Sunday morning."[6] In an unmailed answer to a questionnaire, Gödel described his religion as "baptized Lutheran (but not member of any religious congregation). My belief is *theistic*, not pantheistic, following Leibniz rather than Spinoza."[7]

16.2 Outline of Gödel's proof

The proof uses modal logic, which distinguishes between *necessary* truths and *contingent* truths. In the most common semantics for modal logic, many "possible worlds" are considered. A truth is *necessary* if it is true in all possible worlds. By contrast, a truth is *contingent* if it just happens to be the case, for instance, "more than half of the planet is covered by water". If a statement happens to be true in our world, but is false in another world, then it is a contingent truth. A statement that is true in some world (not necessarily our own) is called a *possible* truth.

Furthermore, the proof uses higher-order (modal) logic because the definition of God employs an explicit quantification over properties.[8]

From axioms 1 through 4, Gödel argued that in *some* possible world there exists God. He used a sort of modal plenitude principle to argue this from the logical consistency of Godlikeness. Note that this property is itself positive, since it is the conjunction of the (infinitely many) positive properties.

Then, Gödel defined *essences*: if x is an object in some world, then the property P is said to be an essence of x if $P(x)$ is true in that world and if P entails all other properties that x has in that world. We also say that x *necessarily exists* if for every essence P the following is true: in every possible world, there is an element y with $P(y)$.

Since necessary existence is positive, it must follow from Godlikeness. Moreover, Godlikeness is an essence of God, since it entails all positive properties, and any nonpositive property is the negation of some positive property, so God cannot have any nonpositive properties. Since any Godlike object is necessarily existent, it follows that any Godlike object in one world is a Godlike object in all worlds, by the definition of necessary existence. Given the existence of a Godlike object in one world, proven above, we may conclude that there is a Godlike object in every possible world, as required.

From these hypotheses, it is also possible to prove that there is only one God in each world by Leibniz's law, the identity of indiscernibles: two or more objects are identical (are one and the same) if they have all their properties in common, and so, there would only be one object in each world that possesses property G. Gödel did not attempt to do so however, as he purposely limited his proof to the issue of existence, rather than uniqueness. This was more to preserve the logical precision of the argument than due to a penchant for polytheism. This uniqueness proof will only work if one supposes that the positiveness of a property is independent of the object to which it is applied, a claim which some have considered to be suspect.

To formalize the argument sketched above, the following definitions and axioms are needed:

- Definition 1: x is God-like if and only if x has as essential properties those and only those properties which are positive
- Definition 2: A is an essence of x if and only if for every property B, x has B necessarily if and only if A entails B
- Definition 3: x necessarily exists if and only if every essence of x is necessarily exemplified

- Axiom 1: Any property entailed by—i.e., strictly implied by—a positive property is positive
- Axiom 2: A property is positive if and only if its negation is not positive
- Axiom 3: The property of being God-like is positive
- Axiom 4: If a property is positive, then it is necessarily positive
- Axiom 5: Necessary existence is a positive property

Axiom 4 assumes that it is possible to single out *positive* properties from among all properties. Gödel comments that "Positive means positive in the moral aesthetic sense (independently of the accidental structure of the world)... It may also mean pure *attribution* as opposed to *privation* (or containing privation)." (Gödel 1995). Axioms 1, 2 and 3 can be summarized by saying that positive properties form a principal ultrafilter.

From these axioms and definitions and a few other axioms from modal logic, the following theorems can be proved:

- Theorem 1: If a property is positive, then it is consistent, i.e., possibly exemplified.

- Theorem 2: The property of being God-like is consistent.

- Theorem 3: If something is God-like, then the property of being God-like is an essence of that thing.

- Theorem 4: Necessarily, the property of being God-like is exemplified.

Symbolically:

1. Ax. $\{P(\varphi) \wedge \Box \, \forall x[\varphi(x) \to \psi(x)]\} \to P(\psi)$
2. Ax. $P(\neg\varphi) \leftrightarrow \neg P(\varphi)$
1. Th. $P(\varphi) \to \Diamond \, \exists x[\varphi(x)]$
1. Df. $G(x) \iff \forall\varphi[P(\varphi) \to \varphi(x)]$
3. Ax. $P(G)$
2. Th. $\Diamond \, \exists x \, G(x)$
2. Df. $\varphi \text{ ess } x \iff \varphi(x) \wedge \forall\psi \, \{\psi(x) \to \Box \, \forall y[\varphi(y) \to \psi(y)]\}$
4. Ax. $P(\varphi) \to \Box \, P(\varphi)$
3. Th. $G(x) \to G \text{ ess } x$
3. Df. $E(x) \iff \forall\varphi[\varphi \text{ ess } x \to \Box \, \exists y \, \varphi(y)]$
5. Ax. $P(E)$
4. Th. $\Box \, \exists x \, G(x)$

There is an ongoing open-source effort to formalize Gödel's proof to a level that is suitable for automated theorem proving or at least computer verification via proof assistants. The effort made headlines in German newspapers. According to the authors of this effort, they were inspired by Melvin Fitting's book.[9]

16.3 See also

- Absolute Infinite

- Existence of God

- Philosophy of religion

- Synthetic proposition

- Theism

- Ontological argument

16.4 Notes

[1] Quoted in Gödel 1995, p. 388. The German original is quoted in Dawson 1997, p. 307. The nested parentheses are in Morgenstern's original diary entry, as quoted by Dawson.

[2] The publication history of the proof in this paragraph is from Gödel 1995, p. 388

[3] Dawson 1997, pp. 6.

[4] Dawson 1997, pp. 210-212.

[5] Wang 1996, p. 317. The ellipsis is Wikipedia's.

[6] Wang 1996, p. 51.

[7] Gödel's answer to a special questionnaire sent him by the sociologist Burke Grandjean. This answer is quoted directly in Wang 1987, p. 18, and indirectly in Wang 1996, p. 112. It's also quoted directly in Dawson 1997, p. 6, who cites Wang 1987. The Grandjean questionnaire is perhaps the most extended autobiographical item in Gödel's papers. Gödel filled it out in pencil and wrote a cover letter, but he never returned it. "Theistic" is italicized in both Wang 1987 and Wang 1996. It is possible that this italicization is Wang's and not Gödel's. The quote follows Wang 1987, with two corrections taken from Wang 1996. Wang 1987 reads "Baptist Lutheran" where Wang 1996 has "baptized Lutheran". "Baptist Lutheran" makes no sense, especially in context, and was presumably a typo or mistranscription. Wang 1987 has "rel. cong.", which in Wang 1996 is expanded to "religious congregation".

[8] Fitting, 2002, p. 139

[9] Knight, David (23 October 2013). "Scientists Use Computer to Mathematically Prove Gödel's God Theorem". *Der Spiegel*. Retrieved 28 October 2013.

16.5 References

- John W. Dawson, Jr (1997). *Logical Dilemmas: The Life and Work of Kurt Godel*. Wellesley, Mass: AK Peters, Ltd. ISBN 1-56881-025-3.

- Melvin Fitting, "Types, Tableaus, and Godel's God" Publisher: Dordrecht Kluwer Academic, 2002, ISBN 1-4020-0604-7, ISBN 978-1-4020-0604-3

- Kurt Gödel (1995). "Ontological Proof". *Collected Works: Unpublished Essays & Lectures, Volume III*. pp. 403–404. Oxford University Press. ISBN 0-19-514722-7

- A. P. Hazen, "On Gödel's Ontological Proof", Australasian Journal of Philosophy, Vol. 76, No 3, pp. 361–377, September 1998

- Jordan Howard Sobel, "Gödel's Ontological Proof" in *On Being and Saying. Essays for Richard Cartwright*, ed. Judith Jarvis Thomson (MIT press, 1987)

- Wang, Hao (1987). *Reflections on Kurt Gödel*. Cambridge, Mass: MIT Press. ISBN 0-262-23127-1.

- Wang, Hao (1996). *A Logical Journey: from Gödel to Philosophy*. Cambridge, Mass: MIT Press. ISBN 0-262-23189-1.

- Small, Christopher. "Reflections on Gödel's Ontological Argument" (PDF). University of Waterloo.

16.6 External links

- Kurt Gödel's Ontological Argument by Christopher Small

- Ontological arguments entry by Graham Oppy in the *Stanford Encyclopedia of Philosophy*

- Annotated bibliography of studies on Gödel's Ontological Argument

- Benzmueller & Paleo 2014, "Automating Goedel's Ontological Proof of God's Existence with Higher-order Automated Theorem Provers"

Chapter 17

Proofs involving the addition of natural numbers

Mathematical proofs for addition of the natural numbers: additive identity, commutativity, and associativity. These proofs are used in the article Addition of natural numbers.

17.1 Definitions

This article will use the Peano axioms for the definitions of addition of the natural numbers, and the successor function $S(a)$. In particular:

For the proof of commutativity, it is useful to define another natural number closely related to the successor function, namely "1". We define 1 to be the successor of 0, in other words,

$1 = S(0)$.

Note that for all natural numbers a,

17.2 Proof of associativity

We prove associativity by first fixing natural numbers a and b and applying induction on the natural number c.

For the base case $c = 0$,

$(a+b)+0 = a+b = a+(b+0)$

Each equation follows by definition [A1]; the first with $a + b$, the second with b.

Now, for the induction. We assume the induction hypothesis, namely we assume that for some natural number c,

$(a+b)+c = a+(b+c)$

Then it follows,

In other words, the induction hypothesis holds for $S(c)$. Therefore, the induction on c is complete.

17.3 Proof of identity element

Definition [A1] states directly that 0 is a right identity. We prove that 0 is a left identity by induction on the natural number a.

For the base case $a = 0$, $0 + 0 = 0$ by definition [A1]. Now we assume the induction hypothesis, that $0 + a = a$. Then

This completes the induction on a.

17.4 Proof of commutativity

We prove commutativity ($a + b = b + a$) by applying induction on the natural number b. First we prove the base cases $b = 0$ and $b = S(0) = 1$ (i.e. we prove that 0 and 1 commute with everything).

The base case $b = 0$ follows immediately from the identity element property (0 is an additive identity), which has been proved above: $a + 0 = a = 0 + a$.

Next we will prove the base case $b = 1$, that 1 commutes with everything, i.e. for all natural numbers a, we have $a + 1 = 1 + a$. We will prove this by induction on a (an induction proof within an induction proof). Clearly, for $a = 0$, we have $0 + 1 = 0 + S(0) = S(0 + 0) = S(0) = 1 = 1 + 0$. Now, suppose $a + 1 = 1 + a$. Then

This completes the induction on a, and so we have proved the base case $b = 1$. Now, suppose that for all natural numbers a, we have $a + b = b + a$. We must show that for all natural numbers a, we have $a + S(b) = S(b) + a$. We have

This completes the induction on b.

17.5 See also

- Binary operation
- Proof
- Ring

17.6 References

- Edmund Landau, Foundations of Analysis, Chelsea Pub Co. ISBN 0-8218-2693-X.

Chapter 18

Analyticity of holomorphic functions

In complex analysis a complex-valued function f of a complex variable z:

- is said to be holomorphic at a point a if it is differentiable at every point within some open disk centered at a, and

- is said to be analytic at a if in some open disk centered at a it can be expanded as a convergent power series

$$f(z) = \sum_{n=0}^{\infty} c_n (z-a)^n$$

(this implies that the radius of convergence is positive).

One of the most important theorems of complex analysis is that **holomorphic functions are analytic**. Among the corollaries of this theorem are

- the identity theorem that two holomorphic functions that agree at every point of an infinite set S with an accumulation point inside the intersection of their domains also agree everywhere in every connected open subset of their domains that contains the set S, and

- the fact that, since power series are infinitely differentiable, so are holomorphic functions (this is in contrast to the case of real differentiable functions), and

- the fact that the radius of convergence is always the distance from the center a to the nearest singularity; if there are no singularities (i.e., if f is an entire function), then the radius of convergence is infinite. Strictly speaking, this is not a corollary of the theorem but rather a by-product of the proof.

- no bump function on the complex plane can be entire. In particular, on any connected open subset of the complex plane, there can be no bump function defined on that set which is holomorphic on the set. This has important ramifications for the study of complex manifolds, as it precludes the use of partitions of unity. In contrast the partition of unity is a tool which can be used on any real manifold.

18.1 Proof

The argument, first given by Cauchy, hinges on Cauchy's integral formula and the power series development of the expression

18.1. PROOF

$$\frac{1}{w-z}.$$

Let D be an open disk centered at a and suppose f is differentiable everywhere within an open neighborhood containing the closure of D. Let C be the positively oriented (i.e., counterclockwise) circle which is the boundary of D and z a point in D. Starting with Cauchy's integral formula, we have

$$f(z) = \frac{1}{2\pi i} \int_C \frac{f(w)}{w-z} \, dw$$

$$= \frac{1}{2\pi i} \int_C \frac{f(w)}{(w-a)-(z-a)} \, dw$$

$$= \frac{1}{2\pi i} \int_C \frac{1}{w-a} \cdot \frac{1}{1-\frac{z-a}{w-a}} f(w) \, dw$$

$$= \frac{1}{2\pi i} \int_C \frac{1}{w-a} \cdot \sum_{n=0}^{\infty} \left(\frac{z-a}{w-a}\right)^n f(w) \, dw$$

$$= \sum_{n=0}^{\infty} \frac{1}{2\pi i} \int_C \frac{(z-a)^n}{(w-a)^{n+1}} f(w) \, dw.$$

Interchange of the integral and infinite sum is justified by observing that $f(w)/(w-a)$ is bounded on C by some positive number M, while for all w in C

$$\left|\frac{z-a}{w-a}\right| \leq r < 1$$

for some positive r as well. We therefore have

$$\left|\frac{(z-a)^n}{(w-a)^{n+1}} f(w)\right| \leq Mr^n,$$

on C, and as the Weierstrass M-test shows the series converges uniformly, the sum and the integral may be interchanged. As the factor $(z-a)^n$ does not depend on the variable of integration w, it may be factored out to yield

$$f(z) = \sum_{n=0}^{\infty} (z-a)^n \frac{1}{2\pi i} \int_C \frac{f(w)}{(w-a)^{n+1}} \, dw,$$

which has the desired form of a power series in z:

$$f(z) = \sum_{n=0}^{\infty} c_n (z-a)^n$$

with coefficients

$$c_n = \frac{1}{2\pi i} \int_C \frac{f(w)}{(w-a)^{n+1}} \, dw.$$

18.2 Remarks

- Since power series can be differentiated term-wise, applying the above argument in the reverse direction and the power series expression for

$$\frac{1}{(w-z)^{n+1}}$$

gives

$$f^{(n)}(a) = \frac{n!}{2\pi i} \int_C \frac{f(w)}{(w-a)^{n+1}} \, dw.$$

This is a Cauchy integral formula for derivatives. Therefore the power series obtained above is the Taylor series of f.

- The argument works if z is any point that is closer to the center a than is any singularity of f. Therefore the radius of convergence of the Taylor series cannot be smaller than the distance from a to the nearest singularity (nor can it be larger, since power series have no singularities in the interiors of their circles of convergence).

- A special case of the identity theorem follows from the preceding remark. If two holomorphic functions agree on a (possibly quite small) open neighborhood U of a, then they coincide on the open disk $B_d(a)$, where d is the distance from a to the nearest singularity.

18.3 External links

- Existence of power series at PlanetMath.org.

Chapter 19

Proofs involving covariant derivatives

This article contains proof of formulas in Riemannian geometry that involve the Christoffel symbols.

19.1 Contracted Bianchi identities

19.1.1 Proof

Start with the Bianchi identity[1]

$$R_{abmn;l} + R_{ablm;n} + R_{abnl;m} = 0$$

Contract both sides of the above equation with a pair of metric tensors:

$$g^{bn}g^{am}(R_{abmn;l} + R_{ablm;n} + R_{abnl;m}) = 0,$$
$$g^{bn}(R^m{}_{bmn;l} - R^m{}_{bml;n} + R^m{}_{bnl;m}) = 0,$$
$$g^{bn}(R_{bn;l} - R_{bl;n} - R_b{}^m{}_{nl;m}) = 0,$$
$$R^n{}_{n;l} - R^n{}_{l;n} - R^{nm}{}_{nl;m} = 0.$$

The first term on the left contracts to yield a Ricci scalar, while the third term contracts to yield a mixed Ricci tensor,

$$R_{;l} - R^n{}_{l;n} - R^m{}_{l;m} = 0.$$

The last two terms are the same (changing dummy index n to m) and can be combined into a single term which shall be moved to the right,

$$R_{;l} = 2R^m{}_{l;m},$$

which is the same as

$$\nabla_m R^m{}_l = \frac{1}{2}\nabla_l R$$

Swapping the index labels l and m yields

$$\nabla_l R^l{}_m = \frac{1}{2}\nabla_m R, Q.E.D. \text{ (return to article)}$$

19.2 The covariant divergence of the Einstein tensor vanishes

19.2.1 Proof

The last equation in Proof 1 above can be expressed as

$$\nabla_l R^l{}_m - \frac{1}{2}\delta^l{}_m \nabla_l R = 0$$

where δ is the Kronecker delta. Since the mixed Kronecker delta is equivalent to the mixed metric tensor,

$$\delta^l{}_m = g^l{}_m,$$

and since the covariant derivative of the metric tensor is zero (so it can be moved in or out of the scope of any such derivative), then

$$\nabla_l R^l{}_m - \frac{1}{2}\nabla_l g^l{}_m R = 0.$$

Factor out the covariant derivative

$$\nabla_l \left(R^l{}_m - \frac{1}{2} g^l{}_m R \right) = 0,$$

then raise the index m throughout

$$\nabla_l \left(R^{lm} - \frac{1}{2} g^{lm} R \right) = 0.$$

The expression in parentheses is the Einstein tensor, so [1]

$$\nabla_l G^{lm} = 0, Q.E.D. \text{ (return to article)}$$

this means that the covariant divergence of the Einstein tensor vanishes.

19.3 See also

- Four-gradient
- Ricci calculus

19.4 References

[1] Synge J.L., Schild A. (1949). *Tensor Calculus.* pp. 87–89–90.

19.5 Books

- Bishop, R.L.; Goldberg, S.I. (1968), *Tensor Analysis on Manifolds* (First Dover 1980 ed.), The Macmillan Company, ISBN 0-486-64039-6

- Danielson, Donald A. (2003). *Vectors and Tensors in Engineering and Physics* (2/e ed.). Westview (Perseus). ISBN 978-0-8133-4080-7.

- Lovelock, David; Hanno Rund (1989) [1975]. *Tensors, Differential Forms, and Variational Principles.* Dover. ISBN 978-0-486-65840-7.

- Synge J.L., Schild A. (1949). *Tensor Calculus.* first Dover Publications 1978 edition. ISBN 978-0-486-63612-2.

- J.R. Tyldesley (1975), *An introduction to Tensor Analysis: For Engineers and Applied Scientists*, Longman, ISBN 0-582-44355-5

- D.C. Kay (1988), *Tensor Calculus*, Schaum's Outlines, McGraw Hill (USA), ISBN 0-07-033484-6

- T. Frankel (2012), *The Geometry of Physics* (3rd ed.), Cambridge University Press, ISBN 978-1107-602601

Chapter 20

Derivation of the Cartesian form for an ellipse

The **derivation of the Cartesian form for an ellipse** is simple and instructive. One simple definition of the ellipse is the "locus of all points of the plane whose distances to two fixed points(called the foci) add to the same constant". See ellipse for other definitions.

Let the foci be points (-c,0) and (c,0). Then the ellipse center is (0, 0). If (x,y) is any point on the ellipse and if d_1 is the distance between (x,y) and (-c,0) and d_2 is the distance between (x,y) and (c,0), i.e.

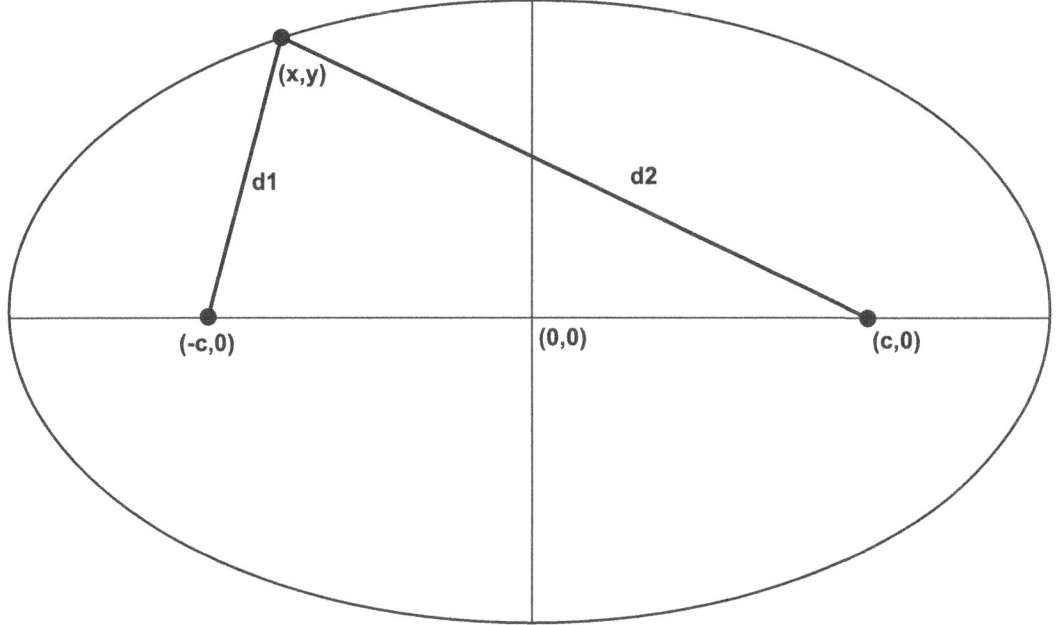

then we can define a

$d_1 + d_2 = 2a$

(**a** here is the semi-major axis, although this is irrelevant for the sake of the proof). From this simple definition we can derive the cartesian equation. Substituting:

$$\sqrt{(x+c)^2 + y^2} + \sqrt{(x-c)^2 + y^2} = 2a$$

To simplify we isolate the radical and square both sides.

$$\sqrt{(x+c)^2 + y^2} = 2a - \sqrt{(x-c)^2 + y^2}$$

$$(x+c)^2 + y^2 = \left(2a - \sqrt{(x-c)^2 + y^2}\right)^2$$

$$(x+c)^2 + y^2 = 4a^2 - 4a\sqrt{(x-c)^2 + y^2} + (x-c)^2 + y^2$$

Solving for the root and simplifying:

$$(x+c)^2 + y^2 - (x-c)^2 - y^2 - 4a^2 = -4a\sqrt{(x-c)^2 + y^2}$$

$$-\frac{1}{4a}((x+c)^2 + y^2 - 4a^2 - (x-c)^2 - y^2) = \sqrt{(x-c)^2 + y^2}$$

Swap sides to return to original format and continue:

$$\sqrt{(x-c)^2 + y^2} = -\frac{1}{4a}((x+c)^2 + y^2 - 4a^2 - (x-c)^2 - y^2)$$

$$\sqrt{(x-c)^2 + y^2} = -\frac{1}{4a}(x^2 + 2xc + c^2 - 4a^2 - x^2 + 2xc - c^2)$$

$$\sqrt{(x-c)^2 + y^2} = -\frac{1}{4a}(4xc - 4a^2)$$

$$\sqrt{(x-c)^2 + y^2} = a - \frac{c}{a}x$$

A final squaring

$$(x-c)^2 + y^2 = a^2 - 2xc + \frac{c^2}{a^2}x^2$$

$$x^2 - 2xc + c^2 + y^2 = a^2 - 2xc + \frac{c^2}{a^2}x^2$$

$$x^2 + c^2 + y^2 = a^2 + \frac{c^2}{a^2}x^2$$

Grouping the x-terms and dividing by $a^2 - c^2$

$$x^2 - \frac{c^2}{a^2}x^2 + y^2 = a^2 - c^2$$

$$x^2\left(1 - \frac{c^2}{a^2}\right) + y^2 = a^2 - c^2$$

Where: $1 = \frac{a^2}{a^2}$

$$x^2\left(\frac{a^2 - c^2}{a^2}\right) + y^2 = a^2 - c^2$$

$$\frac{x^2}{a^2} + \frac{y^2}{a^2 - c^2} = 1$$

If x = 0 then

$$d_1 = d_2 = a = \sqrt{c^2 + b^2}$$

(where b is the semi-minor axis)

Therefore we can substitute

$$b^2 = a^2 - c^2$$

And we have our desired equation:

$$\frac{x^2}{a^2} + \frac{y^2}{b^2} = 1$$

Chapter 21

Derivation of the Routh array

The Routh array is a tabular method permitting one to establish the stability of a system using only the coefficients of the characteristic polynomial. Central to the field of control systems design, the Routh–Hurwitz theorem and Routh array emerge by using the Euclidean algorithm and Sturm's theorem in evaluating Cauchy indices.

21.1 The Cauchy index

Given the system:

$$f(x) = a_0 x^n + a_1 x^{n-1} + \cdots + a_n \quad (1)$$
$$= (x - r_1)(x - r_2) \cdots (x - r_n) \quad (2)$$

Assuming no roots of $f(x) = 0$ lie on the imaginary axis, and letting

N = The number of roots of $f(x) = 0$ with negative real parts, and
P = The number of roots of $f(x) = 0$ with positive real parts

then we have

$$N + P = n \quad (3)$$

Expressing $f(x)$ in polar form, we have

$$f(x) = \rho(x) e^{j\theta(x)} \quad (4)$$

where

$$\rho(x) = \sqrt{\mathfrak{Re}^2[f(x)] + \mathfrak{Im}^2[f(x)]} \quad (5)$$

and

$$\theta(x) = \tan^{-1}\left(\mathfrak{Im}[f(x)]/\mathfrak{Re}[f(x)]\right) \quad (6)$$

from (2) note that

$$\theta(x) = \theta_{r_1}(x) + \theta_{r_2}(x) + \cdots + \theta_{r_n}(x) \quad (7)$$

where

$$\theta_{r_i}(x) = \angle(x - r_i) \quad (8)$$

Now if the ith root of $f(x) = 0$ has a positive real part, then (using the notation y=(RE[y],IM[y]))

$$\begin{aligned}\theta_{r_i}(x)\big|_{x=j\infty} &= \angle(x - r_i)\big|_{x=j\infty} \\ &= \angle(0 - \mathfrak{Re}[r_i], \infty - \mathfrak{Im}[r_i]) \\ &= \angle(-\mathfrak{Re}[r_i], \infty) \\ &= \lim_{\phi \to -\infty} \tan^{-1}\phi = -\frac{\pi}{2} \quad (9)\end{aligned}$$

and

$$\theta_{r_i}(x)\big|_{x=-j\infty} = \angle(-\mathfrak{Re}[r_i], -\infty) = \lim_{\phi \to \infty} \tan^{-1}\phi = \frac{\pi}{2} \quad (10)$$

Similarly, if the ith root of $f(x) = 0$ has a negative real part,

$$\theta_{r_i}(x)\big|_{x=j\infty} = \angle(-\mathfrak{Re}[r_i], \infty) = \lim_{\phi \to \infty} \tan^{-1}\phi = \frac{\pi}{2} \quad (11)$$

and

$$\theta_{r_i}(x)\big|_{x=-j\infty} = \angle(-\mathfrak{Re}[r_i], -\infty) = \lim_{\phi \to -\infty} \tan^{-1}\phi = -\frac{\pi}{2} \quad (12)$$

Therefore, $\theta_{r_i}(x)\Big|_{x=-j\infty}^{x=j\infty} = -\pi$ when the ith root of $f(x)$ has a positive real part, and $\theta_{r_i}(x)\Big|_{x=-j\infty}^{x=j\infty} = \pi$ when the ith root of $f(x)$ has a negative real part. Alternatively,

$$\theta(x)\big|_{x=j\infty} = \angle(x - r_1)\big|_{x=j\infty} + \angle(x - r_2)\big|_{x=j\infty} + \cdots + \angle(x - r_n)\big|_{x=j\infty} = \frac{\pi}{2}N - \frac{\pi}{2}P \quad (13)$$

and

$$\theta(x)\big|_{x=-j\infty} = \angle(x - r_1)\big|_{x=-j\infty} + \angle(x - r_2)\big|_{x=-j\infty} + \cdots + \angle(x - r_n)\big|_{x=-j\infty} = -\frac{\pi}{2}N + \frac{\pi}{2}P \quad (14)$$

So, if we define

$$\Delta = \frac{1}{\pi}\theta(x)\Big|_{-j\infty}^{j\infty} \quad (15)$$

then we have the relationship

$$N - P = \Delta \quad (16)$$

and combining (3) and (16) gives us

$$N = \frac{n+\Delta}{2} \text{ and } P = \frac{n-\Delta}{2} \quad (17)$$

Therefore, given an equation of $f(x)$ of degree n we need only evaluate this function Δ to determine N, the number of roots with negative real parts and P, the number of roots with positive real parts.

Equations (13) and (14) show that at $x = \pm\infty$, $\theta = \theta(x)$ is an integer multiple of $\pi/2$. Note now, in accordance with (6) and Figure 1, the graph of $\tan(\theta)$ vs θ, that varying x over an interval (a,b) where $\theta_a = \theta(x)|_{x=ja}$ and $\theta_b = \theta(x)|_{x=jb}$ are integer multiples of π, this variation causing the function $\theta(x)$ to have increased by π, indicates that in the course of travelling from point a to point b, θ has "jumped" from $+\infty$ to $-\infty$ one more time than it has jumped from $-\infty$ to $+\infty$. Similarly, if we vary x over an interval (a,b) this variation causing $\theta(x)$ to have decreased by π, where again θ is a multiple of π at both $x = ja$ and $x = jb$, implies that $\tan\theta(x) = \Im[f(x)]/\Re[f(x)]$ has jumped from $-\infty$ to $+\infty$ one more time than it has jumped from $+\infty$ to $-\infty$ as x was varied over the said interval.

Thus, $\theta(x)\big|_{-j\infty}^{j\infty}$ is π times the difference between the number of points at which $\Im[f(x)]/\Re[f(x)]$ jumps from $-\infty$ to $+\infty$ and the number of points at which $\Im[f(x)]/\Re[f(x)]$ jumps from $+\infty$ to $-\infty$ as x ranges over the interval $(-j\infty, +j\infty)$ provided that at $x = \pm j\infty$, $\tan[\theta(x)]$ is defined.

In the case where the starting point is on an incongruity (i.e. $\theta_a = \pi/2 \pm i\pi$, $i = 0, 1, 2, ...$) the ending point will be on an incongruity as well, by equation (16) (since N is an integer and P is an integer, Δ will be an integer). In this case, we can achieve this same index (difference in positive and negative jumps) by shifting the axes of the tangent function by $\pi/2$, through adding $\pi/2$ to θ. Thus, our index is now fully defined for any combination of coefficients in $f(x)$ by evaluating $\tan[\theta] = \Im[f(x)]/\Re[f(x)]$ over the interval (a,b) = $(+j\infty, -j\infty)$ when our starting (and thus ending) point is not an incongruity, and by evaluating

$$\tan[\theta'(x)] = \tan[\theta + \pi/2] = -\cot[\theta(x)] = -\Re[f(x)]/\Im[f(x)] \quad (18)$$

over said interval when our starting point is at an incongruity.

This difference, Δ, of negative and positive jumping incongruities encountered while traversing x from $-j\infty$ to $+j\infty$ is called the Cauchy Index of the tangent of the phase angle, the phase angle being $\theta(x)$ or $\theta'(x)$, depending as θ_a is an integer multiple of π or not.

21.2 The Routh criterion

To derive Routh's criterion, first we'll use a different notation to differentiate between the even and odd terms of $f(x)$:

$$f(x) = a_0 x^n + b_0 x^{n-1} + a_1 x^{n-2} + b_1 x^{n-3} + \cdots \quad (19)$$

Now we have:

$$\begin{aligned} f(j\omega) &= a_0(j\omega)^n + b_0(j\omega)^{n-1} + a_1(j\omega)^{n-2} + b_1(j\omega)^{n-3} + \cdots & (20) \\ &= a_0(j\omega)^n + a_1(j\omega)^{n-2} + a_2(j\omega)^{n-4} + \cdots & (21) \\ &\quad + b_0(j\omega)^{n-1} + b_1(j\omega)^{n-3} + b_2(j\omega)^{n-5} + \cdots \end{aligned}$$

Therefore, if n is even,

$$\begin{aligned} f(j\omega) = (-1)^{n/2}\big[a_0\omega^n + a_1\omega^{n-2} + a_2\omega^{n-4} + \cdots\big] & \quad (22) \\ + j(-1)^{(n/2)-1}\big[b_0\omega^{n-1} + b_1\omega^{n-3} + b_2\omega^{n-5} + \cdots\big] \end{aligned}$$

and if n is odd:

$$f(j\omega) = j(-1)^{(n-1)/2}\left[a_0\omega^n + a_1\omega^{n-2} + a_2\omega^{n-4} + \cdots\right] \quad (23)$$
$$+ (-1)^{(n-1)/2}\left[b_0\omega^{n-1} + b_1\omega^{n-3} + b_2\omega^{n-5} + \cdots\right]$$

Now observe that if n is an odd integer, then by (3) $N + P$ is odd. If $N + P$ is an odd integer, then $N - P$ is odd as well. Similarly, this same argument shows that when n is even, $N - P$ will be even. Equation (13) shows that if $N - P$ is even, θ is an integer multiple of π. Therefore, $\tan(\theta)$ is defined for n even, and is thus the proper index to use when n is even, and similarly $\tan(\theta') = \tan(\theta + \pi) = -\cot(\theta)$ is defined for n odd, making it the proper index in this latter case.

Thus, from (6) and (22), for n even:

$$\Delta = I_{-\infty}^{+\infty} \frac{-\Im[f(x)]}{\Re[f(x)]} = I_{-\infty}^{+\infty} \frac{b_0\omega^{n-1} - b_1\omega^{n-3} + \cdots}{a_0\omega^n - a_1\omega^{n-2} + \ldots} \quad (24)$$

and from (18) and (23), for n odd:

$$\Delta = I_{-\infty}^{+\infty} \frac{\Re[f(x)]}{\Im[f(x)]} = I_{-\infty}^{+\infty} \frac{b_0\omega^{n-1} - b_1\omega^{n-3} + \cdots}{a_0\omega^n - a_1\omega^{n-2} + \ldots} \quad (25)$$

Lo and behold we are evaluating the same Cauchy index for both:
$$\Delta = I_{-\infty}^{+\infty} \frac{b_0\omega^{n-1} - b_1\omega^{n-3} + \cdots}{a_0\omega^n - a_1\omega^{n-2} + \cdots} \quad (26)$$

21.3 Sturm's theorem

Sturm gives us a method for evaluating $\Delta = I_{-\infty}^{+\infty} \frac{f_2(x)}{f_1(x)}$. His theorem states as follows:

Given a sequence of polynomials $f_1(x), f_2(x), \ldots, f_m(x)$ where:

1) If $f_k(x) = 0$ then $f_{k-1}(x) \neq 0$, $f_{k+1}(x) \neq 0$, and $\text{sign}[f_{k-1}(x)] = -\text{sign}[f_{k+1}(x)]$

2) $f_m(x) \neq 0$ for $-\infty < x < \infty$

and we define $V(x)$ as the number of changes of sign in the sequence $f_1(x), f_2(x), \ldots, f_m(x)$ for a fixed value of x, then:

$$\Delta = I_{-\infty}^{+\infty} \frac{f_2(x)}{f_1(x)} = V(-\infty) - V(+\infty) \quad (27)$$

A sequence satisfying these requirements is obtained using the Euclidean algorithm, which is as follows:

Starting with $f_1(x)$ and $f_2(x)$, and denoting the remainder of $f_1(x)/f_2(x)$ by $f_3(x)$ and similarly denoting the remainder of $f_2(x)/f_3(x)$ by $f_4(x)$, and so on, we obtain the relationships:

$$f_1(x) = q_1(x)f_2(x) - f_3(x) \quad (28)$$
$$f_2(x) = q_2(x)f_3(x) - f_4(x)$$
$$\ldots$$
$$f_{m-1}(x) = q_{m-1}(x)f_m(x)$$

or in general

21.3. STURM'S THEOREM

$$f_{k-1}(x) = q_{k-1}(x)f_k(x) - f_{k+1}(x)$$

where the last non-zero remainder, $f_m(x)$ will therefore be the highest common factor of $f_1(x), f_2(x), \ldots, f_{m-1}(x)$. It can be observed that the sequence so constructed will satisfy the conditions of Sturm's theorem, and thus an algorithm for determining the stated index has been developed.

It is in applying Sturm's theorem (28) to (26), through the use of the Euclidean algorithm above that the Routh matrix is formed.

We get

$$f_3(\omega) = \frac{a_0}{b_0} f_2(\omega) - f_1(\omega) \quad (29)$$

and identifying the coefficients of this remainder by c_0, $-c_1$, c_2, $-c_3$, and so forth, makes our formed remainder

$$f_3(\omega) = c_0 \omega^{n-2} - c_1 \omega^{n-4} + c_2 \omega^{n-6} - \cdots \quad (30)$$

where

$$c_0 = a_1 - \frac{a_0}{b_0} b_1 = \frac{b_0 a_1 - a_1 b_0}{b_0}; c_1 = a_2 - \frac{a_0}{b_0} b_2 = \frac{b_0 a_2 - a_0 b_2}{b_0}; \ldots \quad (31)$$

Continuing with the Euclidean algorithm on these new coefficients gives us

$$f_4(\omega) = \frac{b_0}{c_0} f_3(\omega) - f_2(\omega) \quad (32)$$

where we again denote the coefficients of the remainder $f_4(\omega)$ by d_0, $-d_1$, d_2, $-d_3$,

making our formed remainder

$$f_4(\omega) = d_0 \omega^{n-3} - d_1 \omega^{n-5} + d_2 \omega^{n-7} - \cdots \quad (33)$$

and giving us

$$d_0 = b_1 - \frac{b_0}{c_0} c_1 = \frac{c_0 b_1 - b_1 c_0}{c_0}; d_1 = b_2 - \frac{b_0}{c_0} c_2 = \frac{c_0 b_2 - b_0 c_2}{c_0}; \ldots \quad (34)$$

The rows of the Routh array are determined exactly by this algorithm when applied to the coefficients of (19). An observation worthy of note is that in the regular case the polynomials $f_1(\omega)$ and $f_2(\omega)$ have as the highest common factor $f_{n+1}(\omega)$ and thus there will be n polynomials in the chain $f_1(x), f_2(x), \ldots, f_m(x)$.

Note now, that in determining the signs of the members of the sequence of polynomials $f_1(x), f_2(x), \ldots, f_m(x)$ that at $\omega = \pm\infty$ the dominating power of ω will be the first term of each of these polynomials, and thus only these coefficients corresponding to the highest powers of ω in $f_1(x), f_2(x), \ldots$, and $f_m(x)$, which are a_0, b_0, c_0, d_0, ... determine the signs of $f_1(x)$, $f_2(x)$, ..., $f_m(x)$ at $\omega = \pm\infty$.

So we get $V(+\infty) = V(a_0, b_0, c_0, d_0, \ldots)$ that is, $V(+\infty)$ is the number of changes of sign in the sequence $a_0 \infty^n$, $b_0 \infty^{n-1}$, $c_0 \infty^{n-2}$, ... which is the number of sign changes in the sequence a_0, b_0, c_0, d_0, ... and $V(-\infty) = V(a_0, -b_0, c_0, -d_0, \ldots)$; that is $V(-\infty)$ is the number of changes of sign in the sequence $a_0(-\infty)^n$, $b_0(-\infty)^{n-1}$, $c_0(-\infty)^{n-2}$, ... which is the number of sign changes in the sequence a_0, $-b_0$, c_0, $-d_0$, ...

Since our chain a_0, b_0, c_0, d_0, ... will have n members it is clear that $V(+\infty) + V(-\infty) = n$ since within $V(a_0, b_0, c_0, d_0, \ldots)$ if going from a_0 to b_0 a sign change has not occurred, within $V(a_0, -b_0, c_0, -d_0, \ldots)$ going from a_0 to $-b_0$ one has, and likewise for all n transitions (there will be no terms equal to zero) giving us n total sign changes.

As $\Delta = V(-\infty) - V(+\infty)$ and $n = V(+\infty) + V(-\infty)$, and from (17) $P = (n - \Delta/2)$, we have that $P = V(+\infty) = V(a_0, b_0, c_0, d_0, \ldots)$ and have derived Routh's theorem -

The number of roots of a real polynomial $f(z)$ which lie in the right half plane $\Re\mathfrak{e}(r_i) > 0$ is equal to the number of changes of sign in the first column of the Routh scheme.

And for the stable case where $P = 0$ then $V(a_0, b_0, c_0, d_0, \ldots) = 0$ by which we have Routh's famous criterion:

In order for all the roots of the polynomial $f(z)$ to have negative real parts, it is necessary and sufficient that all of the elements in the first column of the Routh scheme be different from zero and of the same sign.

21.4 References

- Hurwitz, A., "On the Conditions under which an Equation has only Roots with Negative Real Parts", Rpt. in Selected Papers on Mathematical Trends in Control Theory, Ed. R. T. Ballman et al. New York: Dover 1964

- Routh, E. J., A Treatise on the Stability of a Given State of Motion. London: Macmillan, 1877. Rpt. in Stability of Motion, Ed. A. T. Fuller. London: Taylor & Francis, 1975

- Felix Gantmacher (J.L. Brenner translator) (1959) *Applications of the Theory of Matrices*, pp 177–80, New York: Interscience.

Chapter 22

Deriving the Schwarzschild solution

The Schwarzschild solution is one of the simplest and most useful solutions of the Einstein field equations (see general relativity). It describes spacetime in the vicinity of a non-rotating massive spherically-symmetric object. It is worthwhile deriving this metric in some detail; the following is a reasonably rigorous derivation that is not always seen in the textbooks.

22.1 Assumptions and notation

Working in a coordinate chart with coordinates (r, θ, ϕ, t) labelled 1 to 4 respectively, we begin with the metric in its most general form (10 independent components, each of which is a smooth function of 4 variables). The solution is assumed to be spherically symmetric, static and vacuum. For the purposes of this article, these assumptions may be stated as follows (see the relevant links for precise definitions):

1. A spherically symmetric spacetime is one that is invariant under rotations and taking the mirror image.

2. A static spacetime is one in which all metric components are independent of the time coordinate t (so that $\frac{\partial}{\partial t} g_{\mu\nu} = 0$) and the geometry of the spacetime is unchanged under a time-reversal $t \to -t$.

3. A vacuum solution is one that satisfies the equation $T_{ab} = 0$. From the Einstein field equations (with zero cosmological constant), this implies that $R_{ab} = 0$ since contracting $R_{ab} - \frac{R}{2} g_{ab} = 0$ yields $R = 0$.

4. Metric signature used here is (+,+,+,−).

22.2 Diagonalising the metric

The first simplification to be made is to diagonalise the metric. Under the coordinate transformation, $(r, \theta, \phi, t) \to (r, \theta, \phi, -t)$, all metric components should remain the same. The metric components $g_{\mu 4}$ ($\mu \neq 4$) change under this transformation as:

$$g'_{\mu 4} = \frac{\partial x^\alpha}{\partial x'^\mu} \frac{\partial x^\beta}{\partial x'^4} g_{\alpha\beta} = -g_{\mu 4} \ (\mu \neq 4)$$

But, as we expect $g'_{\mu 4} = g_{\mu 4}$ (metric components remain the same), this means that:

$$g_{\mu 4} = 0 \ (\mu \neq 4)$$

Similarly, the coordinate transformations $(r, \theta, \phi, t) \to (r, \theta, -\phi, t)$ and $(r, \theta, \phi, t) \to (r, -\theta, \phi, t)$ respectively give:

$$g_{\mu 3} = 0 \, (\mu \neq 3)$$
$$g_{\mu 2} = 0 \, (\mu \neq 2)$$

Putting all these together gives:

$$g_{\mu\nu} = 0 \, (\mu \neq \nu)$$

and hence the metric must be of the form:

$$ds^2 = g_{11} \, dr^2 + g_{22} \, d\theta^2 + g_{33} \, d\phi^2 + g_{44} \, dt^2$$

where the four metric components are independent of the time coordinate t (by the static assumption).

22.3 Simplifying the components

On each hypersurface of constant t, constant θ and constant ϕ (i.e., on each radial line), g_{11} should only depend on r (by spherical symmetry). Hence g_{11} is a function of a single variable:

$$g_{11} = A(r)$$

A similar argument applied to g_{44} shows that:

$$g_{44} = B(r)$$

On the hypersurfaces of constant t and constant r, it is required that the metric be that of a 2-sphere:

$$dl^2 = r_0^2(d\theta^2 + \sin^2\theta \, d\phi^2)$$

Choosing one of these hypersurfaces (the one with radius r_0, say), the metric components restricted to this hypersurface (which we denote by $\tilde{g}_{22}$ and $\tilde{g}_{33}$) should be unchanged under rotations through θ and ϕ (again, by spherical symmetry). Comparing the forms of the metric on this hypersurface gives:

$$\tilde{g}_{22}\left(d\theta^2 + \frac{\tilde{g}_{33}}{\tilde{g}_{22}} d\phi^2\right) = r_0^2(d\theta^2 + \sin^2\theta \, d\phi^2)$$

which immediately yields:

$$\tilde{g}_{22} = r_0^2 \text{ and } \tilde{g}_{33} = r_0^2 \sin^2\theta$$

But this is required to hold on each hypersurface; hence,

$$g_{22} = r^2 \text{ and } g_{33} = r^2 \sin^2\theta$$

Thus, the metric can be put in the form:

$$ds^2 = A(r) \, dr^2 + r^2 \, d\theta^2 + r^2 \sin^2\theta \, d\phi^2 + B(r) \, dt^2$$

with A and B as yet undetermined functions of r. Note that if A or B is equal to zero at some point, the metric would be singular at that point.

22.4 Calculating the Christoffel symbols

Using the metric above, we find the Christoffel symbols, where the indices are $(1,2,3,4) = (r,\theta,\phi,t)$. The sign $'$ denotes a total derivative of a function.

$$\Gamma^1_{ik} = \begin{bmatrix} A'/(2A) & 0 & 0 & 0 \\ 0 & -r/A & 0 & 0 \\ 0 & 0 & -r\sin^2\theta/A & 0 \\ 0 & 0 & 0 & -B'/(2A) \end{bmatrix}$$

$$\Gamma^2_{ik} = \begin{bmatrix} 0 & 1/r & 0 & 0 \\ 1/r & 0 & 0 & 0 \\ 0 & 0 & -\sin\theta\cos\theta & 0 \\ 0 & 0 & 0 & 0 \end{bmatrix}$$

$$\Gamma^3_{ik} = \begin{bmatrix} 0 & 0 & 1/r & 0 \\ 0 & 0 & \cot\theta & 0 \\ 1/r & \cot\theta & 0 & 0 \\ 0 & 0 & 0 & 0 \end{bmatrix}$$

$$\Gamma^4_{ik} = \begin{bmatrix} 0 & 0 & 0 & B'/(2B) \\ 0 & 0 & 0 & 0 \\ 0 & 0 & 0 & 0 \\ B'/(2B) & 0 & 0 & 0 \end{bmatrix}$$

22.5 Using the field equations to find A(r) and B(r)

To determine A and B, the vacuum field equations are employed:

$$R_{\alpha\beta} = 0$$

Hence:

$$\Gamma^\rho_{\beta\alpha,\rho} - \Gamma^\rho_{\rho\alpha,\beta} + \Gamma^\rho_{\rho\lambda}\Gamma^\lambda_{\beta\alpha} - \Gamma^\rho_{\beta\lambda}\Gamma^\lambda_{\rho\alpha} = 0$$

where a comma is used to set off the index that is being used for the derivative.

Only four of these equations are nontrivial and upon simplification become:

$4A'B^2 - 2rB''AB + rA'B'B + rB'^2A = 0$

$rA'B + 2A^2B - 2AB - rB'A = 0$

$-2rB''AB + rA'B'B + rB'^2A - 4B'AB = 0$

(The fourth equation is just $\sin^2\theta$ times the second equation)

where the dash means the r derivative of the functions.

Subtracting the first and third equations produces:

$A'B + AB' = 0 \Rightarrow A(r)B(r) = K$

where K is a non-zero real constant. Substituting $A(r)B(r) = K$ into the second equation and tidying up gives:

$rA' = A(1-A)$

which has general solution:

$A(r) = \left(1 + \frac{1}{Sr}\right)^{-1}$

for some non-zero real constant S. Hence, the metric for a static, spherically symmetric vacuum solution is now of the form:

$ds^2 = \left(1 + \frac{1}{Sr}\right)^{-1} dr^2 + r^2(d\theta^2 + \sin^2\theta d\phi^2) + K\left(1 + \frac{1}{Sr}\right) dt^2$

Note that the spacetime represented by the above metric is asymptotically flat, i.e. as $r \to \infty$, the metric approaches that of the Minkowski metric and the spacetime manifold resembles that of Minkowski space.

22.6 Using the Weak-Field Approximation to find K and S

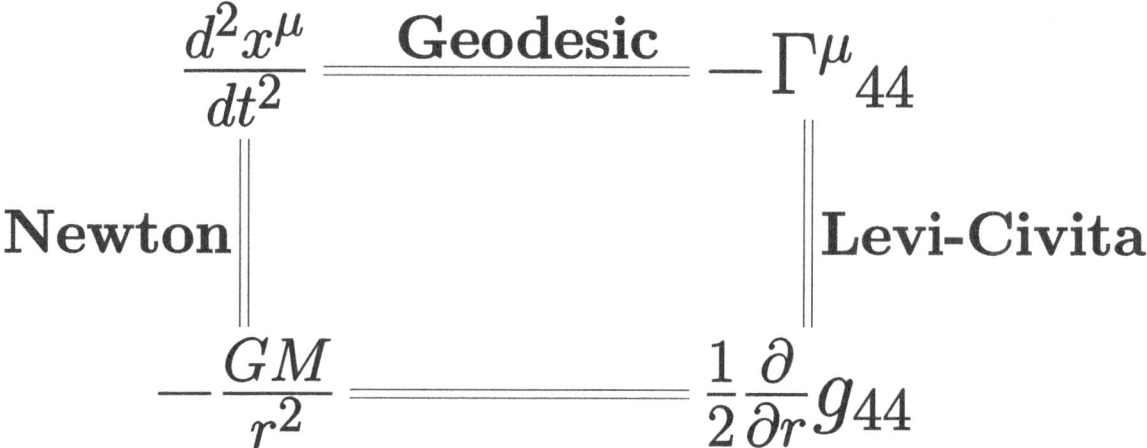

This diagram gives the route to find the Schwarzschild solution by using weak field approximation. The equality on the second row gives g_{44} = -c^2 + 2GM/r, assuming the desired solution degenerates to Minkowski metric when the motion happens far away from the blackhole (r approaches to positive infinity).

The geodesics of the metric (obtained where ds is extremised) must, in some limit (e.g., toward infinite speed of light), agree with the solutions of Newtonian motion (e.g., obtained by Lagrange equations). (The metric must also limit to Minkowski space when the mass it represents vanishes.)

$0 = \delta \int \frac{ds}{dt} dt = \delta \int (KE + PE_g) dt$

(where KE is the kinetic energy and PE_g is the Potential Energy due to gravity) The constants K and S are fully determined by some variant of this approach; from the weak-field approximation one arrives at the result:

$g_{44} = K\left(1 + \frac{1}{Sr}\right) \approx -c^2 + \frac{2Gm}{r} = -c^2\left(1 - \frac{2Gm}{c^2 r}\right)$

where G is the gravitational constant, m is the mass of the gravitational source and c is the speed of light. It is found that:

$K = -c^2$ and $\frac{1}{S} = -\frac{2Gm}{c^2}$

Hence:

$A(r) = \left(1 - \frac{2Gm}{c^2 r}\right)^{-1}$ and $B(r) = -c^2\left(1 - \frac{2Gm}{c^2 r}\right)$

So, the Schwarzschild metric may finally be written in the form:

$ds^2 = \left(1 - \frac{2Gm}{c^2 r}\right)^{-1} dr^2 + r^2(d\theta^2 + \sin^2\theta d\phi^2) - c^2\left(1 - \frac{2Gm}{c^2 r}\right) dt^2$

Note that:

$\frac{2Gm}{c^2} = r_s$

is the definition of the Schwarzschild radius for an object of mass m, so the Schwarzschild metric may be rewritten in the alternative form:

$ds^2 = \left(1 - \frac{r_s}{r}\right)^{-1} dr^2 + r^2(d\theta^2 + \sin^2\theta d\phi^2) - c^2 \left(1 - \frac{r_s}{r}\right) dt^2$

which shows that the metric becomes singular approaching the event horizon (that is, $r \to r_s$). The metric singularity is not a physical one (although there is a real physical singularity at $r = 0$), as can be shown by using a suitable coordinate transformation (e.g. the Kruskal–Szekeres coordinate system).

22.7 Alternative form in isotropic coordinates

The original formulation of the metric uses anisotropic coordinates in which the velocity of light is not the same in the radial and transverse directions. A S Eddington gave alternative forms in isotropic coordinates.[1] For isotropic spherical coordinates r_1, θ, ϕ, coordinates θ and ϕ are unchanged, and then (provided $r \geq \frac{2Gm}{c^2}$)[2]

$r = r_1 \left(1 + \frac{Gm}{2c^2 r_1}\right)^2 \ldots, dr = dr_1 \left(1 - \frac{(Gm)^2}{4c^4 r_1^2}\right) \ldots$, and

$\left(1 - \frac{2Gm}{c^2 r}\right) = \left(1 - \frac{Gm}{2c^2 r_1}\right)^2 / \left(1 + \frac{Gm}{2c^2 r_1}\right)^2 \ldots$

Then for isotropic rectangular coordinates x, y, z,

$x = r_1 \sin(\theta) \cos(\phi) \ldots, y = r_1 \sin(\theta) \sin(\phi) \ldots, z = r_1 \cos(\theta) \ldots$

The metric then becomes, in isotropic rectangular coordinates:

$ds^2 = \left(1 + \frac{Gm}{2c^2 r_1}\right)^4 (dx^2 + dy^2 + dz^2) - c^2 dt^2 \left(1 - \frac{Gm}{2c^2 r_1}\right)^2 / \left(1 + \frac{Gm}{2c^2 r_1}\right)^2 \ldots$

22.8 Dispensing with the static assumption - Birkhoff's theorem

In deriving the Schwarzschild metric, it was assumed that the metric was vacuum, spherically symmetric and static. In fact, the static assumption is stronger than required, as Birkhoff's theorem states that any spherically symmetric vacuum solution of Einstein's field equations is stationary; then one obtains the Schwarzschild solution. Birkhoff's theorem has the consequence that any pulsating star which remains spherically symmetric cannot generate gravitational waves (as the region exterior to the star must remain static).

22.9 See also

- Karl Schwarzschild
- Kerr metric
- Reissner–Nordström metric

22.10 References

[1] A S Eddington, "Mathematical Theory of Relativity", Cambridge UP 1922 (2nd ed.1924, repr.1960), at page 85 and page 93. Symbol usage in the Eddington source for interval s and time-like coordinate t has been converted for compatibility with the usage in the derivation above.

[2] Buchdahl, H. A. (1985). "Isotropic coordinates and Schwarzschild metric". *International Journal of Theoretical Physics* **24** (7): 731–739. Bibcode:1985IJTP...24..731B. doi:10.1007/BF00670880.

Chapter 23

Dual of BCH is an independent source

A certain family of BCH codes have a particularly useful property, which is that treated as linear operators, their dual operators turns their input into an ℓ-wise independent source. That is, the set of vectors from the input vector space are mapped to an ℓ-wise independent source. The proof of this fact below as the following Lemma and Corollary is useful in derandomizing the algorithm for a $1 - 2^{-\ell}$-approximation to MAXEkSAT.

23.1 Lemma

Let $C \subseteq F_2^n$ be a linear code such that $C^\perp$ has distance greater than $\ell + 1$. Then C is an ℓ-wise independent source.

23.2 Proof of Lemma

It is sufficient to show that given any $k \times l$ matrix M, where k is greater than or equal to l, such that the rank of M is l, for all $x \in F_2^k$, xM takes every value in F_2^l the same number of times.

Since M has rank l, we can write M as two matrices of the same size, M_1 and M_2, where M_1 has rank equal to l. This means that xM can be rewritten as $x_1 M_1 + x_2 M_2$ for some x_1 and x_2.

If we consider M written with respect to a basis where the first l rows are the identity matrix, then x_1 has zeros wherever M_2 has nonzero rows, and x_2 has zeros wherever M_1 has nonzero rows.

Now any value y, where $y = xM$, can be written as $x_1 M_1 + x_2 M_2$ for some vectors x_1, x_2.

We can rewrite this as:

$x_1 M_1 = y - x_2 M_2$

Fixing the value of the last $k - l$ coordinates of $x_2 \in F_2^k$ (note that there are exactly 2^{k-l} such choices), we can rewrite this equation again as:

$x_1 M_1 = b$ for some b.

Since M_1 has rank equal to l, there is exactly one solution x_1, so the total number of solutions is exactly 2^{k-l}, proving the lemma.

23.3 Corollary

Recall that BCH_2,m,d is an $[n = 2^m, n - 1 - \lceil d - 2/2 \rceil m, d]_2$ linear code.

Let $C^\perp$ be $\text{BCH}_{2,\log n,\ell+1}$. Then C is an ℓ-wise independent source of size $O(n^{\lfloor \ell/2 \rfloor})$.

23.4 Proof of Corollary

The dimension d of C is just $\lceil (\ell + 1 - 2)/2 \rceil \log n + 1$. So $d = \lceil (\ell - 1)/2 \rceil \log n + 1 = \lfloor \ell/2 \rfloor \log n + 1$. So the cardinality of C considered as a set is just $2^d = O(n^{\lfloor \ell/2 \rfloor})$, proving the Corollary.

23.5 References

Coding Theory notes at University at Buffalo

Coding Theory notes at MIT

Chapter 24

Proofs of Fermat's theorem on sums of two squares

Fermat's theorem on sums of two squares asserts that an odd prime number p can be expressed as

$$p = x^2 + y^2$$

with integer x and y if and only if p is congruent to 1 (mod 4). The statement was announced by Fermat in 1640, but he supplied no proof.

The "only if" clause is easy: a perfect square is congruent to 0 or 1 modulo 4, hence a sum of two squares is congruent to 0, 1, or 2. An odd prime number is congruent to either 1 or 3 modulo 4, and the second possibility has just been ruled out. The first proof that such a representation exists was given by Leonhard Euler in 1747 and was complicated. Since then, many different proofs have been found. Among them, the proof using Minkowski's theorem about convex sets[1] and Don Zagier's short proof based on involutions have appeared.

24.1 Euler's proof by infinite descent

Euler succeeded in proving Fermat's theorem on sums of two squares in 1749, when he was forty-two years old. He communicated this in a letter to Goldbach dated 12 April 1749.[2] The proof relies on infinite descent, and is only briefly sketched in the letter. The full proof consists in five steps and is published in two papers. The first four steps are Propositions 1 to 4 of the first paper[3] and do not correspond exactly to the four steps below. The fifth step below is from the second paper.[4] [5]

1. *The product of two numbers, each of which is a sum of two squares, is itself a sum of two squares.*

   This is a well known property, based on the identity

   $$(a^2 + b^2)(p^2 + q^2) = (ap + bq)^2 + (aq - bp)^2$$

   due to Diophantus of Alexandria.

2. *If a number which is a sum of two squares is divisible by a prime which is a sum of two squares, then the quotient is a sum of two squares.* (This is Euler's first Proposition).

24.1. EULER'S PROOF BY INFINITE DESCENT

Indeed, suppose for example that $a^2 + b^2$ is divisible by $p^2 + q^2$ and that this latter is a prime. Then $p^2 + q^2$ divides

$$(pb - aq)(pb + aq) = p^2b^2 - a^2q^2 = p^2(a^2 + b^2) - a^2(p^2 + q^2).$$

Since $p^2 + q^2$ is a prime, it divides one of the two factors. Suppose that it divides $pb - aq$. Since

$$(a^2 + b^2)(p^2 + q^2) = (ap + bq)^2 + (aq - bp)^2$$

(Diophantus identity) it follows that $p^2 + q^2$ must divide $(ap + bq)^2$. So the equation can be divided by the square of $p^2 + q^2$. Dividing the expression by $(p^2 + q^2)^2$ yields:

$$\frac{a^2 + b^2}{p^2 + q^2} = \left(\frac{ap + bq}{p^2 + q^2}\right)^2 + \left(\frac{aq - bp}{p^2 + q^2}\right)^2$$

and thus expresses the quotient as a sum of two squares, as claimed.

If $p^2 + q^2$ divides $pb + aq$, a similar argument holds by using

$$(a^2 + b^2)(q^2 + p^2) = (aq + bp)^2 + (ap - bq)^2$$

(Diophantus identity).

3. *If a number which can be written as a sum of two squares is divisible by a number which is not a sum of two squares, then the quotient has a factor which is not a sum of two squares.* (This is Euler's second Proposition).

Suppose x divides $a^2 + b^2$ and that the quotient, factored into its prime factors is $p_1 p_2 \cdots p_n$. Then $a^2 + b^2 = x p_1 p_2 \cdots p_n$. If all factors p_i can be written as sums of two squares, then we can divide $a^2 + b^2$ successively by p_1, p_2, etc., and applying the previous step we deduce that each quotient is a sum of two squares. This until we get to x, concluding that x would have to be the sum of two squares. So, by contraposition, if x is not the sum of two squares, then at least one of the primes p_i is not the sum of two squares.

4. *If a and b are relatively prime then every factor of $a^2 + b^2$ is a sum of two squares.* (This is Euler's Proposition 4. The proof sketched below includes the proof of his Proposition 3).

This is the step that uses infinite descent. Let x be a factor of $a^2 + b^2$. We can write

$$a = mx \pm c, \qquad b = nx \pm d$$

where c and d are at most half of x in absolute value. This gives:

$$a^2 + b^2 = m^2x^2 \pm 2mxc + c^2 + n^2x^2 \pm 2nxd + d^2 = Ax + (c^2 + d^2).$$

Therefore, $c^2 + d^2$ must be divisible by x, say $c^2 + d^2 = yx$. If c and d are not relatively prime, then their gcd must be relatively prime to x (else the common factor of their gcd and x would also be a common factor of a and b which we assume are relatively prime). Thus the square of the gcd divides y (as it divides $c^2 + d^2$), giving us an expression of the form $e^2 + f^2 = zx$ for relatively prime e and f, and with z no more than half of x, since

$$zx = e^2 + f^2 \leq c^2 + d^2 \leq \left(\frac{x}{2}\right)^2 + \left(\frac{x}{2}\right)^2 = \frac{1}{2}x^2.$$

If c and d are relatively prime, then we can use them directly instead of switching to e and f.

If x is not the sum of two squares, then by the third step there must be a factor of z which is not the sum of two squares; call it w. This gives an infinite descent, going from x to a smaller number w, both not the sums of two squares but dividing a sum of two squares. Since an infinite descent is impossible, we conclude that x must be expressible as a sum of two squares, as claimed.

5. *Every prime of the form $4n + 1$ is a sum of two squares.* (This is the main result of Euler's second paper).

If $p = 4n + 1$, then by Fermat's Little Theorem each of the numbers $1, 2^{4n}, 3^{4n}, \ldots, (4n)^{4n}$ is congruent to one modulo p. The differences $2^{4n} - 1, 3^{4n} - 2^{4n}, \ldots, (4n)^{4n} - (4n-1)^{4n}$ are therefore all divisible by p. Each of these differences can be factored as

$$a^{4n} - b^{4n} = \left(a^{2n} + b^{2n}\right)\left(a^{2n} - b^{2n}\right).$$

Since p is prime, it must divide one of the two factors. If in any of the $4n - 1$ cases it divides the first factor, then by the previous step we conclude that p is itself a sum of two squares (since a and b differ by 1, they are relatively prime). So it is enough to show that p cannot always divide the second factor. If it divides all $4n - 1$ differences $2^{2n} - 1, 3^{2n} - 2^{2n}, \ldots, (4n)^{2n} - (4n-1)^{2n}$, then it would divide all $4n - 2$ differences of successive terms, all $4n - 3$ differences of the differences, and so forth. Since the k th differences of the sequence $1^k, 2^k, 3^k, \ldots$ are all equal to $k!$ (Finite difference), the $2n$ th differences would all be constant and equal to $(2n)!$, which is certainly not divisible by p. Therefore, p cannot divide all the second factors which proves that p is indeed the sum of two squares.

24.2 Lagrange's proof through quadratic forms

Lagrange completed a proof in 1775[6] based on his general theory of integral quadratic forms. The following is a slight simplification of his argument, due to Gauss, which appears in article 182 of the Disquisitiones Arithmeticae.

A (binary) quadratic form will be taken to be an expression of the form $ax^2 + 2bxy + cy^2$ with a, b, c integers. A number n is said to be *represented by the form* if there exist integers x, y such that $n = ax^2 + 2bxy + cy^2$. Fermat's theorem on sums of two squares is then equivalent to the statement that a prime p is represented by the form $x^2 + y^2$ (i.e., $a = c = 1$, $b = 0$) exactly when p is congruent to 1 modulo 4.

The discriminant of the quadratic form is defined to be $b^2 - ac$ (this is the definition due to Gauss; Lagrange did not require the xy term to have even coefficient, and defined the discriminant as $b^2 - 4ac$). The discriminant of $x^2 + y^2$ is then equal to -1.

Two forms $ax^2 + 2bxy + cy^2$ and $rx'^2 + 2sx'y' + ty'^2$ are *equivalent* if and only if there exist substitutions with integer coefficients

$x = \alpha x' + \beta y'$

$y = \gamma x' + \delta y'$

with $\alpha\delta - \beta\gamma = \pm 1$ such that, when substituted into the first form, yield the second. Equivalent forms are readily seen to have the same discriminant. Moreover, it is clear that equivalent forms will represent exactly the same integers.

Lagrange proved that all positive definite forms of discriminant −1 are equivalent. Thus, to prove Fermat's theorem it is enough to find *any* positive definite form of discriminant −1 that represents p. To do this, it suffices to find an integer m such that p divides $m^2 + 1$. For, finding such an integer, we can consider the form

$$px^2 + 2mxy + \left(\frac{m^2+1}{p}\right)y^2,$$

which has discriminant −1 and represents p by setting $x = 1$ and $y = 0$.

Suppose then that $p = 4n + 1$. Again we invoke Fermat's Little Theorem: for any z relatively prime to p, we know that p divides $z^{p-1} - 1 = z^{4n} - 1 = (z^{2n} - 1)(z^{2n} + 1)$. Moreover, by a theorem of Lagrange, the number of solutions modulo p to a congruence of degree q modulo p is at most q (this follows since the integers modulo p form a field, and a polynomial of degree q has at most q roots). So the congruence $z^{2n} - 1 \equiv 0 \pmod{p}$ has at most $2n$ solutions among the numbers 1, 2, ..., $p - 1 = 4n$. Therefore, there exists some positive integer z strictly smaller than p (and so relatively prime to p) such that p does not divide $z^{2n} - 1$. Since p divides $z^{4n} - 1 = (z^{2n} - 1)(z^{2n} + 1)$, p must divide $z^{2n} + 1$. Setting $m = z^n$ completes the proof.

24.3 Dedekind's two proofs using Gaussian integers

Richard Dedekind gave at least two proofs of Fermat's theorem on sums of two squares, both using the arithmetical properties of the Gaussian integers, which are numbers of the form $a + bi$, where a and b are integers, and i is the square root of −1. One appears in section 27 of his exposition of ideals published in 1877; the second appeared in Supplement XI to Peter Gustav Lejeune Dirichlet's *Vorlesungen über Zahlentheorie*, and was published in 1894.

1. First proof. If p is an odd prime number, then we have $i^{p-1} = (-1)^{\frac{p-1}{2}}$ in the Gaussian integers. Consequently, writing a Gaussian integer $\omega = x + iy$ with $x, y \in \mathbf{Z}$ and applying the Frobenius automorphism in $\mathbf{Z}[i]/(p)$, one finds

$$\omega^p = (x + yi)^p \equiv x^p + y^p i^p \equiv x + (-1)^{\frac{p-1}{2}} yi \pmod{p},$$

since the automorphism fixes the elements of $\mathbf{Z}/(p)$. In the current case, $p = 4n + 1$ for some integer n, and so in the above expression for ω^p, the exponent (p−1)/2 of −1 is even. Hence the right hand side equals ω, so in this case the Frobenius endomorphism of $\mathbf{Z}[i]/(p)$ is the identity. Kummer had already established that if $f \in \{1, 2\}$ is the order of the Frobenius automorphism of $\mathbf{Z}[i]/(p)$, then the ideal (p) in $\mathbf{Z}[i]$ would be a product of $2/f$ distinct prime ideals. (In fact, Kummer had established a much more general result for any extension of $\mathbf{Z}$ obtained by adjoining a primitive m-th root of unity, where m was any positive integer; this is the case $m = 4$ of that result.) Therefore the ideal (p) is the product of two different prime ideals in $\mathbf{Z}[i]$. Since the Gaussian integers are a Euclidean domain for the norm function $N(x + iy) = x^2 + y^2$, every ideal is principal and generated by a nonzero element of the ideal of minimal norm. Since the norm is multiplicative, the norm of a generator α of one of the ideal factors of (p) must be a strict divisor of $N(p) = p^2$, so that we must have $p = N(\alpha) = N(a + bi) = a^2 + b^2$, which gives Fermat's theorem.

2. Second proof. This proof builds on Lagrange's result that if $p = 4n + 1$ is a prime number, then there must be an integer m such that $m^2 + 1$ is divisible by p (we can also see this by Euler's criterion); it also uses the fact that the Gaussian integers are a unique factorization domain (because they are a Euclidean domain). Since $p \in \mathbf{Z}$ does not divide either of the Gaussian integers $m + i$ and $m - i$ (as it does not divide their imaginary parts), but it does divide their product $m^2 + 1$, it follows that p cannot be a prime element in the Gaussian integers. We must therefore have a nontrivial factorization of p in the Gaussian integers, which in view of the norm can have only two factors (since the norm is multiplicative, and $p^2 = N(p)$, there can only be up to two factors of p), so it must be of the form $p = (x + yi)(x - yi)$ for some integers x and y. This immediately yields that $p = x^2 + y^2$.

24.4 Zagier's "one-sentence proof"

If $p = 4k + 1$ is prime, then the set $S = \{(x, y, z) \in \mathbf{N}^3 : x^2 + 4yz = p\}$ (here the set $\mathbf{N}$ of all natural numbers can be taken to include 0 or to exclude 0, and in both cases, x, y and z must be positive for any $(x, y, z) \in S$, as p is an odd prime) is finite and has two involutions: an obvious one $(x, y, z) \to (x, z, y)$, whose fixed points correspond to representations of p as a sum of two squares, and a more complicated one,

$$(x,y,z) \mapsto \begin{cases} (x+2z,\ z,\ y-x-z), & \text{if } x < y - z \\ (2y-x,\ y,\ x-y+z), & \text{if } y - z < x < 2y \\ (x-2y,\ x-y+z,\ y), & \text{if } x > 2y \end{cases}$$

which has exactly one fixed point, $(1, 1, k)$; however, the number of fixed points of an involution of a finite set S has the same parity as the cardinality of S, so this number is odd (hence, not zero) for the first involution as well, proving that p is a sum of two squares.

This proof, due to Zagier, is a simplification of an earlier proof by Heath-Brown, which in turn was inspired by a proof of Liouville. The technique of the proof is a combinatorial analogue of the topological principle that the Euler characteristics of a topological space with an involution and of its fixed point set have the same parity and is reminiscent of the use of *sign-reversing involutions* in the proofs of combinatorial bijections.

24.5 References

- Richard Dedekind, *The theory of algebraic integers*.

- Harold M. Edwards, *Fermat's Last Theorem. A genetic introduction to algebraic number theory*. Graduate Texts in Mathematics no. 50, Springer-Verlag, NY, 1977.

- C. F. Gauss, *Disquisitiones Arithmeticae* (English Edition). Transl. by Arthur A. Clarke. Springer-Verlag, 1986.

- Goldman, Jay R. (1998), *The Queen of Mathematics: A historically motivated guide to Number Theory*, A K Peters, ISBN 1-56881-006-7

- D. R. Heath-Brown, *Fermat's two squares theorem*. Invariant, 11 (1984) pp. 3–5.

- John Stillwell, Introduction to *Theory of Algebraic Integers* by Richard Dedekind. Cambridge Mathematical Library, Cambridge University Press, 1996.

- Don Zagier, *A one-sentence proof that every prime $p \equiv 1 \mod 4$ is a sum of two squares*. Amer. Math. Monthly 97 (1990), no. 2, 144, doi:10.2307/2323918

24.6 Notes

[1] See Goldman's book, §22.5

[2] Euler à Goldbach, lettre CXXV

[3] De numerus qui sunt aggregata quorum quadratorum. (Novi commentarii academiae scientiarum Petropolitanae 4 (1752/3), 1758, 3-40)

[4] Demonstratio theorematis FERMATIANI omnem numerum primum formae 4n+1 esse summam duorum quadratorum. (Novi commentarii academiae scientiarum Petropolitanae 5 (1754/5), 1760, 3-13)

[5] The summary is taken from Edwards book, pages 45-48; italics in the original.

[6] Nouv. Mém. Acad. Berlin, année 1771, 125; ibid. année 1773, 275; ibid année 1775, 351.

24.7 External links

- Two more proofs at PlanetMath.org
- "A one-sentence proof of the theorem". Archived from the original on 5 February 2012.
- reprint of Heath-Brown's proof, with commentary

Chapter 25

Furstenberg's proof of the infinitude of primes

In number theory, **Hillel Furstenberg's proof of the infinitude of primes** is a celebrated topological proof that the integers contain infinitely many prime numbers. When examined closely, the proof is less a statement about topology than a statement about certain properties of arithmetic sequences.[1] Unlike Euclid's classical proof, Furstenberg's proof is a proof by contradiction. The proof was published in 1955 in the *American Mathematical Monthly* while Furstenberg was still an undergraduate student at Yeshiva University.

25.1 Furstenberg's proof

Define a topology on the integers $\mathbf{Z}$, called the evenly spaced integer topology, by declaring a subset $U \subseteq \mathbf{Z}$ to be an open set if and only if it is either the empty set, $\emptyset$, or it is a union of arithmetic sequences $S(a, b)$ (for $a \neq 0$), where

$$S(a,b) = \{an + b \mid n \in \mathbb{Z}\} = a\mathbb{Z} + b.$$

In other words, U is open if and only if every $x \in U$ admits some non-zero integer a such that $S(a, x) \subseteq U$. The axioms for a topology are easily verified:

- By definition, $\emptyset$ is open; $\mathbf{Z}$ is just the sequence $S(1, 0)$, and so is open as well.

- Any union of open sets is open: for any collection of open sets Ui and x in their union U, any of the numbers ai for which $S(ai, x) \subseteq Ui$ also shows that $S(ai, x) \subseteq U$.

- The intersection of two (and hence finitely many) open sets is open: let U_1 and U_2 be open sets and let $x \in U_1 \cap U_2$ (with numbers a_1 and a_2 establishing membership). Set a to be the lowest common multiple of a_1 and a_2. Then $S(a, x) \subseteq S(ai, x) \subseteq U_i$.

This topology has two notable properties:

1. Since any non-empty open set contains an infinite sequence, a finite set cannot be open; put another way, the complement of a finite set cannot be a closed set.

2. The basis sets $S(a, b)$ are both open and closed: they are open by definition, and we can write $S(a, b)$ as the complement of an open set as follows:

$$S(a,b) = \mathbb{Z} \setminus \bigcup_{j=1}^{a-1} S(a, b+j).$$

The only integers that are not integer multiples of prime numbers are -1 and $+1$, i.e.

$$\mathbb{Z} \setminus \{-1, +1\} = \bigcup_{p \text{ prime}} S(p, 0).$$

By the first property, the set on the left-hand side cannot be closed. On the other hand, by the second property, the sets $S(p, 0)$ are closed. So, if there were only finitely many prime numbers, then the set on the right-hand side would be a finite union of closed sets, and hence closed. This would be a contradiction, so there must be infinitely many prime numbers.

25.2 Notes

[1] Mercer, Idris D. (2009). "On Furstenberg's Proof of the Infinitude of Primes" (PDF). *American Mathematical Monthly* **116**: 355–356. doi:10.4169/193009709X470218.

25.3 References

- Aigner, Martin; Ziegler, Günter M. (1998). "Proofs from The Book". Berlin, New York: Springer-Verlag.

- Furstenberg, Harry (1955). "On the infinitude of primes". *American Mathematical Monthly* (Mathematical Association of America) **62** (5): 353. doi:10.2307/2307043. JSTOR 2307043. MR 0068566

25.4 External links

- Furstenberg's proof that there are infinitely many prime numbers at Everything2
- Fürstenberg's proof of the infinitude of primes at PlanetMath.org.

Chapter 26

Proofs involving the Moore–Penrose pseudoinverse

Main article: Moore–Penrose pseudoinverse

Let A be an m-by-n matrix over a field $\mathbb{K}$, where $\mathbb{K}$, is either the field $\mathbb{R}$, of real numbers or the field $\mathbb{C}$, of complex numbers. Then there is a unique n-by-m matrix A^+ over $\mathbb{K}$, such that:

1. $A A^+ A = A$

2. $A^+ A A^+ = A^+$

3. $(AA^+)^* = AA^+$

4. $(A^+A)^* = A^+A$

A^+ is called the Moore-Penrose pseudoinverse of A. Notice that A is also the Moore-Penrose pseudoinverse of A^+. That is, $(A^+)^+ = A$.

26.1 Useful lemmas

These results are used in the proofs below. In the following lemmas, A is a matrix with complex elements and n columns, B is a matrix with complex elements and n rows.

26.1.1 Lemma 1: $A^*A = 0 \Rightarrow A = 0$

The assumption says that all elements of A^*A are zero. Therefore

$$0 = \operatorname{Tr}(A^*A) = \sum_{j=1}^{n}(A^*A)_{jj} = \sum_{j=1}^{n}\sum_{i=1}^{m}(A^*)_{ji}A_{ij} = \sum_{i=1}^{m}\sum_{j=1}^{n}|A_{ij}|^2$$

Therefore all A_{ij} equal 0 i.e. $A=0$.

26.1.2 Lemma 2: $A^*AB = 0 \Rightarrow AB = 0$

26.1.3 Lemma 3: $ABB^* = 0 \Rightarrow AB = 0$

This is proved in a manner similar to the argument of Lemma 2 (or by simply taking the Hermitian conjugate).

26.2 Existence and uniqueness

26.2.1 Proof of uniqueness

Suppose that B and C are two n-by-m matrices over $\mathbb{K}$ satisfying the Moore-Penrose criteria. Observe then that

$AB = (AB)^* = B^*A^* = B^*(ACA)^* = B^*A^*C^*A^* = (AB)^*(AC)^* = ABAC = AC$.

Analogously we conclude that $BA=CA$. The proof is completed by observing that then

$B = BAB = BAC = CAC = C$.

26.2.2 Proof of existence

The proof proceeds in stages.

1-by-1 matrices

For any $x \in \mathbb{K}$, we define $x^+ := \begin{cases} x^{-1}, & \text{if } x \neq 0 \\ 0, & \text{if } x = 0 \end{cases}$

It is easy to see that x^+ is a pseudoinverse of x (interpreted as a 1-by-1 matrix).

Square diagonal matrices

Let D be an n-by-n matrix over K with zeros off the diagonal. We define D^+ as an n-by-n matrix over K with $(D^+)_{ij} := (D_{ij})^+$. We write simply D^+_{ij} for $(D^+)_{ij} = (D_{ij})^+$.

Notice that D^+ is also a matrix with zeros off the diagonal.

We now show that D^+ is a pseudoinverse of D:

1. $(DD^+D)_{ij} = D_{ij}D^+_{ij}D_{ij} = D_{ij} \Rightarrow DD^+D = D$

2. $(D^+DD^+)_{ij} = D^+_{ij}D_{ij}D^+_{ij} = D^+_{ij} \Rightarrow D^+DD^+ = D^+$

3. $(DD^+)^*_{ij} = \overline{(DD^+)_{ji}} = \overline{D_{ji}D^+_{ji}} = (D_{ji}D^+_{ji})^* = D_{ji}D^+_{ji} = D_{ij}D^+_{ij} \Rightarrow (DD^+)^* = DD^+$

4. $(D^+D)^*_{ij} = \overline{(D^+D)_{ji}} = \overline{D^+_{ji}D_{ji}} = (D^+_{ji}D_{ji})^* = D^+_{ji}D_{ji} = D^+_{ij}D_{ij} \Rightarrow (D^+D)^* = D^+D$

General diagonal matrices

Arbitrary matrices

The singular value decomposition theorem states that there exists a factorization of the form

$$A = U\Sigma V^*$$

where:

> U is an m-by-m unitary matrix over K.
>
> Σ is an m-by-n matrix over K with nonnegative numbers on the diagonal and zeros off the diagonal.
>
> V is an n-by-n unitary matrix over K.[1]

Define A^+ as $V\Sigma^+ U^*$.

We now show that A^+ is a pseudoinverse of A:

1. $AA^+A = U\Sigma V^* V\Sigma^+ U^* U\Sigma V^* = U\Sigma\Sigma^+\Sigma V^* = U\Sigma V^* = A$
2. $A^+AA^+ = V\Sigma^+ U^* U\Sigma V^* V\Sigma^+ U^* = V\Sigma^+\Sigma\Sigma^+ U^* = V\Sigma^+ U^* = A^+$
3. $(AA^+)^* = (U\Sigma V^* V\Sigma^+ U^*)^* = (U\Sigma\Sigma^+ U^*)^* = U(\Sigma\Sigma^+)^* U^* = U(\Sigma\Sigma^+)U^* = U\Sigma V^* V\Sigma^+ U^* = AA^+$
4. $(A^+A)^* = (V\Sigma^+ U^* U\Sigma V^*)^* = (V\Sigma^+\Sigma V^*)^* = V(\Sigma^+\Sigma)^* V^* = V(\Sigma^+\Sigma)V^* = V\Sigma^+ U^* U\Sigma V^* = A^+A$

26.3 Basic properties

26.3.1 $A^{*+}=A^{+*}$

The proof works by showing that A^{+*} satisfies the four criteria for the pseudoinverse of A^*. Since this amounts to just substitution, it is not shown here.

The proof of this relation is given as Exercise 1.18c in.[2]

26.3.2 Identities

$A^+ = A^+ A^{+*} A^*$

$A^+ = A^+AA^+$ and $AA^+ = (AA^+)^*$ imply that $A^+ = A^+(A\,A^+)^* = A^+A^{+*}A^*$.

$A^+ = A^* A^{+*} A^+$

$A^+ = A^+AA^+$ and $A^+A = (A^+A)^*$ imply that $A^+ = (A^+A)^*A^+ = A^*A^{+*}A^+$.

$A = A^{+*} A^* A$

$A = A\,A^+\,A$ and $A\,A^+ = (A\,A^+)^*$ imply that $A = (A\,A^+)^*\,A = A^{+*}\,A^*\,A$.

$A = A\,A^*\,A^{+*}$

$A = A\,A^+\,A$ and $A^+\,A = (A^+\,A)^*$ imply that $A = A\,(A^+\,A)^* = A\,A^*\,A^{+*}$.

$A^* = A^*\,A\,A^+$

This is the conjugate transpose of $A = A^{+*}\,A^*\,A$ above.

$A* = A^+ A A*$

This is the conjugate transpose of $A = A A* A^{+*}$ above.

26.4 Reduction to the Hermitian case

The results of this section show that the computation of the pseudoinverse is reducible to its construction in the Hermitian case. It suffices to show that the putative constructions satisfy the defining criteria.

26.4.1 $A^+ = A* (A A*)^+$

This relation is given as exercise 18(d) in,[2] for the reader to prove, "for every matrix A". Write $D = A*(A A*)^+$. Observe that

Similarly, $(AA*)^+ AA*(AA*)^+ = (AA*)^+$ implies that $A*(AA*)^+ AA*(AA*)^+ = A*(AA*)^+$ i.e. $DAD = D$.

Additionally, $AD = AA*(AA*)^+$ so $AD = (AD)*$.

Finally, $DA = A*(AA*)^+A$ implies that $(DA)* = A*((AA*)^+)*A = A*((AA*)^+)A = DA$.

Therefore $D = A^+$.

26.4.2 $A^+ = (A* A)^+ A*$

This is proved in an analogous manner to the case above, using Lemma 2 instead of Lemma 3.

26.5 Products

For the first three proofs, we consider products $C = AB$.

26.5.1 A has orthonormal columns

If A has orthonormal columns i.e. $A*A = I$ then $A^+ = A*$. Write $D = B^+ A^+ = B^+ A*$. We show that D satisfies the Moore-Penrose criteria.

$CDC = ABB^+ A*AB = ABB^+ B = AB = C$.

$DCD = B^+ A*ABB^+ A* = B^+ BB^+ A* = B^+ A* = D$

$(CD)* = D*B*A* = A(B^+)*B*A* = A(BB^+)*A* = ABB^+ A* = CD$

$(DC)* = B*A*D* = B*A*A(B^+)* = (B^+B)* = B^+B = B^+A*AB = DC$

Therefore $D = C^+$

26.5.2 B has orthonormal rows

If B has orthonormal rows i.e. $BB* = I$ then $B^+ = B*$. Write $D = B^+ A^+ = B*A^+$. We show that D satisfies the Moore-Penrose criteria.

$CDC = ABB*A^+ AB = AA^+ AB = AB = C$.

$DCD = B*A^+ ABB*A^+ = B*A^+ AA^+ = B*A^+ = D$

$(CD)* = D*B*A* = (A^+)*BB*A* = (A^+)*A* = (AA^+)* = AA^+ = ABB*A^+ = CD$

$(DC)^* = B^*A^*D^* = B^*A^*(A^+)^*B = B^*(A^+A)^*B = B^*A^+AB = DC$

Therefore $D = C^+$

26.5.3 *A* has full column rank and *B* has full row rank

Since A has full column rank, A^*A is invertible so $(A^*A)^+ = (A^*A)^{-1}$. Similarly, since B has full row rank, BB^* is invertible so $(BB^*)^+ = (BB^*)^{-1}$.

Write $D = B^+A^+ = B^*(BB^*)^{-1}(A^*A)^{-1}A^*$. We show that D satisfies the Moore-Penrose criteria.

$CDC = ABB^*(BB^*)^{-1}(A^*A)^{-1}A^*AB = AB = C$.

$DCD = B^*(BB^*)^{-1}(A^*A)^{-1}A^*ABB^*(BB^*)^{-1}(A^*A)^{-1}A^* = B^*(BB^*)^{-1}(A^*A)^{-1}A^* = D$

$CD = ABB^*(BB^*)^{-1}(A^*A)^{-1}A^* = A(A^*A)^{-1}A^* = (A(A^*A)^{-1}A^*)^* \Rightarrow (CD)^* = CD$.

$DC = B^*(BB^*)^{-1}(A^*A)^{-1}A^*AB = B^*(BB^*)^{-1}B = (B^*(BB^*)^{-1}B)^* \Rightarrow (DC)^* = DC$.

Therefore $D = C^+$

26.5.4 Conjugate transpose

Here, $B = A^*$, and thus $C = AA^*$ and $D = A^{+*}A^+$. We show that indeed D satisfies the four Moore-Penrose criteria.

$$CDC = AA^*A^{+*}A^+AA^* = A(A^+A)^*A^+AA^* = AA^+AA^+AA^* = AA^+AA^* = AA^* = C$$

$$DCD = A^{+*}A^+AA^*A^{+*}A^+ = A^{+*}A^+A(A^+A)^*A^+ = A^{+*}A^+AA^+AA^+ = A^{+*}A^+AA^+ = A^{+*}A^+AA^+ = A^{+*}A^+ = D$$

$$(CD)^* = (AA^*A^{+*}A^+)^* = A^{+*}A^+AA^* = A^{+*}(A^+A)^*A^* = A^{+*}A^*A^{+*}A^* = (AA^+)^*(AA^+)^* = AA^+AA^+ = A(A^+A)^*A^+ =$$

$$= AA^*A^{+*}A^+ = CD$$

$$(DC)^* = (A^{+*}A^+AA^*)^* = AA^*A^{+*}A^+ = A(A^+A)^*A^+ = AA^+AA^+ = (AA^+)^*(AA^+)^* = A^{+*}A^*A^{+*}A^* = A^{+*}(A^+A)^*A^* =$$

$$= A^{+*}A^+AA^* = DC$$

Therefore $D = C^+$. In other words:

$$(AA^*)^+ = A^{+*}A^+$$

and, since $(A^*)^* = A$

$$(A^*A)^+ = A^+A^{+*}$$

26.6 Projectors and subspaces

Define $P = AA^+$ and $Q = A^+A$. Observe that $P^2 = AA^+AA^+ = AA^+ = P$. Similarly $Q^2 = Q$, and finally, $P = P^*$ and $Q = Q^*$. Thus P and Q are orthogonal projection operators. Orthogonality follows from the relations $P = P^*$ and $Q = Q^*$. Indeed, consider the operator P: any vector decomposes as

$x = Px + (I-P)x$

and for all vectors x and y satisfying $Px=x$ and $(I-P)y = y$, we have

$x^*y = (Px)^*(I-P)y = x^*P^*(I-P)y = x^*P(I-P)y = 0$.

It follows that $PA = AA^+A = A$ and $A^+P = A^+AA^+ = A^+$. Similarly, $QA^+ = A^+$ and $AQ = A$. The orthogonal components are now readily identified.

If y belongs to the range of A then for some x, $y = Ax$ and $Py = PAx = Ax = y$. Conversely, if $Py = y$ then $y = AA^+y$ so that y belongs to the range of A. It follows that P is the orthogonal projector onto the range of A. $I - P$ is then the orthogonal projector onto the orthogonal complement of the range of A, which equals the kernel of A^*.

A similar argument using the relation $QA^* = A^*$ establishes that Q is the orthogonal projector onto the range of A^* and $(I-Q)$ is the orthogonal projector onto the kernel of A.

Using the relations $P(A^+)^* = P^*(A^+)^* = (A^+P)^* = (A^+)^*$ and $P = P^* = (A^+)^*A^*$ it follows that the range of P equals the range of $(A^+)^*$, which in turn implies that the range of $I-P$ equals the kernel of A^+. Similarly $QA^+ = A^+$ implies that the range of Q equals the range of A^+. Therefore we find,

$\text{Ker}(A^+) = \text{Ker}(A^*)$.
$\text{Im}(A^+) = \text{Im}(A^*)$.

26.7 Additional properties

26.7.1 Least-squares minimization

In the general case, it is shown here for any $m \times n$ matrix A that $\|Ax - b\|_2 \geq \|Az - b\|_2$ where $z = A^+b$. This lower bound need not be zero as the system $Ax = b$ may not have a solution (e.g. when the matrix A does not have full rank or the system is overdetermined).

To prove this, we first note that (stating the complex case), using the fact that $P = AA^+$ satisfies $PA = A$ and $P = P^*$, we have

$$\begin{aligned} A^*(Az - b) &= A^*(AA^+b - b) \\ &= A^*(Pb - b) \\ &= A^*P^*b - A^*b \\ &= (PA)^*b - A^*b \\ &= 0 \end{aligned}$$

so that

$$\begin{aligned} \|Ax - b\|_2^2 &= \|Az - b\|_2^2 + (A(x - z))^*(Az - b) + \text{c.c.} + \|A(x - z)\|_2^2 \\ &= \|Az - b\|_2^2 + (x - z)^*A^*(Az - b) + \text{c.c.} + \|A(x - z)\|_2^2 \\ &= \|Az - b\|_2^2 + \|A(x - z)\|_2^2 \\ &\geq \|Az - b\|_2^2 \end{aligned}$$

as claimed.

If A is injective i.e. one-to-one (which implies $m \geq n$), then the bound is attained uniquely at z.

26.7.2 Minimum-norm solution to a linear system

The proof above also shows that if the system $Ax = b$ is satisfiable i.e. has a solution, then necessarily $z = A^+b$ is a solution (not necessarily unique). We show here that z is the smallest such solution (its Euclidean norm is uniquely minimum).

To see this, note first, with $Q = A^+A$, that $Qz = A^+AA^+b = A^+b = z$ and that $Q^* = Q$. Therefore, assuming that $Ax = b$, we have

$$\begin{aligned} z^*(x-z) &= (Qz)^*(x-z) \\ &= z^*Q(x-z) \\ &= z^*(A^+Ax - z) \\ &= z^*(A^+b - z) \\ &= 0. \end{aligned}$$

Thus

$$\begin{aligned} \|x\|_2^2 &= \|z\|_2^2 + 2z^*(x-z) + \|x-z\|_2^2 \\ &= \|z\|_2^2 + \|x-z\|_2^2 \\ &\geq \|z\|_2^2 \end{aligned}$$

with equality if and only if $x = z$, as was to be shown.

26.8 References

[1] Some authors use slightly different dimensions for the factors. The two definitions are equivalent.

[2] Ben-Israel, Adi; Thomas N.E. Greville (2003). *Generalized Inverses*. Springer-Verlag. ISBN 0-387-00293-6.

Chapter 27

Sharp-P-completeness of 01-permanent

The correct title of this article is **#P-completeness of 01-permanent**. The substitution or omission of the # is because of technical restrictions.

The **#P-completeness of 01-permanent**, sometimes known as **Valiant's theorem**,[1] is a mathematical proof about the permanent of matrices, considered a seminal result in computational complexity theory.[2][3] In a 1979 scholarly paper, Leslie Valiant proved[4] that the computational problem of computing the permanent of a matrix is #P-hard, even if the matrix is restricted to have entries that are all 0 or 1. In this restricted case, computing the permanent is even #P-complete, because it corresponds to the #P problem of counting the number of permutation matrices one can get by changing ones into zeroes.

Valiant's 1979 paper also introduced #P as a complexity class.[5]

27.1 Significance

One reason for interest in the computational complexity of the permanent is that it provides an example of a problem where constructing a single solution can be done efficiently but where counting all solutions is hard.[6] As Papadimitriou writes in his book *Computational Complexity*:

Specifically, computing the permanent (shown to be difficult by Valiant's results) is closely connected with finding a perfect matching in a bipartite graph, which is solvable in polynomial time by the Hopcroft–Karp algorithm.[7][8] For a bipartite graph with $2n$ vertices partitioned into two parts with n vertices each, the number of perfect matchings equals the permanent of its biadjacency matrix and the square of the number of perfect matchings is equal to the permanent of its adjacency matrix.[9] Since any 0–1 matrix is the biadjacency matrix of some bipartite graph, Valiant's theorem implies[9] that the problem of counting the number of perfect matchings in a bipartite graph is #P-complete, and in conjunction with Toda's theorem this implies that it is hard for the entire polynomial hierarchy.[10][11]

The computational complexity of the permanent also has some significance in other aspects of complexity theory: it is not known whether NC equals P (informally, whether every polynomially-solvable problem can be solved by a polylogarithmic-time parallel algorithm) and Ketan Mulmuley has suggested an approach to resolving this question that relies on writing the permanent as the determinant of a matrix.[12]

Hartmann [13] proved a generalization of Valiant's theorem concerning the complexity of computing immanants of matrices that generalize both the determinant and the permanent.

27.2 Ben-Dor and Halevi's proof

Below, the proof that computing the permanent of a 01-matrix is #P-complete is described. It mainly follows the proof by Ben-Dor & Halevi (1993).[14]

27.2.1 Overview

Any square matrix $A = (a_{ij})$ can be viewed as the adjacency matrix of a directed graph, with a_{ij} representing the weight of the edge from vertex i to vertex j. Then, the permanent of A is equal to the sum of the weights of all cycle-covers of the graph; this is a graph-theoretic interpretation of the permanent.

#SAT, a function problem related to the Boolean satisfiability problem, is the problem of counting the number of satisfying assignments of a given Boolean formula. It is a #P-complete problem (by definition), as any NP machine can be encoded into a Boolean formula by a process similar to that in Cook's theorem, such that the number of satisfying assignments of the Boolean formula is equal to the number of accepting paths of the NP machine. Any formula in SAT can be rewritten as a formula in 3-CNF form preserving the number of satisfying assignments, and so #SAT and #3SAT are equivalent and #3SAT is #P-complete as well.

In order to prove that 01-Permanent is #P-hard, it is therefore sufficient to show that the number of satisfying assignments for a 3-CNF formula can be expressed succinctly as a function of the permanent of a matrix that contains only the values 0 and 1. This is usually accomplished in two steps:

1. Given a 3-CNF formula φ, construct a directed integer-weighted graph G_ϕ, such that the sum of the weights of cycle covers of G_ϕ (or equivalently, the permanent of its adjacency matrix) is equal to the number of satisfying assignments of φ. This establishes that Permanent is #P-hard.

2. Through a series of reductions, reduce Permanent to 01-Permanent, the problem of computing the permanent of a matrix all entries 0 or 1. This establishes that 01-permanent is #P-hard as well.

27.2.2 Constructing the integer graph

Given a 3CNF-formula ϕ with m clauses and n variables, one can construct a weighted, directed graph G_ϕ such that

1. each satisfying assignment for ϕ will have a corresponding set of cycle covers in G_ϕ where the sum of the weights of cycle covers in this set will be 12^m; and

2. all other cycle covers in G_ϕ will have weights summing to 0.

Thus if $(\#\phi)$ is the number of satisfying assignments for ϕ, the permanent of this graph will be $12^m \cdot (\#\phi)$. (Valiant's original proof constructs a graph with entries in $\{-1, 0, 1, 2, 3\}$ whose permanent is $4^{t(\phi)} \cdot (\#\phi)$ where $t(\phi)$ is "twice the number of occurrences of literals in ϕ " $- m$.)

The graph construction makes use of a component that is treated as a "black box." To keep the explanation simple, the properties of this component are given without actually defining the structure of the component.

To specify $G\varphi$, one first constructs a variable node in $G\varphi$ for each of the n variables in φ. Additionally, for each of the m clauses in φ, one constructs a clause component Cj in $G\varphi$ that functions as a sort of "black box." All that needs to be noted about Cj is that it has three input edges and three output edges. The input edges come either from variable nodes or from previous clause components (e.g., Co for some $o < j$) and the output edges go either to variable nodes or to later clause components (e.g., Co for some $o > j$). The first input and output edges correspond with the first variable of the clause j, and so on. Thus far, all of the nodes that will appear in the graph $G\varphi$ have been specified.

Next, one would consider the edges. For each variable x_i of ϕ, one makes a true cycle (T-cycle) and a false cycle (F-cycle) in G_ϕ. To create the T-cycle, one starts at the variable node for x_i and draw an edge to the clause component C_j that corresponds to the first clause in which x_i appears. If x_i is the first variable in the clause of ϕ corresponding to C_j, this edge will be the first input edge of C_j, and so on. Thence, draw an edge to the next clause component corresponding to the next clause of ϕ in which x_i appears, connecting it from the appropriate output edge of C_j to the appropriate input edge of the next clause component, and so on. After the last clause in which x_i appears, we connect the appropriate output edge of the corresponding clause component back to x_i's variable node. Of course, this completes the cycle. To create the F-cycle, one would follow the same procedure, but connect x_i's variable node to those clause components in which ~x_i appears, and finally back to x_i's variable node. All of these edges outside the clause components are termed *external edges*, all of which have weight 1. Inside the clause components, the edges are termed *internal edges*. Every external edge is part of a T-cycle or an F-cycle (but not both—that would force inconsistency).

Note that the graph G_ϕ is of size linear in $|\phi|$, so the construction can be done in polytime (assuming that the clause components do not cause trouble).

Notable properties of the graph

A useful property of G_ϕ is that its cycle covers correspond to variable assignments for ϕ. For a cycle cover **Z** of G_ϕ, one can say that **Z** induces an assignment of values for the variables in ϕ just in case **Z** contains all of the external edges in x_i's T-cycle and none of the external edges in x_i's F-cycle for all variables x_i that the assignment makes true, and vice versa for all variables x_i that the assignment makes false. Although any given cycle cover **Z** need not induce an assignment for ϕ, any one that does induces exactly one assignment, and the same assignment induced depends only on the external edges of **Z**. The term **Z** is considered an incomplete cycle cover at this stage, because one talks only about its external edges, M. In the section below, one considers M-completions to show that one has a set of cycle covers corresponding to each M that have the necessary properties.

The sort of **Z**'s that don't induce assignments are the ones with cycles that "jump" inside the clause components. That is, if for every C_j, at least one of C_j's input edges is in **Z**, and every output edge of the clause components is in **Z** when the corresponding input edge is in **Z**, then **Z** is proper with respect to each clause component, and **Z** will produce a satisfying assignment for ϕ. This is because proper **Z**'s contain either the complete T-cycle or the complete F-cycle of every variable x_i in ϕ as well as each including edges going into and coming out of each clause component. Thus, these **Z**'s assign either true or false (but never both) to each x_i and ensure that each clause is satisfied. Further, the sets of cycle covers corresponding to all such **Z**'s have weight 12^m, and any other **Z**'s have weight 0. The reasons for this depend on the construction of the clause components, and are outlined below.

The clause component

To understand the relevant properties of the clause components C_j, one needs the notion of an M-completion. A cycle cover **Z** induces a satisfying assignment just in case its external edges satisfy certain properties. For any cycle cover of G_ϕ, consider only its external edges, the subset M. Let M be a set of external edges. A set of internal edges L is an M-completion just in case $M \cup L$ is a cycle cover of G_ϕ. Further, denote the set of all M-completions by L^M and the set of all resulting cycle covers of G_ϕ by Z^M.

Recall that construction of G_ϕ was such that each external edge had weight 1, so the weight of Z^M, the cycle covers resulting from any M, depends only on the internal edges involved. We add here the premise that the construction of the clause components is such that the sum over possible M-completions of the weight of the internal edges in each clause component, where M is proper relative to the clause component, is 12. Otherwise the weight of the internal edges is 0. Since there are m clause components, and the selection of sets of internal edges, L, within each clause component is independent of the selection of sets of internal edges in other clause components, so one can multiply everything to get the weight of Z^M. So, the weight of each Z^M, where M induces a satisfying assignment, is 12^m. Further, where M does not induce a satisfying assignment, M is not proper with respect to some C_j, so the product of the weights of internal edges in Z^M will be 0.

The clause component is a weighted, directed graph with 7 nodes with edges weighted and nodes arranged to yield the properties specified above, and is given in Appendix A of Ben-Dor and Halevi (1993). Note that the internal edges here have weights drawn from the set $\{-1, 0, 1, 2, 3\}$; not all edges have 0–1 weights.

Finally, since the sum of weights of all the sets of cycle covers inducing any particular satisfying assignment is 12^m, and the sum of weights of all other sets of cycle covers is 0, one has $\text{Perm}(G\varphi) = 12^m \cdot (\#\varphi)$. The following section reduces computing Perm(G_ϕ) to the permanent of a 01 matrix.

27.2.3 01-Matrix

The above section has shown that Permanent is #P-hard. Through a series of reductions, any permanent can be reduced to the permanent of a matrix with entries only 0 or 1. This will prove that 01-Permanent is #P-hard as well.

Reduction to a non-negative matrix

Using modular arithmetic, convert an integer matrix A into an equivalent non-negative matrix A' so that the permanent of A can be computed easily from the permanent of A', as follows:

Let A be an $n \times n$ integer matrix where no entry has a magnitude larger than μ.

- Compute $Q = 2n! \cdot \mu^n + 1$. The choice of Q is due to the fact that $|\text{Perm}(A)| \le n! \cdot \mu^n$
- Compute $A' = A \bmod Q$
- Compute $P = \text{Perm}(A') \bmod Q$
- If $P < Q/2$ then Perm(A) = P. Otherwise Perm(A) = $P - Q$

The transformation of A into A' is polynomial in n and $\log(\mu)$, since the number of bits required to represent Q is polynomial in n and $\log(\mu)$

An example of the transformation and why it works is given below.

$$A = \begin{bmatrix} 2 & -2 \\ -2 & 1 \end{bmatrix}$$

Perm(A) = $2 \cdot 1 + (-2) \cdot (-2) = 6$

Here, $n = 2$, $\mu = 2$, and $\mu^n = 4$, so $Q = 17$. Thus

$$A' = A \bmod 17 = \begin{bmatrix} 2 & 15 \\ 15 & 1 \end{bmatrix}.$$

Note how the elements are non-negative because of the modular arithmetic. It is simple to compute the permanent

Perm(A') = $2 \cdot 1 + 15 \cdot 15 = 227$

so $P = 227 \bmod 17 = 6$. Then $P < Q/2$, so Perm(A) = $P = 6$.

Reduction to powers of 2

Note that any number can be decomposed into a sum of powers of 2. For example,

$13 = 2^3 + 2^2 + 2^0$

This fact is used to convert a non-negative matrix into an equivalent matrix whose entries are all powers of 2. The reduction can be expressed in terms of graphs equivalent to the matrices.

Let G be a n-node weighted directed graph with non-negative weights, where largest weight is W. Every edge e with weight w is converted into an equivalent edge with weights in powers of 2 as follows:

27.2. BEN-DOR AND HALEVI'S PROOF

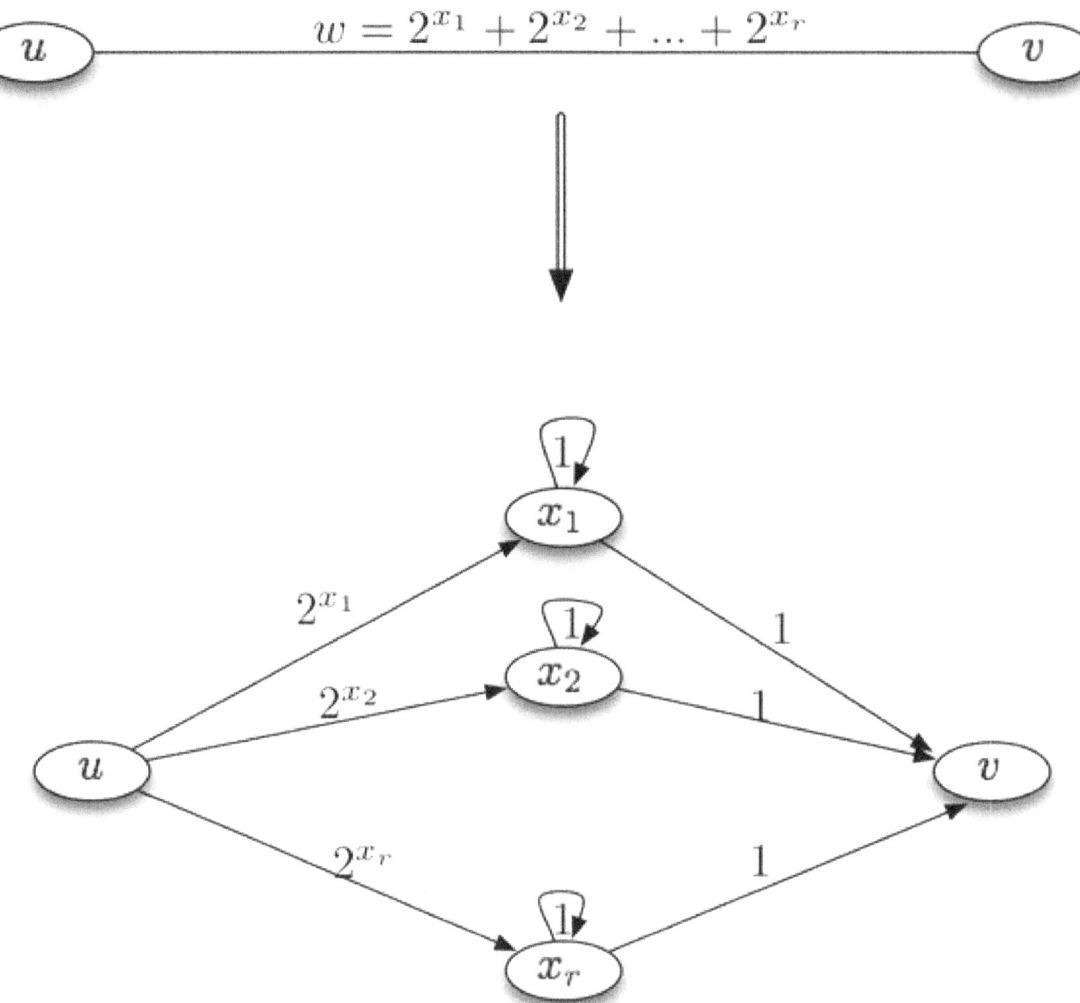

Figure 1: Construction of 2Power from NonNeg

$$w = 2^{x_1} + 2^{x_2} + \cdots + 2^{x_r}, \quad 0 \leq x_1 \leq x_2 \leq \cdots \leq x_r \leq \log(w)$$

This can be seen graphically in the Figure 1. The subgraph that replaces the existing edge contains r nodes and $3r$ edges.

To prove that this produces an equivalent graph G' that has the same permanent as the original, one must show the correspondence between the cycle covers of G and G'.

Consider some cycle-cover R in G.

- If an edge e is not in R, then to cover all the nodes in the new sub graph, one must use the self-loops. Since all self-loops have a weight of 1, the weight of cycle-covers in R and R' match.

- If e is in R, then in all the corresponding cycle-covers in G', there must be a path from u to v, where u and v are the nodes of edge e. From the construction, one can see that there are r different paths and sum of all these paths equal to the weight of the edge in the original graph G. So the weight of corresponding cycle-covers in G and G' match.

Note that the size of G' is polynomial in n and $\log W$.

Reduction to 0–1

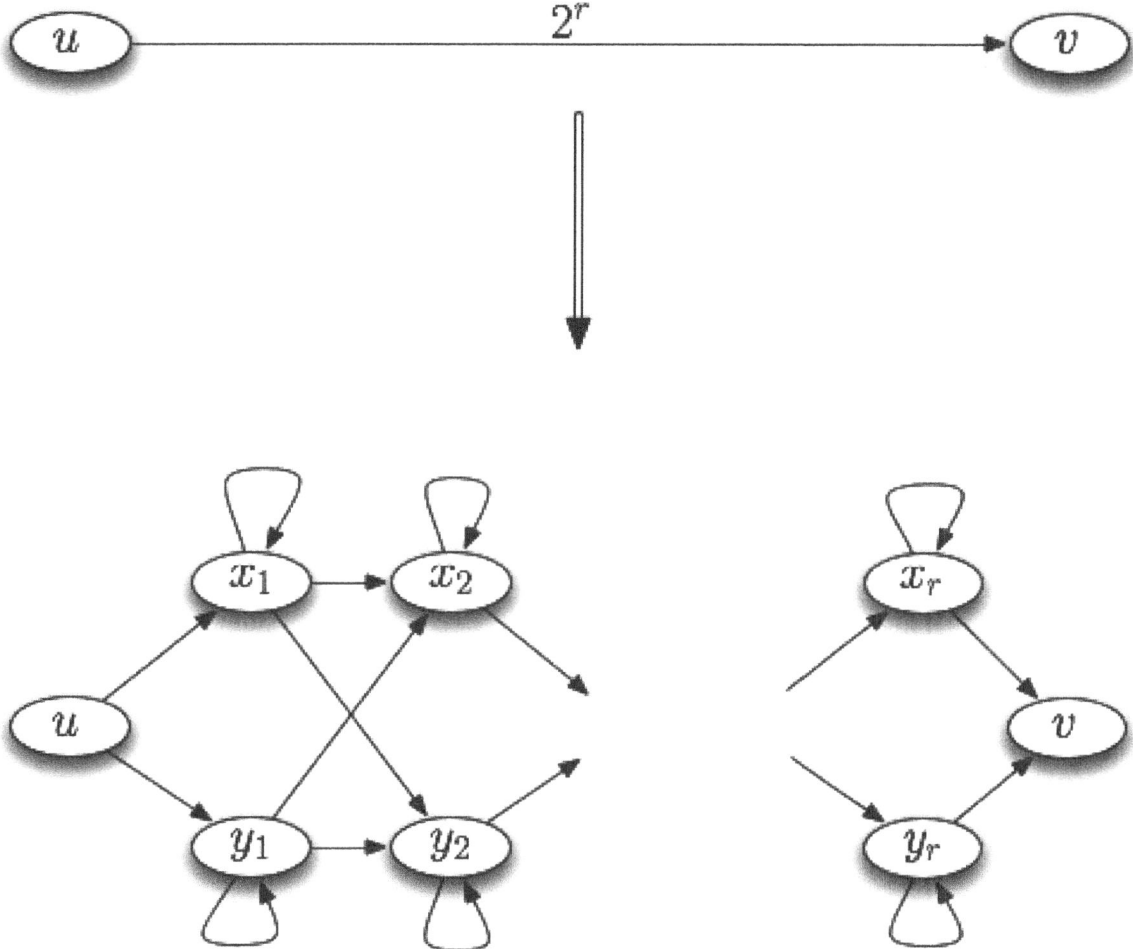

Figure 2: Construction of a 01-matrix from 2Power

The objective here is to reduce a matrix whose entries are powers of 2 into an equivalent matrix containing only zeros and ones (i.e. a directed graph where each edge has a weight of 1).

Let G be a n-node directed graph where all the weights on edges are powers of two. Construct a graph, G', where the weight of each edge is 1 and Perm(G) = Perm(G'). The size of this new graph, G', is polynomial in n and p where the maximal weight of any edge in graph G is 2^p.

This reduction is done locally at each edge in G that has a weight larger than 1. Let $e = (u,v)$ be an edge in G with a weight $w = 2^r > 1$. It is replaced by a subgraph J_e that is made up of $2r$ nodes and $6r$ edges as seen in Figure 2. Each edge in J_e has a weight of 1. Thus, the resulting graph G' contains only edges with a weight of 1.

Consider some cycle-cover R in G.

- If an original edge e from graph G is not in R, one cannot create a path through the new subgraph J_e. The only way to form a cycle cover over J_e in such a case is for each node in the subgraph to take its self-loop. As each edge has a weight of one, the weight of the resulting cycle cover is equal to that of the original cycle cover.

- However, if the edge in G is a part of the cycle cover then in any cycle cover of G' there must be a path from u to v in the subgraph. At each step down the subgraph there are two choices one can make to form such a path. One must make this choice r times, resulting in 2^r possible paths from u to v. Thus, there are 2^r possible cycle covers and since each path has a weight of 1, the sum of the weights of all these cycle covers equals the weight of the original cycle cover.

27.3 Aaronson's proof

Quantum computer scientist Scott Aaronson[15] has proved #P-hardness of permanent using quantum methods.

27.4 References

[1] Christos H. Papadimitriou. *Computational Complexity*. Addison-Wesley, 1994. ISBN 0-201-53082-1. Page 443

[2] Allen Kent, James G. Williams, Rosalind Kent and Carolyn M. Hall (editors). *Encyclopedia of microcomputers*.Marcel Dekker, 1999. ISBN 978-0-8247-2722-2; p. 34

[3] Jin-Yi Cai, A. Pavan and D. Sivakumar, *On the Hardness of Permanent.* In: STACS, '99: 16th Annual Symposium on Theoretical Aspects of Computer Science, Trier, Germany, March 4–6, 1999 Proceedings. pp. 90–99. Springer-Verlag, New York, LLC Pub. Date: October 2007. ISBN 978-3-540-65691-3; p. 90.

[4] Leslie G. Valiant (1979). "The Complexity of Computing the Permanent". *Theoretical Computer Science* (Elsevier) **8** (2): 189–201. doi:10.1016/0304-3975(79)90044-6.

[5] Lance Fortnow. *My Favorite Ten Complexity Theorems of the Past Decade.* Foundations of Software Technology and Theoretical Computer Science: Proceedings of the 14th Conference, Madras, India, December 15–17, 1994. P. S. Thiagarajan (editor), pp. 256–275, Springer-Verlag, New York, 2007. ISBN 978-3-540-58715-6; p. 265

[6] Bürgisser, Peter (2000). *Completeness and reduction in algebraic complexity theory*. Algorithms and Computation in Mathematics **7**. Berlin: Springer-Verlag. p. 2. ISBN 3-540-66752-0. Zbl 0948.68082.

[7] John E. Hopcroft, Richard M. Karp: *An $n^{5/2}$ Algorithm for Maximum Matchings in Bipartite Graphs.* SIAM J. Comput. 2(4), 225–231 (1973)

[8] Cormen, Thomas H.; Leiserson, Charles E., Rivest, Ronald L., Stein, Clifford (2001) [1990]. "26.5: The relabel-to-front algorithm". *Introduction to Algorithms* (2nd ed.). MIT Press and McGraw-Hill. pp. 696–697. ISBN 0-262-03293-7.

[9] Dexter Kozen. *The Design and Analysis of Algorithms.* Springer-Verlag, New York, 1991. ISBN 978-0-387-97687-7; pp. 141–142

[10] Seinosuke Toda. PP is as Hard as the Polynomial-Time Hierarchy. SIAM Journal on Computing, Volume 20 (1991), Issue 5, pp. 865–877.

[11] 1998 Gödel Prize. Seinosuke Toda

[12] Ketan Mulmuley. Lower Bounds in a Parallel Model without Bit Operations. SIAM Journal on Computing, Volume 28 (1999), Issue 4, pp. 1460–1509.

[13] W. Hartmann. On the complexity of immanants. Linear and Multilinear Algebra 18 (1985), no. 2, pp. 127–140.

[14] Ben-Dor, Amir; Halevi, Shai (1993). "Proceedings of the 2nd Israel Symposium on the Theory and Computing Systems" (PDF). pp. 108–117. |contribution= ignored (help).

[15] S. Aaronson, A Linear-Optical Proof that the Permanent is #P-Hard

Chapter 28

Proof of Fermat's Last Theorem for specific exponents

Several proofs for Fermat's Last Theorem for specific exponents have been developed.

28.1 Mathematical preliminaries

Fermat's Last Theorem states that no three positive integers (a, b, c) can satisfy the equation $a^n + b^n = c^n$ for any integer value of n greater than two. If n equals two, the equation has infinitely many solutions, the Pythagorean triples.

A solution (a, b, c) for a given n is equivalent to a solution for all the factors of n. For illustration, let n be factored into g and h, $n = gh$. Then (a^g, b^g, c^g) is a solution for the exponent h

$$(a^g)^h + (b^g)^h = (c^g)^h$$

Conversely, to prove that Fermat's equation has *no* solutions for $n > 2$, it suffices to prove that it has no solutions for $n = 4$ and for all odd primes p.

For any such odd exponent p, every positive-integer solution of the equation $a^p + b^p = c^p$ corresponds to a general integer solution to the equation $a^p + b^p + c^p = 0$. For example, if $(3, 5, 8)$ solves the first equation, then $(3, 5, -8)$ solves the second. Conversely, any solution of the second equation corresponds to a solution to the first. The second equation is sometimes useful because it makes the symmetry between the three variables a, b and c more apparent.

28.1.1 Primitive solutions

If two of the three numbers (a, b, c) can be divided by a fourth number d, then all three numbers are divisible by d. For example, if a and c are divisible by $d = 13$, then b is also divisible by 13. This follows from the equation

$$b^n = c^n - a^n$$

If the right-hand side of the equation is divisible by 13, then the left-hand side is also divisible by 13. Let g represent the greatest common divisor of a, b, and c. Then (a, b, c) may be written as $a = gx$, $b = gy$, and $c = gz$ where the three numbers (x, y, z) are pairwise coprime. In other words, the greatest common divisor (GCD) of each pair equals one

$$GCD(x, y) = GCD(x, z) = GCD(y, z) = 1$$

If (a, b, c) is a solution of Fermat's equation, then so is (x, y, z), since the equation

$$a^n + b^n = c^n = g^n x^n + g^n y^n = g^n z^n$$

implies the equation

$$x^n + y^n = z^n.$$

A pairwise coprime solution (x, y, z) is called a *primitive solution*. Since every solution to Fermat's equation can be reduced to a primitive solution by dividing by their greatest common divisor g, Fermat's Last Theorem can be proven by demonstrating that no primitive solutions exist.

28.1.2 Even and odd

Integers can be divided into even and odd, those that are divisible by two and those that are not. The even integers are ...−4, −2, 0, 2, 4, whereas the odd integers are −3, −1, 1, 3,... The property of whether an integer is even (or not) is known as its parity. If two numbers are both even or both odd, they have the same parity. By contrast, if one is even and the other odd, they have different parity.

The addition, subtraction and multiplication of even and odd integers obey simple rules. The addition or subtraction of two even numbers or of two odd numbers always produces an even number, e.g., 4 + 6 = 10 and 3 + 5 = 8. Conversely, the addition or subtraction of an odd and even number is always odd, e.g., 3 + 8 = 11. The multiplication of two odd numbers is always odd, but the multiplication of an even number with any number is always even. An odd number raised to a power is always odd and an even number raised to power is always even.

In any primitive solution (x, y, z) to the equation $x^n + y^n = z^n$, one number is even and the other two numbers are odd. They cannot all be even, for then they would not be coprime; they could all be divided by two. However, they cannot be all odd, since the sum of two odd numbers $x^n + y^n$ is never an odd number z^n. Therefore, at least one number must be even and at least one number must be odd. It follows that the third number is also odd, because the sum of an even and an odd number is itself odd.

28.1.3 Prime factorization

The fundamental theorem of arithmetic states that any natural number can be written in only one way (uniquely) as the product of prime numbers. For example, 42 equals the product of prime numbers 2×3×7, and no other product of prime numbers equals 42, aside from trivial re-arrangements such as 7×3×2. This unique factorization property is the basis on which much of number theory is built.

One consequence of this unique factorization property is that if a pth power of a number equals a product such as

$$x^p = uv$$

and if u and v are coprime (share no prime factors), then u and v are themselves the pth power of two other numbers, $u = r^p$ and $v = s^p$.

As described below, however, some number systems do not have unique factorization. This fact led to the failure of Lamé's 1847 general proof of Fermat's Last Theorem.

28.1.4 Two cases

Since the time of Sophie Germain, Fermat's Last Theorem has been separated into two cases that are proven separately. The first case (case I) is to show that there are no primitive solutions (x,y,z) to the equation $x^p + y^p = z^p$ under the condition that p does not divide the product xyz. The second case (case II) corresponds to the condition that p does divide the product xyz. Since x, y, and z are pairwise coprime, p divides only one of the three numbers.

Main article: Sophie Germain's theorem

28.2 $n = 4$

Only one mathematical proof by Fermat has survived, in which Fermat uses the technique of infinite descent to show that the area of a right triangle with integer sides can never equal the square of an integer.[1] This result is known as Fermat's right triangle theorem. As shown below, his proof is equivalent to demonstrating that the equation

$$x^4 - y^4 = z^2$$

has no primitive solutions in integers (no pairwise coprime solutions). In turn, this is equivalent to proving Fermat's Last Theorem for the case n=4, since the equation $a^4 + b^4 = c^4$ can be written as $c^4 - b^4 = (a^2)^2$. Alternative proofs of the case $n = 4$ were developed later[2] by Frénicle de Bessy,[3] Euler,[4] Kausler,[5] Barlow,[6] Legendre,[7] Schopis,[8] Terquem,[9] Bertrand,[10] Lebesgue,[11] Pepin,[12] Tafelmacher,[13] Hilbert,[14] Bendz,[15] Gambioli,[16] Kronecker,[17] Bang,[18] Sommer,[19] Bottari,[20] Rychlik,[21] Nutzhorn,[22] Carmichael,[23] Hancock,[24] Vrănceanu,[25] Grant and Perella,[26] Barbara,[27] and Dolan.[28] For one proof by infinite descent, see Infinite descent#Non-solvability of $r^2 + s^4 = t^4$.

28.2.1 Application to right triangles

Fermat's proof demonstrates that no right triangle with integer sides can have an area that is a square.[29] Let the right triangle have sides (u, v, w), where the area equals $uv/2$ and, by the Pythagorean theorem, $u^2 + v^2 = w^2$. If the area were equal to the square of an integer s

$$uv/2 = s^2$$

then

$$(u + v)^2 = w^2 + 4s^2$$
$$(u - v)^2 = w^2 - 4s^2$$

Multiplying these equations together yields

$$(u^2 - v^2)^2 = w^4 - 2^4 s^4$$

But as Fermat proved, there can be no integer solution to the equation

$$x^4 - y^4 = z^2$$

of which this is a special case with $z = (u^2 - v^2)$, $x = w$ and $y = 2s$.

The first step of Fermat's proof is to factor the left-hand side[30]

$$(x^2 + y^2)(x^2 - y^2) = z^2$$

Since x and y are coprime (this can be assumed because otherwise the factors could be cancelled), the greatest common divisor of $x^2 + y^2$ and $x^2 - y^2$ is either 2 (case A) or 1 (case B). The theorem is proven separately for these two cases.

28.2.2 Proof for Case A

In this case, both x and y are odd and z is even. Since (y^2, z, x^2) form a primitive Pythagorean triple, they can be written

$$z = 2de$$
$$y^2 = d^2 - e^2$$
$$x^2 = d^2 + e^2$$

where d and e are coprime and $d > e > 0$. Thus,

$$x^2 y^2 = d^4 - e^4$$

which produces another solution (d, e, xy) that is smaller $(0 < d < x)$. By the above argument that solutions cannot be shrunk indefinitely, this proves that the original solution was impossible.

28.2.3 Proof for Case B

In this case, the two factors are coprime. Since their product is a square z^2, they must each be a square

$$x^2 + y^2 = s^2$$
$$x^2 - y^2 = t^2$$

The numbers s and t are both odd, since $s^2 + t^2 = 2x^2$, an even number, and since x and y cannot both be even. Therefore, the sum and difference of s and t are likewise even numbers, so we define integers u and v as

$$u = (s + t)/2$$
$$v = (s - t)/2$$

Since s and t are coprime, so are u and v; only one of them can be even. Since $y^2 = 2uv$, exactly one of them is even. For illustration, let u be even; then the numbers may be written as $u=2m^2$ and $v=k^2$. Since (u, v, x) form a primitive Pythagorean triple

$$(s^2 + t^2)/2 = u^2 + v^2 = x^2$$

they can be expressed in terms of smaller integers d and e using Euclid's formula

$$u = 2de$$
$$v = d^2 - e^2$$
$$x = d^2 + e^2$$

Since $u = 2m^2 = 2de$, and since d and e are coprime, they must be squares themselves, $d = g^2$ and $e = h^2$. This gives the equation

$$v = d^2 - e^2 = g^4 - h^4 = k^2$$

The solution (g, h, k) is another solution to the original equation, but smaller $(0 < g < d < x)$. Applying the same procedure to (g, h, k) would produce another solution, still smaller, and so on. But this is impossible, since natural numbers cannot be shrunk indefinitely. Therefore, the original solution (x, y, z) was impossible.

28.3 $n = 3$

Fermat sent the letters which had the problem as the case of $n = 3$ in 1636, 1640, 1657.[31] Euler sent the letter which he had the proof of the case of $n = 3$ to Goldbach at 4 August 1753.[32] Euler had the complete and pure elemental proof in 1760.[33] The case $n = 3$ was proven by Euler in 1770.[34][35][36][37] Independent proofs were published by several other mathematicians,[38] including Kausler,[5] Legendre,[7][39] Calzolari,[40] Lamé,[41] Tait,[42] Günther,[43] Gambioli,[16] Krey,[44] Rychlik,[21] Stockhaus,[45] Carmichael,[46] van der Corput,[47] Thue,[48] and Duarte.[49]

As Fermat did for the case $n = 4$, Euler used the technique of infinite descent.[50] The proof assumes a solution (x, y, z) to the equation $x^3 + y^3 + z^3 = 0$, where the three non-zero integers x, y, and z are pairwise coprime and not all positive. One of the three must be even, whereas the other two are odd. Without loss of generality, z may be assumed to be even.

Since x and y are both odd, they cannot be equal. If $x = y$, then $2x^3 = -z^3$, which implies that x is even, a contradiction.

Since x and y are both odd, their sum and difference are both even numbers

$$2u = x + y$$
$$2v = x - y$$

where the non-zero integers u and v are coprime and have different parity (one is even, the other odd). Since $x = u + v$ and $y = u - v$, it follows that

$$-z^3 = (u + v)^3 + (u - v)^3 = 2u(u^2 + 3v^2)$$

Since u and v have opposite parity, $u^2 + 3v^2$ is always an odd number. Therefore, since z is even, u is even and v is odd. Since u and v are coprime, the greatest common divisor of $2u$ and $u^2 + 3v^2$ is either 1 (case A) or 3 (case B).

28.3.1 Proof for Case A

In this case, the two factors of $-z^3$ are coprime. This implies that three does not divide u and that the two factors are cubes of two smaller numbers, r and s

$$2u = r^3$$
$$u^2 + 3v^2 = s^3$$

Since $u^2 + 3v^2$ is odd, so is s. A crucial lemma shows that if s is odd and if it satisfies an equation $s^3 = u^2 + 3v^2$, then it can be written in terms of two coprime integers e and f

$$s = e^2 + 3f^2$$

so that

$$u = e(e^2 - 9f^2)$$
$$v = 3f(e^2 - f^2)$$

Since u is even and v odd, then e is even and f is odd. Since

$$r^3 = 2u = 2e(e - 3f)(e + 3f)$$

The factors $2e$, $(e-3f)$, and $(e+3f)$ are coprime since 3 cannot divide e: If e were divisible by 3, then 3 would divide u, violating the designation of u and v as coprime. Since the three factors on the right-hand side are coprime, they must individually equal cubes of smaller integers

$$-2e = k^3$$
$$e - 3f = l^3$$
$$e + 3f = m^3$$

which yields a smaller solution $k^3 + l^3 + m^3 = 0$. Therefore, by the argument of infinite descent, the original solution (x, y, z) was impossible.

28.3.2 Proof for Case B

In this case, the greatest common divisor of $2u$ and $u^2 + 3v^2$ is 3. That implies that 3 divides u, and one may express $u = 3w$ in terms of a smaller integer, w. Since u is divisible by 4, so is w; hence, w is also even. Since u and v are coprime, so are v and w. Therefore, neither 3 nor 4 divide v.

Substituting u by w in the equation for z^3 yields

$$-z^3 = 6w(9w^2 + 3v^2) = 18w(3w^2 + v^2)$$

Because v and w are coprime, and because 3 does not divide v, then $18w$ and $3w^2 + v^2$ are also coprime. Therefore, since their product is a cube, they are each the cube of smaller integers, r and s

$$18w = r^3$$
$$3w^2 + v^2 = s^3$$

By the lemma above, since s is odd and equal to a number of the form $3w^2 + v^2$, it too can be expressed in terms of smaller coprime numbers, e and f.

$$s = e^2 + 3f^2$$

A short calculation shows that

$$v = e(e^2 - 9f^2)$$
$$w = 3f(e^2 - f^2)$$

Thus, e is odd and f is even, because v is odd. The expression for $18w$ then becomes

$$r^3 = 18w = 54f(e^2 - f^2) = 54f(e+f)(e-f) = 3^3 \times 2f(e+f)(e-f).$$

Since 3^3 divides r^3 we have that 3 divides r, so $(r/3)^3$ is an integer that equals $2f(e+f)(e-f)$. Since e and f are coprime, so are the three factors $2e$, $e+f$, and $e-f$; therefore, they are each the cube of smaller integers, k, l, and m.

$$-2e = k^3$$
$$e + f = l^3$$
$$e - f = m^3$$

which yields a smaller solution $k^3 + l^3 + m^3 = 0$. Therefore, by the argument of infinite descent, the original solution (x, y, z) was impossible.

28.4 $n = 5$

Fermat's Last Theorem for $n = 5$ states that no three coprime integers x, y and z can satisfy the equation

$$x^5 + y^5 + z^5 = 0$$

This was proven[51] neither independently nor collaboratively by Dirichlet and Legendre around 1825.[52][32] Alternative proofs were developed[53] by Gauss,[54] Lebesgue,[55] Lamé,[56] Gambioli,[16][57] Werebrusow,[58] Rychlik,[59] van der Corput,[47] and Terjanian.[60]

Dirichlet's proof for $n = 5$ is divided into the two cases (cases I and II) defined by Sophie Germain. In case I, the exponent 5 does not divide the product xyz. In case II, 5 does divide xyz.

1. Case I for $n = 5$ can be proven immediately by Sophie Germain's theorem(1823) if the auxiliary prime $\theta = 11$.

2. Case II is divided into the two cases (cases II(i) and II(ii)) by Dirichlet in 1825. Case II(i) is the case which one of x, y, z is divided by either 5 and 2. Case II(ii) is the case which one of x, y, z is divided by 5 and another one of x, y, z is divided by 2. In July 1825, Dirichlet proved the case II(i) for $n = 5$. In September 1825, Legendre proved the case II(ii) for $n = 5$. After Legendre's proof, Dirichlet completed the proof for the case II(ii) for $n = 5$ by the extended argument for the case II(i).[32]

28.4.1 Proof for Case A

Case A for $n = 5$ can be proven immediately by Sophie Germain's theorem if the auxiliary prime $\theta = 11$. A more methodical proof is as follows. By Fermat's little theorem,

$$x^5 \equiv x \pmod{5}$$
$$y^5 \equiv y \pmod{5}$$
$$z^5 \equiv z \pmod{5}$$

and therefore

$$x + y + z \equiv 0 \pmod{5}$$

This equation forces two of the three numbers x, y, and z to be equivalent modulo 5, which can be seen as follows: Since they are indivisible by 5, x, y and z cannot equal 0 modulo 5, and must equal one of four possibilities: ±1 or ±2. If they were all different, two would be opposites and their sum modulo 5 would be zero (implying contrary to the assumption of this case that the other one would be 0 modulo 5).

Without loss of generality, x and y can be designated as the two equivalent numbers modulo 5. That equivalence implies that

$$x^5 \equiv y^5 \pmod{25} \text{ (note change in modulo)}$$
$$-z^5 \equiv x^5 + y^5 \equiv 2 x^5 \pmod{25}$$

However, the equation $x \equiv y \pmod{5}$ also implies that

$$-z \equiv x + y \equiv 2x \pmod{5}$$
$$-z^5 \equiv 2^5 x^5 \equiv 32 x^5 \pmod{25}$$

Combining the two results and dividing both sides by x^5 yields a contradiction

$$2 \equiv 32 \pmod{25}$$

Thus, case A for $n = 5$ has been proven.

28.4.2 Proof for Case B

28.5 *n* = 7

The case $n = 7$ was proven[61] by Gabriel Lamé in 1839.[62] His rather complicated proof was simplified in 1840 by Victor-Amédée Lebesgue,[63] and still simpler proofs[64] were published by Angelo Genocchi in 1864, 1874 and 1876.[65] Alternative proofs were developed by Théophile Pépin[66] and Edmond Maillet.[67]

28.6 *n* = 6, 10, and 14

Fermat's Last Theorem has also been proven for the exponents $n = 6$, 10, and 14. Proofs for $n = 6$ have been published by Kausler,[5] Thue,[68] Tafelmacher,[69] Lind,[70] Kapferer,[71] Swift,[72] and Breusch.[73] Similarly, Dirichlet[74] and Terjanian[75] each proved the case $n = 14$, while Kapferer[71] and Breusch[73] each proved the case $n = 10$. Strictly speaking, these proofs are unnecessary, since these cases follow from the proofs for $n = 3$, 5, and 7, respectively. Nevertheless, the reasoning of these even-exponent proofs differs from their odd-exponent counterparts. Dirichlet's proof for $n = 14$ was published in 1832, before Lamé's 1839 proof for $n = 7$.

28.7 Notes

[1] Freeman L. "Fermat's One Proof". Retrieved 2009-05-23.

[2] Ribenboim, pp. 15–24.

[3] Frénicle de Bessy, *Traité des Triangles Rectangles en Nombres*, vol. I, 1676, Paris. Reprinted in *Mém. Acad. Roy. Sci.*, **5**, 1666–1699 (1729).

[4] Euler L (1738). "Theorematum quorundam arithmeticorum demonstrationes". *Comm. Acad. Sci. Petrop.* **10**: 125–146.. Reprinted *Opera omnia*, ser. I, "Commentationes Arithmeticae", vol. I, pp. 38–58, Leipzig:Teubner (1915).

[5] Kausler CF (1802). "Nova demonstratio theorematis nec summam, nec differentiam duorum cuborum cubum esse posse". *Novi Acta Acad. Petrop.* **13**: 245–253.

[6] Barlow P (1811). *An Elementary Investigation of Theory of Numbers*. St. Paul's Church-Yard, London: J. Johnson. pp. 144–145.

[7] Legendre AM (1830). *Théorie des Nombres (Volume II)* (3rd ed.). Paris: Firmin Didot Frères. Reprinted in 1955 by A. Blanchard (Paris).

[8] Schopis (1825). *Einige Sätze aus der unbestimmten Analytik*. Gummbinnen: Programm.

[9] Terquem O (1846). "Théorèmes sur les puissances des nombres". *Nouv. Ann. Math.* **5**: 70–87.

[10] Bertrand J (1851). *Traité Élémentaire d'Algèbre*. Paris: Hachette. pp. 217–230, 395.

[11] Lebesgue VA (1853). "Résolution des équations biquadratiques $z^2 = x^4 \pm 2^m y^4$, $z^2 = 2^m x^4 - y^4$, $2^m z^2 = x^4 \pm y^4$". *J. Math. Pures Appl.* **18**: 73–86.
Lebesgue VA (1859). *Exercices d'Analyse Numérique*. Paris: Leiber et Faraguet. pp. 83–84, 89.
Lebesgue VA (1862). *Introduction à la Théorie des Nombres*. Paris: Mallet-Bachelier. pp. 71–73.

[12] Pepin T (1883). "Étude sur l'équation indéterminée $ax^4 + by^4 = cz^2$". *Atti Accad. Naz. Lincei* **36**: 34–70.

[13] Tafelmacher WLA (1893). "Sobre la ecuación $x^4 + y^4 = z^4$". *Ann. Univ. Chile* **84**: 307–320.

[14] Hilbert D (1897). "Die Theorie der algebraischen Zahlkörper". *Jahresbericht der Deutschen Mathematiker-Vereinigung* **4**: 175–546. Reprinted in 1965 in *Gesammelte Abhandlungen, vol. I* by New York:Chelsea.

[15] Bendz TR (1901). *Öfver diophantiska ekvationen* $x^n + y^n = z^n$. Uppsala: Almqvist & Wiksells Boktrycken.

[16] Gambioli D (1901). "Memoria bibliographica sull'ultimo teorema di Fermat". *Period. Mat.* **16**: 145–192.

[17] Kronecker L (1901). *Vorlesungen über Zahlentheorie, vol. I*. Leipzig: Teubner. pp. 35–38. Reprinted by New York: Springer-Verlag in 1978.

[18] Bang A (1905). "Nyt Bevis for at Ligningen $x^4 - y^4 = z^4$, ikke kan have rationale Løsinger". *Nyt Tidsskrift Mat.* **16B**: 35–36.

[19] Sommer J (1907). *Vorlesungen über Zahlentheorie*. Leipzig: Teubner.

[20] Bottari A. "Soluzione intere dell'equazione pitagorica e applicazione alla dimostrazione di alcune teoremi dellla teoria dei numeri". *Period. Mat.* **23**: 104–110.

[21] Rychlik K (1910). "On Fermat's last theorem for $n = 4$ and $n = 3$ (in Bohemian)". *Časopis Pěst. Mat.* **39**: 65–86.

[22] Nutzhorn F (1912). "Den ubestemte Ligning $x^4 + y^4 = z^4$". *Nyt Tidsskrift Mat.* **23B**: 33–38.

[23] Carmichael RD (1913). "On the impossibility of certain Diophantine equations and systems of equations". *Amer. Math. Monthly* **20** (7): 213–221. doi:10.2307/2974106. JSTOR 2974106.

[24] Hancock H (1931). *Foundations of the Theory of Algebraic Numbers, vol. I*. New York: Macmillan.

[25] Vrănceanu G (1966). "Asupra teorema lui Fermat pentru $n=4$". *Gaz. Mat. Ser. A* **71**: 334–335. Reprinted in 1977 in *Opera matematica*, vol. 4, pp. 202–205, București:Edit. Acad. Rep. Soc. Romana.

[26] Grant, Mike, and Perella, Malcolm, "Descending to the irrational", *Mathematical Gazette* 83, July 1999, pp.263-267.

[27] Barbara, Roy, "Fermat's last theorem in the case n=4", *Mathematical Gazette* 91, July 2007, 260-262.

[28] Dolan, Stan, "Fermat's method of *descente infinie*", *Mathematical Gazette* 95, July 2011, 269-271.

[29] Fermat P. "Ad Problema XX commentarii in ultimam questionem Arithmeticorum Diophanti. Area trianguli rectanguli in numeris non potest esse quadratus", *Oeuvres*, vol. I, p. 340 (Latin), vol. III, pp. 271–272 (French). Paris:Gauthier-Villars, 1891, 1896.

[30] Ribenboim, pp. 11–14.

[31] Dickson (2005, p. 546)

[32] O'Connor & Robertson (1996)

[33] Bergmann (1966)

[34] Euler L (1770) *Vollständige Anleitung zur Algebra*, Roy.Acad. Sci., St. Petersburg.

[35] Freeman L. "Fermat's Last Theorem: Proof for $n = 3$". Retrieved 2009-05-23.

[36] J. J. Mačys (2007). "On Euler's hypothetical proof". *Mathematical Notes* **82** (3–4): 352–356. doi:10.1134/S0001434607090088. MR 2364600.

[37] Euler (1822, pp. 399, 401-402)

[38] Ribenboim, pp. 33, 37–41.

[39] Legendre AM (1823). "Recherches sur quelques objets d'analyse indéterminée, et particulièrement sur le théorème de Fermat". *Mém. Acad. Roy. Sci. Institut France* **6**: 1–60. Reprinted in 1825 as the "Second Supplément" for a printing of the 2nd edition of *Essai sur la Théorie des Nombres*, Courcier (Paris). Also reprinted in 1909 in *Sphinx-Oedipe*, **4**, 97–128.

[40] Calzolari L (1855). *Tentativo per dimostrare il teorema di Fermat sull'equazione indeterminata* xn + yn = zn. Ferrara.

[41] Lamé G (1865). "Étude des binômes cubiques $x^3 \pm y^3$". *C. R. Acad. Sci. Paris* **61**: 921–924, 961–965.

[42] Tait PG (1872). "Mathematical Notes". *Proc. Roy. Soc. Edinburgh* **7**: 144.

[43] Günther S (1878). "Über die unbestimmte Gleichung $x^3 + y^3 = z^3$". *Sitzungsberichte Böhm. Ges. Wiss.*: 112–120.

[44] Krey H (1909). "Neuer Beweis eines arithmetischen Satzes". *Math. Naturwiss. Blätter* **6**: 179–180.

[45] Stockhaus H (1910). *Beitrag zum Beweis des Fermatschen Satzes*. Leipzig: Brandstetter.

[46] Carmichael RD (1915). *Diophantine Analysis*. New York: Wiley.

[47] van der Corput JG (1915). "Quelques formes quadratiques et quelques équations indéterminées". *Nieuw Archief Wisk.* **11**: 45–75.

[48] Thue A (1917). "Et bevis for at ligningen $A^3 + B^3 = C^3$ er unmulig i hele tal fra nul forskjellige tal A, B og C". *Arch. Mat. Naturv.* **34** (15). Reprinted in *Selected Mathematical Papers* (1977), Oslo:Universitetsforlaget, pp. 555–559.

[49] Duarte FJ (1944). "Sobre la ecuación $x^3 + y^3 + z^3 = 0$". *Ciencias Fis. Mat. Naturales (Caracas)* **8**: 971–979.

[50] Ribenboim, pp. 24–49.

[51] Freeman L. "Fermat's Last Theorem: Proof for $n = 5$". Retrieved 2009-05-23.

[52] Ribenboim, p. 49.

[53] Ribenboim, pp. 55–57.

[54] Gauss CF (1875, posthumous). "Neue Theorie der Zerlegung der Cuben". *Zur Theorie der complexen Zahlen, Werke, vol. II* (2nd ed.). Königl. Ges. Wiss. Göttingen. pp. 387–391. Check date values in: |date= (help)

[55] Lebesgue VA (1843). "Théorèmes nouveaux sur l'équation indéterminée $x^5 + y^5 = az^5$". *J. Math. Pures Appl.* **8**: 49–70.

[56] Lamé G (1847). "Mémoire sur la résolution en nombres complexes de l'équation $A^5 + B^5 + C^5 = 0$". *J. Math. Pures Appl.* **12**: 137–171.

[57] Gambioli D (1903/4). "Intorno all'ultimo teorema di Fermat". *Il Pitagora* **10**: 11–13, 41–42. Check date values in: |date= (help)

[58] Werebrusow AS (1905). "On the equation $x^5 + y^5 = Az^5$ *(in Russian)*". *Moskov. Math. Samml.* **25**: 466–473.

[59] Rychlik K (1910). "On Fermat's last theorem for $n = 5$ *(in Bohemian)*". *Časopis Pěst. Mat.* **39**: 185–195, 305–317.

[60] Terjanian G (1987). "Sur une question de V. A. Lebesgue". *Ann. Inst. Fourier* **37** (3): 19–37. doi:10.5802/aif.1096.

[61] Ribenboim, pp. 57–63.

[62] Lamé G (1839). "Mémoire sur le dernier théorème de Fermat". *C. R. Acad. Sci. Paris* **9**: 45–46.
Lamé G (1840). "Mémoire d'analyse indéterminée démontrant que l'équation $x^7 + y^7 = z^7$ est impossible en nombres entiers". *J. Math. Pures Appl.* **5**: 195–211.

[63] Lebesgue VA (1840). "Démonstration de l'impossibilité de résoudre l'équation $x^7 + y^7 + z^7 = 0$ en nombres entiers". *J. Math. Pures Appl.* **5**: 276–279, 348–349.

[64] Freeman L. "Fermat's Last Theorem: Proof for $n = 7$". Retrieved 2009-05-23.

[65] Genocchi A (1864). "Intorno all'equazioni $x^7 + y^7 + z^7 = 0$". *Annali Mat.* **6**: 287–288.
Genocchi A (1874). "Sur l'impossibilité de quelques égalités doubles". *C. R. Acad. Sci. Paris* **78**: 433–436.
Genocchi A (1876). "Généralisation du théorème de Lamé sur l'impossibilité de l'équation $x^7 + y^7 + z^7 = 0$". *C. R. Acad. Sci. Paris* **82**: 910–913.

[66] Pepin T (1876). "Impossibilité de l'équation $x^7 + y^7 + z^7 = 0$". *C. R. Acad. Sci. Paris* **82**: 676–679, 743–747.

[67] Maillet E (1897). "Sur l'équation indéterminée $ax^{\lambda t} + by^{\lambda t} = cz^{\lambda t}$". *Assoc. Française Avanc. Sci., St. Etienne (sér. II)* **26**: 156–168.

[68] Thue A (1896). "Über die Auflösbarkeit einiger unbestimmter Gleichungen". *Det Kongel. Norske Videnskabers Selskabs Skrifter* **7**. Reprinted in *Selected Mathematical Papers*, pp. 19–30, Oslo:Universitetsforlaget (1977).

[69] Tafelmacher WLA (1897). "La ecuación $x^3 + y^3 = z^2$: Una demostración nueva del teorema de fermat para el caso de las sestas potencias". *Ann. Univ. Chile, Santiago* **97**: 63–80.

[70] Lind B (1909). "Einige zahlentheoretische Sätze". *Arch. Math. Phys.* **15**: 368–369.

[71] Kapferer H (1913). "Beweis des Fermatschen Satzes für die Exponenten 6 und 10". *Archiv Math. Phys.* **21**: 143–146.

[72] Swift E (1914). "Solution to Problem 206". *Amer. Math. Monthly* **21**: 238–239.

[73] Breusch R (1960). "A simple proof of Fermat's last theorem for $n = 6, n = 10$". *Math. Mag.* **33** (5): 279–281. doi:10.2307/3029800. JSTOR 3029800.

[74] Dirichlet PGL (1832). "Démonstration du théorème de Fermat pour le cas des 14e puissances". *J. Reine Angew. Math.* **9**: 390–393. Reprinted in *Werke*, vol. I, pp. 189–194, Berlin:G. Reimer (1889); reprinted New York:Chelsea (1969).

[75] Terjanian G (1974). "L'équation $x^{14} + y^{14} = z^{14}$ en nombres entiers". *Bull. Sci. Math. (sér. 2)* **98**: 91–95.

28.8 References

- Aczel, Amir (1996-09-30). *Fermat's Last Theorem: Unlocking the Secret of an Ancient Mathematical Problem*. Four Walls Eight Windows. ISBN 978-1-56858-077-7.

- Dickson LE (1919). *History of the Theory of Numbers. Volume II. Diophantine Analysis*. New York: Chelsea Publishing. pp. 545–550, 615–621, 731–776.

  - Dickson, LE (2005) [1920], *History of the theory of numbers. Vol. II: Diophantine analysis*, New York: Dover Publications, ISBN 978-0-486-44233-4, MR 0245500

- Edwards, HM (2008-05-23). *Fermat's Last Theorem: A Genetic Introduction to Algebraic Number Theory*. Graduate Texts in Mathematics **50** (3rd printing 2000 ed.). New York: Springer-Verlag. ISBN 978-0-387-95002-0.

- Mordell LJ (1921). *Three Lectures on Fermat's Last Theorem*. Cambridge: Cambridge University Press.

- Ribenboim P (2000). *Fermat's Last Theorem for Amateurs*. New York: Springer-Verlag. ISBN 978-0-387-98508-4.

- Singh S (October 1998). *Fermat's Enigma*. New York: Anchor Books. ISBN 978-0-385-49362-8.

- Stark H (1978). *An Introduction to Number Theory*. MIT Press. ISBN 0-262-69060-8.

28.9 Further reading

- Bell, Eric T. (1998-08-06) [1961]. *The Last Problem*. New York: The Mathematical Association of America. ISBN 978-0-88385-451-8.

- Benson, Donald C. (2001-04-05). *The Moment of Proof: Mathematical Epiphanies*. Oxford University Press. ISBN 978-0-19-513919-8.

- Bergmann, G. (1966), "Über Eulers Beweis des großen Fermatschen Satzes für den Exponenten 3.", *Mathematische Annalen* (Springer) **164** (2): 159–175, doi:10.1007/BF01429054, Zbl 0138.25101

- Brudner, Harvey J. (1994). *Fermat and the Missing Numbers*. WLC, Inc. ISBN 978-0-9644785-0-3.

- Faltings G (July 1995). "The Proof of Fermat's Last Theorem by R. Taylor and A. Wiles" (PDF). *Notices of the AMS* **42** (7): pp. 743–746. ISSN 0002-9920.

- Euler, L. (1822), *Elements of Algebra* (3rd ed.), London: Longman, pp. 399, 401–402

- Mozzochi, Charles (2000-12-07). *The Fermat Diary*. American Mathematical Society. ISBN 978-0-8218-2670-6.

- Ribenboim P (1979). *13 Lectures on Fermat's Last Theorem*. New York: Springer Verlag. ISBN 978-0-387-90432-0.

- van der Poorten, Alf (1996-03-06). *Notes on Fermat's Last Theorem*. WileyBlackwell. ISBN 978-0-471-06261-5.

28.10 External links

- Elkies, Noam D. "Tables of Fermat "near-misses" - approximate solutions of $x^n + y^n = z^n$".

- Freeman, Larry (2005). "Fermat's Last Theorem Blog". A blog that covers the history of Fermat's Last Theorem from Pierre Fermat to Andrew Wiles.

- Ribet, Ken (1995). "Galois representations and modular forms" (PDF). Discusses various material which is related to the proof of Fermat's Last Theorem: elliptic curves, modular forms, Galois representations and their deformations, Frey's construction, and the conjectures of Serre and of Taniyama–Shimura.

- Shay, David (2003). "Fermat's Last Theorem". Retrieved 2004-08-05. The story, the history and the mystery.

- "The bluffer's guide to Fermat's Last Theorem".

- Weisstein, Eric W., "Fermat's Last Theorem", *MathWorld*.

- O'Connor, John J.; Robertson, Edmund F. (1996), *Fermat's last theorem*, MacTutor History of Mathematical Topics - University of St Andrews.

- "The Proof". The title of one edition of the PBS television series NOVA, discusses Andrew Wiles's effort to prove Fermat's Last Theorem.

- "The Whole Story". Edited version of ~2,000-word essay published in Prometheus magazine, describing Andrew Wiles's successful journey.

- "Documentary Movie on Fermat's Last Theorem (1996)". Simon Singh and John Lynch's film tells the enthralling and emotional story of Andrew Wiles.

Portrait of Pierre de Fermat.

Leonhard Euler by Jakob Emanuel Handmann.

Portrait of Peter Gustav Lejeune Dirichlet.

Caricature of Adrien-Marie Legendre (the only surviving portrait of him).

Chapter 29

Proof of the Euler product formula for the Riemann zeta function

Leonhard Euler proved the **Euler product formula for the Riemann zeta function** in his thesis *Variae observationes circa series infinitas* (*Various Observations about Infinite Series*), published by St Petersburg Academy in 1737.[1][2]

29.1 The Euler product formula

The Euler product formula for the Riemann zeta function reads

$$\sum_{n=1}^{\infty} \frac{1}{n^s} = \prod_{p \text{ prime}} \frac{1}{1-p^{-s}}$$

where the left hand side equals the Riemann zeta function:

$$\zeta(s) = \sum_{n=1}^{\infty} \frac{1}{n^s} = 1 + \frac{1}{2^s} + \frac{1}{3^s} + \frac{1}{4^s} + \frac{1}{5^s} + \ldots$$

and the product on the right hand side extends over all prime numbers p:

$$\prod_{p \text{ prime}} \frac{1}{1-p^{-s}} = \frac{1}{1-2^{-s}} \cdot \frac{1}{1-3^{-s}} \cdot \frac{1}{1-5^{-s}} \cdot \frac{1}{1-7^{-s}} \cdots \frac{1}{1-p^{-s}} \cdots$$

29.2 Proof of the Euler product formula

This sketch of a proof only makes use of simple algebra commonly taught in high school. This was originally the method by which Euler discovered the formula. There is a certain sieving property that we can use to our advantage:

$$\zeta(s) = 1 + \frac{1}{2^s} + \frac{1}{3^s} + \frac{1}{4^s} + \frac{1}{5^s} + \ldots$$

$$\frac{1}{2^s}\zeta(s) = \frac{1}{2^s} + \frac{1}{4^s} + \frac{1}{6^s} + \frac{1}{8^s} + \frac{1}{10^s} + \ldots$$

The method of Eratosthenes used to sieve out prime numbers is employed in this proof.

Subtracting the second equation from the first we remove all elements that have a factor of 2:

$$\left(1 - \frac{1}{2^s}\right)\zeta(s) = 1 + \frac{1}{3^s} + \frac{1}{5^s} + \frac{1}{7^s} + \frac{1}{9^s} + \frac{1}{11^s} + \frac{1}{13^s} + \ldots$$

Repeating for the next term:

$$\frac{1}{3^s}\left(1 - \frac{1}{2^s}\right)\zeta(s) = \frac{1}{3^s} + \frac{1}{9^s} + \frac{1}{15^s} + \frac{1}{21^s} + \frac{1}{27^s} + \frac{1}{33^s} + \ldots$$

Subtracting again we get:

$$\left(1 - \frac{1}{3^s}\right)\left(1 - \frac{1}{2^s}\right)\zeta(s) = 1 + \frac{1}{5^s} + \frac{1}{7^s} + \frac{1}{11^s} + \frac{1}{13^s} + \frac{1}{17^s} + \ldots$$

where all elements having a factor of 3 or 2 (or both) are removed.

It can be seen that the right side is being sieved. Repeating infinitely we get:

$$\ldots \left(1 - \frac{1}{11^s}\right)\left(1 - \frac{1}{7^s}\right)\left(1 - \frac{1}{5^s}\right)\left(1 - \frac{1}{3^s}\right)\left(1 - \frac{1}{2^s}\right)\zeta(s) = 1$$

Dividing both sides by everything but the $\zeta(s)$ we obtain:

$$\zeta(s) = \frac{1}{\left(1 - \frac{1}{2^s}\right)\left(1 - \frac{1}{3^s}\right)\left(1 - \frac{1}{5^s}\right)\left(1 - \frac{1}{7^s}\right)\left(1 - \frac{1}{11^s}\right)\ldots}$$

This can be written more concisely as an infinite product over all primes p:

$$\zeta(s) = \prod_{p\text{ prime}} \frac{1}{1 - p^{-s}}$$

To make this proof rigorous, we need only observe that when $\Re(s) > 1$, the sieved right-hand side approaches 1, which follows immediately from the convergence of the Dirichlet series for $\zeta(z)$.

29.3 The case $s = 1$

An interesting result can be found for $\zeta(1)$

$$\ldots \left(1 - \frac{1}{11}\right)\left(1 - \frac{1}{7}\right)\left(1 - \frac{1}{5}\right)\left(1 - \frac{1}{3}\right)\left(1 - \frac{1}{2}\right)\zeta(1) = 1$$

which can also be written as,

$$\ldots \left(\frac{10}{11}\right)\left(\frac{6}{7}\right)\left(\frac{4}{5}\right)\left(\frac{2}{3}\right)\left(\frac{1}{2}\right)\zeta(1) = 1$$

which is,

$$\left(\frac{\ldots \cdot 10 \cdot 6 \cdot 4 \cdot 2 \cdot 1}{\ldots \cdot 11 \cdot 7 \cdot 5 \cdot 3 \cdot 2}\right)\zeta(1) = 1$$

as, $\zeta(1) = 1 + \frac{1}{2} + \frac{1}{3} + \frac{1}{4} + \frac{1}{5} + \ldots$
thus,

$$1 + \frac{1}{2} + \frac{1}{3} + \frac{1}{4} + \frac{1}{5} + \ldots = \frac{2 \cdot 3 \cdot 5 \cdot 7 \cdot 11 \cdot \ldots}{1 \cdot 2 \cdot 4 \cdot 6 \cdot 10 \cdot \ldots}$$

We know that the left-hand side of the equation diverges to infinity, therefore the numerator on the right-hand side (the primorial) must also be infinite for divergence. This proves that there are infinitely many prime numbers.

29.4 Another proof

Each factor (for a given prime p) in the product above can be expanded to a geometric series consisting of the reciprocal of p raised to multiples of s, as follows

$$\frac{1}{1-p^{-s}} = 1 + \frac{1}{p^s} + \frac{1}{p^{2s}} + \frac{1}{p^{3s}} + \ldots + \frac{1}{p^{ks}} + \ldots$$

When $\Re(s) > 1$, we have $|p^{-s}| < 1$ and this series converges absolutely. Hence we may take a finite number of factors, multiply them together, and rearrange terms. Taking all the primes p up to some prime number limit q, we have

$$\left| \zeta(s) - \prod_{p \leq q} \left(\frac{1}{1-p^{-s}} \right) \right| < \sum_{n=q+1}^{\infty} \frac{1}{n^\sigma}$$

where σ is the real part of s. By the fundamental theorem of arithmetic, the partial product when expanded out gives a sum consisting of those terms n^{-s} where n is a product of primes less than or equal to q. The inequality results from the fact that therefore only integers larger than q can fail to appear in this expanded out partial product. Since the difference between the partial product and ζ(s) goes to zero when σ > 1, we have convergence in this region.

29.5 References

- John Derbyshire, *Prime Obsession: Bernhard Riemann and The Greatest Unsolved Problem in Mathematics*, Joseph Henry Press, 2003, ISBN 978-0-309-08549-6

29.6 Notes

[1] O'Connor, J.J. and Robertson, E.F. (February 1996). "A history of calculus". University of St Andrews. Retrieved 2007-08-07.

[2] John Derbyshire (2003), chapter 7, "The Golden Key, and an Improved Prime Number Theorem"

Chapter 30

Proofs involving ordinary least squares

The purpose of this page is to provide supplementary materials for the Ordinary least squares article, reducing the load of the main article with mathematics and improving its accessibility, while at the same time retaining the completeness of exposition.

30.1 Least squares estimator for β

Using matrix notation, the sum of squared residuals is given by

$$S(b) = (y - Xb)'(y - Xb)$$

Where $'$ denotes the matrix transpose.

Since this is a quadratic expression and $S(b) \geq 0$, the global minimum will be found by differentiating it with respect to b:

$$0 = \frac{dS}{db'}(\hat{\beta}) = \frac{d}{db'}\left(y'y - b'X'y - y'Xb + b'X'Xb\right)\bigg|_{b=\hat{\beta}} = -2X'y + 2X'X\hat{\beta}$$

By assumption matrix X has full column rank, and therefore $X'X$ is invertible and the least squares estimator for β is given by

$$\hat{\beta} = (X'X)^{-1}X'y$$

30.2 Unbiasedness and Variance of $\hat{\beta}$

Plug $y = X\beta + \varepsilon$ into the formula for $\hat{\beta}$ and then use the Law of iterated expectation:

$$\begin{aligned}
E[\hat{\beta}] &= E\left[(X'X)^{-1}X'(X\beta + \varepsilon)\right] \\
&= \beta + E\left[(X'X)^{-1}X'\varepsilon\right] \\
&= \beta + E\left[E\left[(X'X)^{-1}X'\varepsilon|X\right]\right] \\
&= \beta + E\left[(X'X)^{-1}X'E[\varepsilon|X]\right] \quad = \beta,
\end{aligned}$$

where $E[\varepsilon|X] = 0$ by assumptions of the model.

For the variance, let $\sigma^2 I$ denote the covariance matrix of ε. Then,

$$\begin{aligned}
E[(\hat{\beta} - \beta)(\hat{\beta} - \beta)^T] &= E\left[((X'X)^{-1}X'\varepsilon)((X'X)^{-1}X'\varepsilon)^T\right] \\
&= E\left[(X'X)^{-1}X'\varepsilon\varepsilon'X(X'X)^{-1}\right] \\
&= E\left[(X'X)^{-1}X'\sigma^2 X(X'X)^{-1}\right] \\
&= E\left[\sigma^2(X'X)^{-1}X'X(X'X)^{-1}\right] \\
&= \sigma^2(X'X)^{-1},
\end{aligned}$$

where we used the fact that $\hat{\beta} - \beta$ is just an affine transformation of ε by the matrix $(X'X)^{-1}X'$ (see article on the multivariate normal distribution under the affine transformation section).

For a simple linear regression model, where $\beta = [\beta_0, \beta_1]^T$ (β_0 is the y-intercept and β_1 is the slope), one obtains

$$\begin{aligned}
\sigma^2(X'X)^{-1} &= \sigma^2 \left(\sum x_i x_i'\right)^{-1} \\
&= \sigma^2 \left(\sum (1, x_i)'(1, x_i)\right)^{-1} \\
&= \sigma^2 \left(\sum \begin{pmatrix} 1 & x_i \\ x_i & x_i^2 \end{pmatrix}\right)^{-1} \\
&= \sigma^2 \begin{pmatrix} n & \sum x_i \\ \sum x_i & \sum x_i^2 \end{pmatrix}^{-1} \\
&= \sigma^2 \cdot \frac{1}{n\sum x_i^2 - (\sum x_i)^2} \begin{pmatrix} \sum x_i^2 & -\sum x_i \\ -\sum x_i & n \end{pmatrix} \\
&= \sigma^2 \cdot \frac{1}{n\sum_{i=1}^n (x_i - \bar{x})^2} \begin{pmatrix} \sum x_i^2 & -\sum x_i \\ -\sum x_i & n \end{pmatrix}
\end{aligned}$$

$$Var(\beta_1) = \frac{\sigma^2}{\sum_{i=1}^n (x_i - \bar{x})^2}.$$

30.3 Expected value of $\hat{\sigma}^2$

First we will plug in the expression for y into the estimator, and use the fact that $X'M = MX = 0$ (matrix M projects onto the space orthogonal to X):

$$\hat{\sigma}^2 = \tfrac{1}{n} y'My = \tfrac{1}{n}(X\beta + \varepsilon)'M(X\beta + \varepsilon) = \tfrac{1}{n}\varepsilon'M\varepsilon$$

Now we can recognize $\varepsilon'M\varepsilon$ as a 1×1 matrix, such matrix is equal to its own trace. This is useful because by properties of trace operator, **tr**(AB)=**tr**(BA), and we can use this to separate disturbance ε from matrix M which is a function of regressors X:

$$E\,\hat{\sigma}^2 = \tfrac{1}{n} E\left[\operatorname{tr}(\varepsilon'M\varepsilon)\right] = \tfrac{1}{n}\operatorname{tr}\left(E[M\varepsilon\varepsilon']\right)$$

Using the Law of iterated expectation this can be written as

$$E\,\hat{\sigma}^2 = \tfrac{1}{n}\operatorname{tr}\left(E\left[M\,E[\varepsilon\varepsilon'|X]\right]\right) = \tfrac{1}{n}\operatorname{tr}\left(E[\sigma^2 MI]\right) = \tfrac{1}{n}\sigma^2 E\left[\operatorname{tr} M\right]$$

Recall that $M = I - P$ where P is the projection onto linear space spanned by columns of matrix X. By properties of a projection matrix, it has $p = \text{rank}(X)$ eigenvalues equal to 1, and all other eigenvalues are equal to 0. Trace of a matrix is equal to the sum of its characteristic values, thus $\text{tr}(P)=p$, and $\text{tr}(M) = n - p$. Therefore

$$\text{E}\,\hat{\sigma}^2 = \frac{n-p}{n}\sigma^2$$

Note: in the later section "Maximum likelihood" we show that under the additional assumption that errors are distributed normally, the estimator $\hat{\sigma}^2$ is proportional to a chi-squared distribution with $n - p$ degrees of freedom, from which the formula for expected value would immediately follow. However the result we have shown in this section is valid regardless of the distribution of the errors, and thus has importance on its own.

30.4 Consistency and asymptotic normality of $\hat{\beta}$

Estimator $\hat{\beta}$ can be written as

$$\hat{\beta} = \left(\tfrac{1}{n}X'X\right)^{-1}\tfrac{1}{n}X'y = \beta + \left(\tfrac{1}{n}X'X\right)^{-1}\tfrac{1}{n}X'\varepsilon = \beta + \left(\frac{1}{n}\sum_{i=1}^{n}x_i x_i'\right)^{-1}\left(\frac{1}{n}\sum_{i=1}^{n}x_i \varepsilon_i\right)$$

We can use the law of large numbers to establish that

$$\frac{1}{n}\sum_{i=1}^{n}x_i x_i' \xrightarrow{p} \text{E}[x_i x_i'] = \frac{Q_{xx}}{n}, \qquad \frac{1}{n}\sum_{i=1}^{n}x_i \varepsilon_i \xrightarrow{p} \text{E}[x_i \varepsilon_i] = 0$$

By Slutsky's theorem and continuous mapping theorem these results can be combined to establish consistency of estimator $\hat{\beta}$:

$$\hat{\beta} \xrightarrow{p} \beta + Q_{xx}^{-1} \cdot 0 = \beta$$

The central limit theorem tells us that

$$\tfrac{1}{\sqrt{n}}\sum_{i=1}^{n} x_i \varepsilon_i \xrightarrow{d} \mathcal{N}(0, V), \text{ where } V = \text{Var}[x_i \varepsilon_i] = \text{E}[\varepsilon_i^2 x_i x_i'] = \text{E}\left[\,\text{E}[\varepsilon_i^2|x_i]\,x_i x_i'\,\right] = \sigma^2 \tfrac{Q_{xx}}{n}$$

Applying Slutsky's theorem again we'll have

$$\sqrt{n}(\hat{\beta} - \beta) = \left(\frac{1}{n}\sum_{i=1}^{n}x_i x_i'\right)^{-1}\left(\frac{1}{\sqrt{n}}\sum_{i=1}^{n}x_i \varepsilon_i\right) \xrightarrow{d} Q_{xx}^{-1}n \cdot \mathcal{N}\left(0, \sigma^2\frac{Q_{xx}}{n}\right) = \mathcal{N}(0, \sigma^2 Q_{xx}^{-1} n)$$

30.5 Maximum likelihood approach

Maximum likelihood estimation is a generic technique for estimating the unknown parameters in a statistical model by constructing a log-likelihood function corresponding to the joint distribution of the data, then maximizing this function over all possible parameter values. In order to apply this method, we have to make an assumption about the distribution of y given X so that the log-likelihood function can be constructed. The connection of maximum likelihood estimation to OLS arises when this distribution is modeled as a multivariate normal.

30.5. MAXIMUM LIKELIHOOD APPROACH

Specifically, assume that the errors ε have multivariate normal distribution with mean 0 and variance matrix $\sigma^2 I$. Then the distribution of y conditionally on X is

$$y|X \sim \mathcal{N}(X\beta, \sigma^2 I)$$

and the log-likelihood function of the data will be

$$\mathcal{L}(\beta, \sigma^2|X) = \ln\left(\frac{1}{(2\pi)^{n/2}(\sigma^2)^{n/2}} e^{-\frac{1}{2}(y-X\beta)'(\sigma^2 I)^{-1}(y-X\beta)}\right)$$

$$= -\frac{n}{2}\ln 2\pi - \frac{n}{2}\ln\sigma^2 - \frac{1}{2\sigma^2}(y-X\beta)'(y-X\beta)$$

Differentiating this expression with respect to β and σ^2 we'll find the ML estimates of these parameters:

$$\frac{\partial \mathcal{L}}{\partial \beta'} = -\frac{1}{2\sigma^2}\left(-2X'y + 2X'X\beta\right) = 0 \quad \Rightarrow \quad \hat{\beta} = (X'X)^{-1}X'y$$

$$\frac{\partial \mathcal{L}}{\partial \sigma^2} = -\frac{n}{2}\frac{1}{\sigma^2} + \frac{1}{2\sigma^4}(y-X\beta)'(y-X\beta) = 0 \quad \Rightarrow \quad \hat{\sigma}^2 = \frac{1}{n}(y-X\hat{\beta})'(y-X\hat{\beta}) = \frac{1}{n}S(\hat{\beta})$$

We can check that this is indeed a maximum by looking at the Hessian matrix of the log-likelihood function.

30.5.1 Finite sample distribution

Since we have assumed in this section that the distribution of error terms is known to be normal, it becomes possible to derive the explicit expressions for the distributions of estimators $\hat{\beta}$ and $\hat{\sigma}^2$:

$$\hat{\beta} = (X'X)^{-1}X'y = (X'X)^{-1}X'(X\beta + \varepsilon) = \beta + (X'X)^{-1}X'\mathcal{N}(0, \sigma^2 I)$$

so that by the affine transformation properties of multivariate normal distribution

$$\hat{\beta}|X \sim \mathcal{N}(\beta, \sigma^2(X'X)^{-1}).$$

Similarly the distribution of $\hat{\sigma}^2$ follows from

$$\hat{\sigma}^2 = \tfrac{1}{n}(y - X(X'X)^{-1}X'y)'(y - X(X'X)^{-1}X'y)$$
$$= \tfrac{1}{n}(My)'My$$
$$= \tfrac{1}{n}(X\beta + \varepsilon)'M(X\beta + \varepsilon)$$
$$= \tfrac{1}{n}\varepsilon'M\varepsilon,$$

where $M = I - X(X'X)^{-1}X'$ is the symmetric projection matrix onto subspace orthogonal to X, and thus $MX = X'M = 0$. We have argued before that this matrix has rank of $n-p$, and thus by properties of chi-squared distribution,

$$\tfrac{n}{\sigma^2}\hat{\sigma}^2|X = (\varepsilon/\sigma)'M(\varepsilon/\sigma) \sim \chi^2_{n-p}$$

Moreover, the estimators $\hat{\beta}$ and $\hat{\sigma}^2$ turn out to be independent (conditional on X), a fact which is fundamental for construction of the classical t- and F-tests. The independence can be easily seen from following: the estimator $\hat{\beta}$ represents coefficients of vector decomposition of $\hat{y} = X\hat{\beta} = Py = X\beta + P\varepsilon$ by the basis of columns of X, as such $\hat{\beta}$ is a function of $P\varepsilon$. At the same time, the estimator $\hat{\sigma}^2$ is a norm of vector $M\varepsilon$ divided by n, and thus this estimator is a function of $M\varepsilon$. Now, random variables *(Pε, Mε)* are jointly normal as a linear transformation of ε, and they are also uncorrelated because $PM = 0$. By properties of multivariate normal distribution, this means that $P\varepsilon$ and $M\varepsilon$ are independent, and therefore estimators $\hat{\beta}$ and $\hat{\sigma}^2$ will be independent as well.

Chapter 31

Proofs involving the Laplace–Beltrami operator

Main article: Laplace operator

31.1 −div is adjoint to d

The claim is made that −div is adjoint to d:

$$\int_M df(X)\,\omega = -\int_M f\,\mathrm{div}\,X\,\omega$$

Proof of the above statement:

$$\int_M (f\mathrm{div}(X) + X(f))\omega = \int_M (f\mathcal{L}_X + \mathcal{L}_X(f))\omega$$

$$= \int_M \mathcal{L}_X f\omega = \int_M \mathrm{d}\iota_X f\omega = \int_{\partial M} \iota_X f\omega$$

If f has compact support, then the last integral vanishes, and we have the desired result.

31.2 Laplace–de Rham operator

One may prove that the Laplace–de Rham operator is equivalent to the definition of the Laplace–Beltrami operator, when acting on a scalar function f. This proof reads as:

$$\Delta f = \mathrm{d}\delta f + \delta\,\mathrm{d}f = \delta\,\mathrm{d}f = \delta\,\partial_i f\,\mathrm{d}x^i$$

$$= -*\mathrm{d}*\partial_i f\,\mathrm{d}x^i = -*\mathrm{d}(\varepsilon_{iJ}\sqrt{|g|}\partial^i f\,\mathrm{d}x^J)$$

$$= - * \varepsilon_{iJ} \, \partial_j(\sqrt{|g|}\partial^i f) \, \mathrm{d}x^j \wedge \mathrm{d}x^J = - * \frac{1}{\sqrt{|g|}} \partial_i(\sqrt{|g|} \, \partial^i f) \mathrm{vol}_n$$

$$= -\frac{1}{\sqrt{|g|}} \partial_i(\sqrt{|g|} \, \partial^i f),$$

where vol_n; is the volume form and ε is the completely antisymmetric Levi-Civita symbol. Note that in the above, the italic lower-case index i is a single index, whereas the upper-case Roman J stands for all of the remaining $(n-1)$ indices. Notice that the Laplace–de Rham operator is actually minus the Laplace–Beltrami operator; this minus sign follows from the conventional definition of the properties of the codifferential. Unfortunately, Δ is used to denote both; reader beware.

31.3 Properties

Given scalar functions f and h, and a real number a, the Laplacian has the property:

$$\Delta(fh) = f\,\Delta h + 2\partial_i f\, \partial^i h + h\,\Delta f.$$

31.3.1 Proof

$$\Delta(fh) = \delta\,\mathrm{d}fh = \delta(f\,\mathrm{d}h + h\,\mathrm{d}f) = *\mathrm{d}(f*\mathrm{d}h) + *\mathrm{d}(h*\mathrm{d}f)$$

$$= *(f\,\mathrm{d}*\mathrm{d}h + \mathrm{d}f \wedge *\mathrm{d}h + \mathrm{d}h \wedge *\mathrm{d}f + h\,\mathrm{d}*\mathrm{d}f)$$

$$= f*\mathrm{d}*\mathrm{d}h + *(\mathrm{d}f \wedge *\mathrm{d}h + \mathrm{d}h \wedge *\mathrm{d}f) + h*\mathrm{d}*\mathrm{d}f$$

$$= f\,\Delta h$$

$$+ *(\partial_i f\,\mathrm{d}x^i \wedge \varepsilon_{jJ}\sqrt{|g|}\partial^j h\,\mathrm{d}x^J + \partial_i h\,\mathrm{d}x^i \wedge \varepsilon_{jJ}\sqrt{|g|}\partial^j f\,\mathrm{d}x^J)$$

$$+ h\,\Delta f$$

$$= f\,\Delta h + (\partial_i f\,\partial^i h + \partial_i h\,\partial^i f)*\mathrm{vol}_n + h\,\Delta f$$

$$= f\,\Delta h + 2\partial_i f\,\partial^i h + h\,\Delta f$$

where f and h are scalar functions.

Chapter 32

Proofs of convergence of random variables

This article is supplemental for "Convergence of random variables" and provides proofs for selected results.

Several results will be established using the **portmanteau lemma**: A sequence $\{Xn\}$ converges in distribution to X if and only if any of the following conditions are met:

1. $E[f(Xn)] \to E[f(X)]$ for all bounded, continuous functions f;
2. $E[f(Xn)] \to E[f(X)]$ for all bounded, Lipschitz functions f;
3. $\limsup\{\Pr(Xn \in C)\} \leq \Pr(X \in C)$ for all closed sets C;

32.1 Convergence almost surely implies convergence in probability

$$X_n \xrightarrow{as} X \quad \Rightarrow \quad X_n \xrightarrow{p} X$$

Proof: If $\{Xn\}$ converges to X almost surely, it means that the set of points $\{\omega: \lim Xn(\omega) \neq X(\omega)\}$ has measure zero; denote this set O. Now fix $\varepsilon > 0$ and consider a sequence of sets

$$A_n = \bigcup_{m \geq n} \{|X_m - X| > \varepsilon\}$$

This sequence of sets is decreasing: $An \supseteq An_{+1} \supseteq ...$, and it decreases towards the set

$$A_\infty = \bigcap_{n \geq 1} A_n.$$

For this decreasing sequence of events, their probabilities are also a decreasing sequence, and it decreases towards the $\Pr(A\infty)$; we shall show now that this number is equal to zero. Now any point ω in the complement of O is such that $\lim Xn(\omega) = X(\omega)$, which implies that $|Xn(\omega) - X(\omega)| < \varepsilon$ for all n greater than a certain number N. Therefore, for all $n \geq N$ the point ω will not belong to the set An, and consequently it will not belong to $A\infty$. This means that $A\infty$ is disjoint with O, or equivalently, $A\infty$ is a subset of O and therefore $\Pr(A\infty) = 0$.

Finally, consider

$$\Pr(|X_n - X| > \varepsilon) \leq \Pr(A_n) \xrightarrow[n \to \infty]{} 0,$$

which by definition means that Xn converges in probability to X.

32.2 Convergence in probability does not imply almost sure convergence in the discrete case

If X_n are independent random variables assuming value one with probability $1/n$ and zero otherwise, then X_n converges to zero in probability but not almost surely. This can be verified using the Borel–Cantelli lemmas.

32.3 Convergence in probability implies convergence in distribution

$$X_n \xrightarrow{p} X \quad \Rightarrow \quad X_n \xrightarrow{d} X,$$

32.3.1 Proof for the case of scalar random variables

Lemma. Let X, Y be random variables, a a real number and $\varepsilon > 0$. Then

$$\Pr(Y \leq a) \leq \Pr(X \leq a + \varepsilon) + \Pr(|Y - X| > \varepsilon).$$

(or $\{Y \leq a\} \subset \{X \leq a + \varepsilon\} \cup \{|Y - X| > \varepsilon\}$.)

Proof of lemma:

$$\begin{aligned}
\Pr(Y \leq a) &= \Pr(Y \leq a,\ X \leq a + \varepsilon) + \Pr(Y \leq a,\ X > a + \varepsilon) \\
&\leq \Pr(X \leq a + \varepsilon) + \Pr(Y - X \leq a - X,\ a - X < -\varepsilon) \\
&\leq \Pr(X \leq a + \varepsilon) + \Pr(Y - X < -\varepsilon) \\
&\leq \Pr(X \leq a + \varepsilon) + \Pr(Y - X < -\varepsilon) + \Pr(Y - X > \varepsilon) \\
&= \Pr(X \leq a + \varepsilon) + \Pr(|Y - X| > \varepsilon)
\end{aligned}$$

Proof of the theorem: Recall that in order to prove convergence in distribution, one must show that the sequence of cumulative distribution functions converges to the FX at every point where FX is continuous. Let a be such a point. For every $\varepsilon > 0$, due to the preceding lemma, we have:

$$\Pr(X_n \leq a) \leq \Pr(X \leq a + \varepsilon) + \Pr(|X_n - X| > \varepsilon)$$
$$\Pr(X \leq a - \varepsilon) \leq \Pr(X_n \leq a) + \Pr(|X_n - X| > \varepsilon)$$

So, we have

$$\Pr(X \leq a - \varepsilon) - \Pr(|X_n - X| > \varepsilon) \leq \Pr(X \leq a + \varepsilon) + \Pr(|X_n - X| > \varepsilon).$$

Taking the limit as $n \to \infty$, we obtain:

$$F_X(a - \varepsilon) \leq F_X(a + \varepsilon),$$

where $FX(a) = \Pr(X \leq a)$ is the cumulative distribution function of X. This function is continuous at a by assumption, and therefore both $FX(a-\varepsilon)$ and $FX(a+\varepsilon)$ converge to $FX(a)$ as $\varepsilon \to 0^+$. Taking this limit, we obtain

$$\lim_{n \to \infty} \Pr(X_n \leq a) = \Pr(X \leq a),$$

which means that $\{Xn\}$ converges to X in distribution.

32.3.2 Proof for the generic case

We see that $|X_n - X|$ converges in probability to zero, and also X converges to X in distribution trivially. Applying the property proved later on this page we conclude that X_n converges to X in distribution.

32.4 Convergence in distribution to a constant implies convergence in probability

$$X_n \xrightarrow{d} c \quad \Rightarrow \quad X_n \xrightarrow{p} c, \text{ provided } c \text{ is a constant.}$$

Proof: Fix $\varepsilon > 0$. Let $B_\varepsilon(c)$ be the open ball of radius ε around point c, and $B_\varepsilon^c(c)$ its complement. Then

$$\Pr(|X_n - c| \geq \varepsilon) = \Pr(X_n \in B_\varepsilon^c(c)).$$

By the portmanteau lemma (part C), if X_n converges in distribution to c, then the limsup of the latter probability must be less than or equal to $\Pr(c \in B_\varepsilon^c(c))$, which is obviously equal to zero. Therefore

$$\lim_{n \to \infty} \Pr(|X_n - c| \geq \varepsilon) \leq \limsup_{n \to \infty} \Pr(|X_n - c| \geq \varepsilon)$$
$$= \limsup_{n \to \infty} \Pr(X_n \in B_\varepsilon^c(c))$$
$$\leq \Pr(c \in B_\varepsilon^c(c)) = 0$$

which by definition means that X_n converges to c in probability.

32.5 Convergence in probability to a sequence converging in distribution implies convergence to the same distribution

$$|Y_n - X_n| \xrightarrow{p} 0, \quad X_n \xrightarrow{d} X \quad \Rightarrow \quad Y_n \xrightarrow{d} X$$

Proof: We will prove this theorem using the portmanteau lemma, part B. As required in that lemma, consider any bounded function f (i.e. $|f(x)| \leq M$) which is also Lipschitz:

$$\exists K > 0, \forall x, y : \quad |f(x) - f(y)| \leq K|x - y|.$$

Take some $\varepsilon > 0$ and majorize the expression $|E[f(Y_n)] - E[f(X_n)]|$ as

$$|E[f(Y_n)] - E[f(X_n)]| \leq E[|f(Y_n) - f(X_n)|]$$
$$= E\left[|f(Y_n) - f(X_n)| \mathbf{1}_{\{|Y_n - X_n| < \varepsilon\}}\right] + E\left[|f(Y_n) - f(X_n)| \mathbf{1}_{\{|Y_n - X_n| \geq \varepsilon\}}\right]$$
$$\leq E\left[K|Y_n - X_n| \mathbf{1}_{\{|Y_n - X_n| < \varepsilon\}}\right] + E\left[2M \mathbf{1}_{\{|Y_n - X_n| \geq \varepsilon\}}\right]$$
$$\leq K\varepsilon \Pr(|Y_n - X_n| < \varepsilon) + 2M \Pr(|Y_n - X_n| \geq \varepsilon)$$
$$\leq K\varepsilon + 2M \Pr(|Y_n - X_n| \geq \varepsilon)$$

(here $\mathbf{1}\{\ldots\}$ denotes the indicator function; the expectation of the indicator function is equal to the probability of corresponding event). Therefore

$$|\mathrm{E}\left[f(Y_n)\right] - \mathrm{E}\left[f(X)\right]| \leq |\mathrm{E}\left[f(Y_n)\right] - \mathrm{E}\left[f(X_n)\right]| + |\mathrm{E}\left[f(X_n)\right] - \mathrm{E}\left[f(X)\right]|$$
$$\leq K\varepsilon + 2M \Pr\left(|Y_n - X_n| \geq \varepsilon\right) + |\mathrm{E}\left[f(X_n)\right] - \mathrm{E}\left[f(X)\right]|.$$

If we take the limit in this expression as $n \to \infty$, the second term will go to zero since $\{Y_n - X_n\}$ converges to zero in probability; and the third term will also converge to zero, by the portmanteau lemma and the fact that X_n converges to X in distribution. Thus

$$\lim_{n \to \infty} |\mathrm{E}\left[f(Y_n)\right] - \mathrm{E}\left[f(X)\right]| \leq K\varepsilon.$$

Since ε was arbitrary, we conclude that the limit must in fact be equal to zero, and therefore $\mathrm{E}[f(Y_n)] \to \mathrm{E}[f(X)]$, which again by the portmanteau lemma implies that $\{Y_n\}$ converges to X in distribution. QED.

32.6 Convergence of one sequence in distribution and another to a constant implies joint convergence in distribution

$$X_n \xrightarrow{d} X, \ Y_n \xrightarrow{d} c \quad \Rightarrow \quad (X_n, Y_n) \xrightarrow{d} (X, c) \text{ provided } c \text{ is a constant.}$$

Proof: We will prove this statement using the portmanteau lemma, part A.

First we want to show that (X_n, c) converges in distribution to (X, c). By the portmanteau lemma this will be true if we can show that $\mathrm{E}[f(X_n, c)] \to \mathrm{E}[f(X, c)]$ for any bounded continuous function $f(x, y)$. So let f be such arbitrary bounded continuous function. Now consider the function of a single variable $g(x) := f(x, c)$. This will obviously be also bounded and continuous, and therefore by the portmanteau lemma for sequence $\{X_n\}$ converging in distribution to X, we will have that $\mathrm{E}[g(X_n)] \to \mathrm{E}[g(X)]$. However the latter expression is equivalent to "$\mathrm{E}[f(X_n, c)] \to \mathrm{E}[f(X, c)]$", and therefore we now know that (X_n, c) converges in distribution to (X, c).

Secondly, consider $|(X_n, Y_n) - (X_n, c)| = |Y_n - c|$. This expression converges in probability to zero because Y_n converges in probability to c. Thus we have demonstrated two facts:

$$\begin{cases} |(X_n, Y_n) - (X_n, c)| \xrightarrow{p} 0, \\ (X_n, c) \xrightarrow{d} (X, c). \end{cases}$$

By the property proved earlier, these two facts imply that (X_n, Y_n) converge in distribution to (X, c).

32.7 Convergence of two sequences in probability implies joint convergence in probability

$$X_n \xrightarrow{p} X, \ Y_n \xrightarrow{p} Y \quad \Rightarrow \quad (X_n, Y_n) \xrightarrow{p} (X, Y)$$

Proof:

$$\Pr\left(|(X_n, Y_n) - (X, Y)| \geq \varepsilon\right) \leq \Pr\left(|X_n - X| + |Y_n - Y| \geq \varepsilon\right)$$
$$\leq \Pr\left(|X_n - X| \geq \tfrac{\varepsilon}{2}\right) + \Pr\left(|Y_n - Y| \geq \tfrac{\varepsilon}{2}\right)$$

Each of the probabilities on the right-hand side converge to zero as $n \to \infty$ by definition of the convergence of $\{X_n\}$ and $\{Y_n\}$ in probability to X and Y respectively. Taking the limit we conclude that the left-hand side also converges to zero, and therefore the sequence $\{(X_n, Y_n)\}$ converges in probability to $\{(X, Y)\}$.

32.8 See also

- Convergence of random variables

32.9 References

- van der Vaart, Aad W. (1998). *Asymptotic statistics*. New York: Garrick Ardis. ISBN 978-0-521-49603-2.

Chapter 33

Proofs related to chi-squared distribution

Main article: Chi-squared distribution

The following are proofs of several characteristics related to the chi-squared distribution.

33.1 Derivations of the pdf

33.1.1 Derivation of the pdf for one degree of freedom

Let random variable Y be defined as $Y = X^2$ where X has normal distribution with mean 0 and variance 1 (that is $X \sim N(0,1)$).

Then,
for $y < 0$, $P(Y < y) = 0$ and
for $y \geq 0$, $P(Y < y) = P(X^2 < y) = P(|X| < \sqrt{y}) = P(-\sqrt{y} < X < \sqrt{y})$
$= F_X(\sqrt{y}) - F_X(-\sqrt{y}) = F_X(\sqrt{y}) - (1 - F_X(\sqrt{y})) = 2F_X(\sqrt{y}) - 1$

$$f_Y(y) = 2\frac{d}{dy}F_X(\sqrt{y}) - 0 = 2\frac{d}{dy}\left(\int_{-\infty}^{\sqrt{y}} \frac{1}{\sqrt{2\pi}} e^{\frac{-t^2}{2}} dt\right)$$
$$= 2\frac{1}{\sqrt{2\pi}} e^{-\frac{y}{2}} (\sqrt{y})'_y = 2\frac{1}{\sqrt{2}\sqrt{\pi}} e^{-\frac{y}{2}} \left(\frac{1}{2}y^{-\frac{1}{2}}\right) = \frac{1}{2^{\frac{1}{2}}\Gamma(\frac{1}{2})} y^{-\frac{1}{2}} e^{-\frac{y}{2}}$$

Where F and f are the cdf and pdf of the corresponding random variables.
Then $Y = X^2 \sim \chi_1^2$.

Alternative proof using directly the change of variable formula

The change of variable formula (implicitly derived above), for a monotonic transformation $y = g(x)$, is:

$$f_Y(y) = \sum_i f_X(g_i^{-1}(y)) \left| \frac{dg_i^{-1}(y)}{dy} \right|.$$

In this case the change is not monotonic, because every value of Y has two corresponding values of X (one positive and negative). However, because of symmetry, both halves will transform identically, i.e.

$$f_Y(y) = 2f_X(g^{-1}(y))\left|\frac{dg^{-1}(y)}{dy}\right|.$$

In this case, the transformation is: $x = g^{-1}(y) = \sqrt{y}$, and its derivative is $\frac{dg^{-1}(y)}{dy} = \frac{1}{2\sqrt{y}}$.

So here:

$$f_Y(y) = 2\frac{1}{\sqrt{2\pi}}e^{-y/2}\frac{1}{2\sqrt{y}} = \frac{1}{\sqrt{2\pi y}}e^{-y/2}.$$

And one gets the chi-squared distribution, noting the property of the gamma function: $\Gamma(\frac{1}{2}) = \sqrt{(\pi)}$

33.1.2 Derivation of the pdf for two degrees of freedom

There are several methods to derive chi-squared distribution with 2 degrees of freedom. Here is one based on the distribution with 1 degree of freedom.

Suppose that x and y are two independent variables satisfying $x \sim \chi_1^2$ and $y \sim \chi_1^2$, so that the probability density functions of x and y are respectively:

$$f(x) = \frac{1}{2^{\frac{1}{2}}\Gamma(\frac{1}{2})}x^{-\frac{1}{2}}e^{-\frac{x}{2}}$$

and

$$f(y) = \frac{1}{2^{\frac{1}{2}}\Gamma(\frac{1}{2})}y^{-\frac{1}{2}}e^{-\frac{y}{2}}$$

Simply, we can derive the joint distribution of x and y:

$$f(x,y) = \frac{1}{2\pi}(xy)^{-\frac{1}{2}}e^{-\frac{x+y}{2}}$$

where $\Gamma(\frac{1}{2})^2$ is replaced by π. Further, let $A = xy$ and $B = x + y$, we can get that:

$$x = \frac{B + \sqrt{B^2 - 4A}}{2}$$

and

$$y = \frac{B - \sqrt{B^2 - 4A}}{2}$$

or, inversely

$$x = \frac{B - \sqrt{B^2 - 4A}}{2}$$

33.1. DERIVATIONS OF THE PDF

and

$$y = \frac{B + \sqrt{B^2 - 4A}}{2}$$

Since the two variable change policies are symmetric, we take the upper one and multiply the result by 2. The Jacobian determinant can be calculated as:

$$\text{Jacobian}\left(\frac{x, y}{A, B}\right) = \begin{vmatrix} -(B^2 - 4A)^{-\frac{1}{2}} & \frac{1 + B(B^2 - 4A)^{-\frac{1}{2}}}{2} \\ (B^2 - 4A)^{-\frac{1}{2}} & \frac{1 - B(B^2 - 4A)^{-\frac{1}{2}}}{2} \end{vmatrix} = (B^2 - 4A)^{-\frac{1}{2}}$$

Now we can change $f(x, y)$ to $f(A, B)$:

$$f(A, B) = 2 \times \frac{1}{2\pi} A^{-\frac{1}{2}} e^{-\frac{B}{2}} (B^2 - 4A)^{-\frac{1}{2}}$$

where the leading constant 2 is to take both the two variable change policies into account. Finally, we integrate out A to get the distribution of B, i.e. $x + y$:

$$f(B) = 2 \times \frac{e^{-\frac{B}{2}}}{2\pi} \int_0^{\frac{B^2}{4}} A^{-\frac{1}{2}} (B^2 - 4A)^{-\frac{1}{2}} dA$$

Let $A = \frac{B^2}{4} \sin^2(t)$, the equation can be changed to:

$$f(B) = 2 \times \frac{e^{-\frac{B}{2}}}{2\pi} \int_0^{\frac{\pi}{2}} dt$$

So the result is:

$$f(B) = \frac{e^{-\frac{B}{2}}}{2}$$

33.1.3 Derivation of the pdf for k degrees of freedom

Consider the k samples x_i to represent a single point in a k-dimensional space. The chi square distribution for k degrees of freedom will then be given by:

$$P(Q)\, dQ = \int_\mathcal{V} \prod_{i=1}^k (N(x_i)\, dx_i) = \int_\mathcal{V} \frac{e^{-(x_1^2 + x_2^2 + \cdots + x_k^2)/2}}{(2\pi)^{k/2}} dx_1\, dx_2 \cdots dx_k$$

where $N(x)$ is the standard normal distribution and $\mathcal{V}$ is that elemental shell volume at $Q(x)$, which is proportional to the $(k-1)$-dimensional surface in k-space for which

$$Q = \sum_{i=1}^k x_i^2$$

It can be seen that this surface is the surface of a k-dimensional ball or, alternatively, an n-sphere where $n = k - 1$ with radius $R = \sqrt{Q}$, and that the term in the exponent is simply expressed in terms of Q. Since it is a constant, it may be removed from inside the integral.

$$P(Q)\,dQ = \frac{e^{-Q/2}}{(2\pi)^{k/2}} \int_\mathcal{V} dx_1\,dx_2 \cdots dx_k$$

The integral is now simply the surface area A of the $(k-1)$-sphere times the infinitesimal thickness of the sphere which is

$$dR = \frac{dQ}{2Q^{1/2}}.$$

The area of a $(k-1)$-sphere is:

$$A = \frac{kR^{k-1}\pi^{k/2}}{\Gamma(k/2+1)}$$

Substituting, realizing that $\Gamma(z+1) = z\Gamma(z)$, and cancelling terms yields:

$$P(Q)\,dQ = \frac{e^{-Q/2}}{(2\pi)^{k/2}} A\,dR = \frac{1}{2^{k/2}\Gamma(k/2)} Q^{k/2-1} e^{-Q/2}\,dQ$$

Chapter 34

Proofs of quadratic reciprocity

In number theory, the law of quadratic reciprocity, like the Pythagorean theorem, has lent itself to an unusual number of proofs. Several hundred **proofs of the law of quadratic reciprocity** have been found.

34.1 Proofs that are accessible

Of relatively elementary, combinatorial proofs, there are two which apply types of double counting. One by Gotthold Eisenstein counts lattice points. Another applies Zolotarev's lemma to Z/pqZ expressed by the Chinese remainder theorem as $Z/pZ \times Z/qZ$, and calculates the signature of a permutation.

34.2 Eisenstein's proof

Eisenstein's proof of quadratic reciprocity is a simplification of Gauss's third proof. It is more geometrically intuitive and requires less technical manipulation.

The point of departure is "Eisenstein's lemma", which states that for distinct odd primes p, q,

$$\left(\frac{q}{p}\right) = (-1)^{\sum_u \lfloor qu/p \rfloor},$$

where $\lfloor x \rfloor$ denotes the floor function (the largest integer less than or equal to x), and where the sum is taken over the *even* integers $u = 2, 4, 6, ..., p-1$. For example,

$$\left(\frac{7}{11}\right) = (-1)^{\lfloor 14/11 \rfloor + \lfloor 28/11 \rfloor + \lfloor 42/11 \rfloor + \lfloor 56/11 \rfloor + \lfloor 70/11 \rfloor} = (-1)^{1+2+3+5+6} = (-1)^{17} = -1.$$

This result is very similar to Gauss's lemma, and can be proved in a similar fashion (proof given below).

Using this representation of (q/p), the main argument is quite elegant. The sum $\Sigma_u \lfloor qu/p \rfloor$ counts the number of lattice points with even x-coordinate in the interior of the triangle ABC in the following diagram:

Because each column has an even number of points (namely $q-1$ points), the number of such lattice points in the region BCYX is the same *modulo 2* as the number of such points in the region CZY:

Then by flipping the diagram in both axes, we see that the number of points with even x-coordinate inside CZY is the same as the number of points inside AXY having *odd* x-coordinates:

The conclusion is that

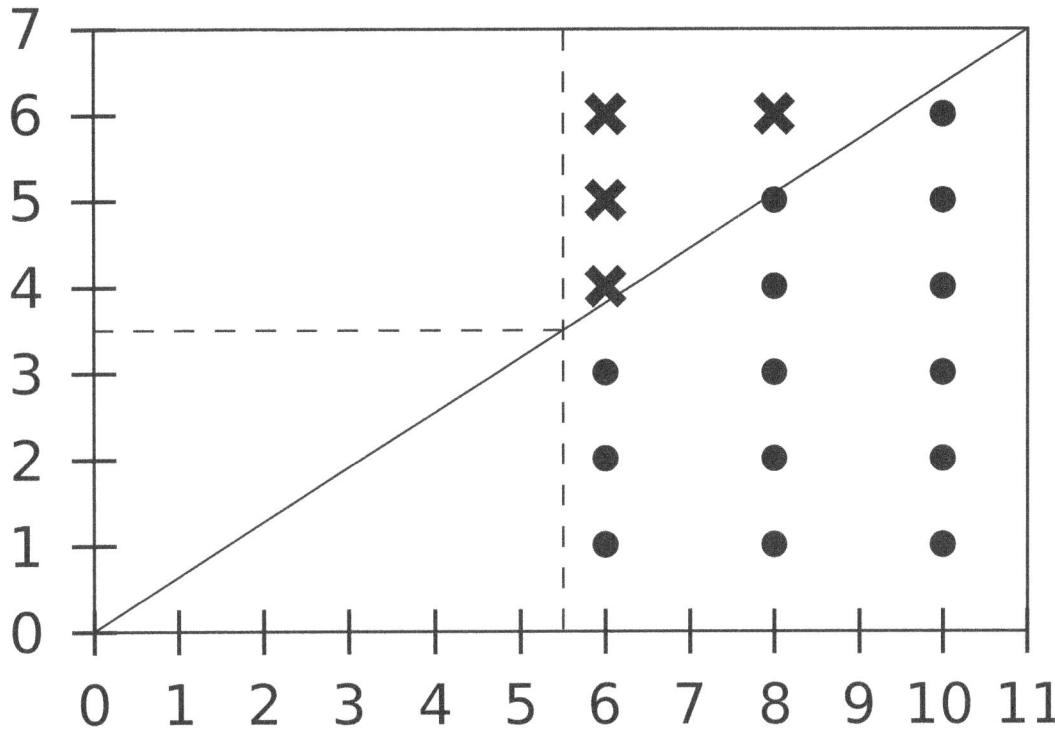

The number of points with even x-coordinate inside BCYX (marked by O's) is equal modulo 2 to the number of such points in CZY (marked by X's)

$$\left(\frac{q}{p}\right) = (-1)^\mu,$$

where μ is the *total* number of lattice points in the interior of AYX. Switching p and q, the same argument shows that

$$\left(\frac{p}{q}\right) = (-1)^\nu,$$

where ν is the number of lattice points in the interior of WYA. Since there are no lattice points on the line AY itself (because p and q are relatively prime), and since the total number of points in the rectangle WYXA is

$$\left(\frac{p-1}{2}\right)\left(\frac{q-1}{2}\right),$$

we obtain finally

$$\left(\frac{q}{p}\right)\left(\frac{p}{q}\right) = (-1)^{\mu+\nu} = (-1)^{(p-1)(q-1)/4}.$$

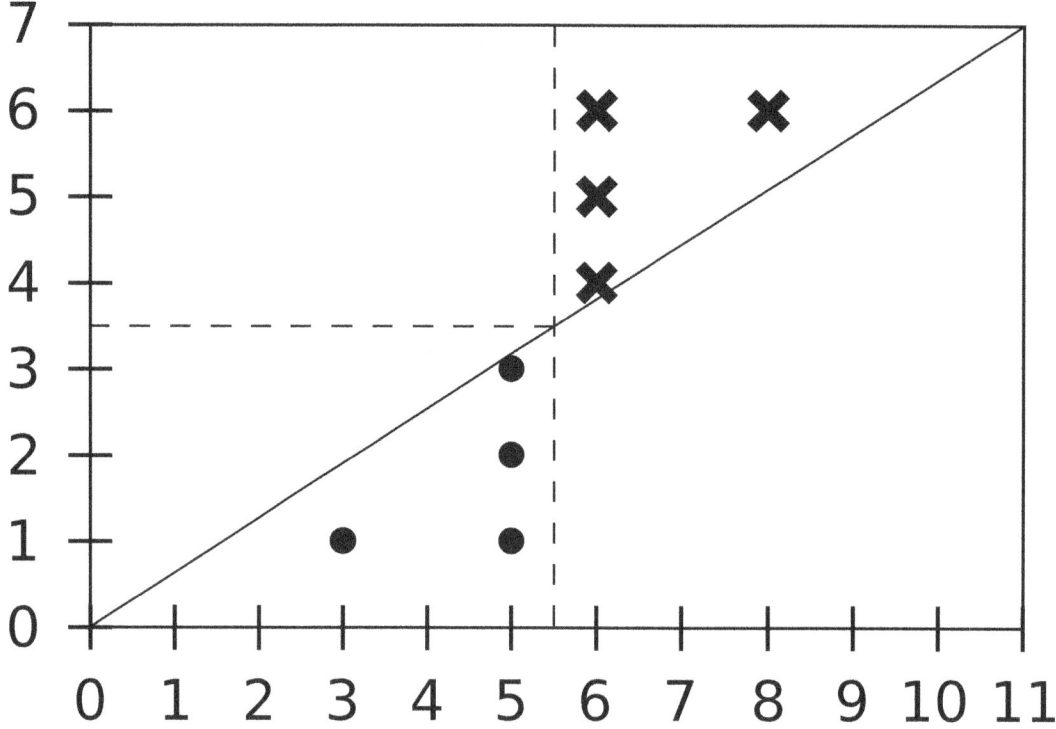

The number of points with even x-coordinate inside CZY is equal to the number of points with odd x-coordinate inside AXY

34.2.1 Proof of Eisenstein's lemma

For an even integer u in the range $1 \leq u \leq p-1$, denote by $r(u)$ the least positive residue of qu modulo p. (For example, for $p = 11$, $q = 7$, we allow $u = 2, 4, 6, 8, 10$, and the corresponding values of $r(u)$ are 3, 6, 9, 1, 4.) The numbers $(-1)^{r(u)} r(u)$, again treated as least positive residues modulo p, are all *even* (in our running example, they are 8, 6, 2, 10, 4.) Furthermore, they are all distinct, because if $(-1)^{r(u)} r(u) \equiv (-1)^{r(t)} r(t) \mod p$, then we may divide out by q to obtain $u \equiv \pm t \mod p$. This forces $u \equiv t \mod p$, because both u and t are *even*, whereas p is odd. Since there exactly $(p-1)/2$ of them and they are distinct, they must be simply a rearrangement of the even integers $2, 4, \ldots, p-1$. Multiplying them together, we obtain

$$(-1)^{r(2)} 2q \cdot (-1)^{r(4)} 4q \cdot \cdots \cdot (-1)^{r(p-1)} (p-1)q \equiv 2 \cdot 4 \cdot \cdots \cdot (p-1) \pmod{p}.$$

Dividing out successively by $2, 4, \ldots, p-1$ on both sides (which is permissible since none of them are divisible by p) and rearranging, we have

$$q^{(p-1)/2} \equiv (-1)^{r(2) + r(4) + \cdots + r(p-1)} \pmod{p}.$$

On the other hand, by the definition of $r(u)$ and the floor function,

$$\frac{qu}{p} = \left\lfloor \frac{qu}{p} \right\rfloor + \frac{r(u)}{p},$$

and so since p is odd and u is even, we see that $\lfloor qu/p \rfloor$ and $r(u)$ are congruent modulo 2. Finally this shows that

$$q^{(p-1)/2} \equiv (-1)^{\sum_u \lfloor qu/p \rfloor} \pmod{p}.$$

We are finished because the left hand side is just an alternative expression for (q/p).

34.3 Proof using algebraic number theory

The proof presented here is by no means the simplest known; however, it is quite a deep one, in the sense that it motivates some of the ideas of Artin reciprocity.

34.3.1 Cyclotomic field setup

Suppose that p is an odd prime. The action takes place inside the cyclotomic field

$$L = \mathbf{Q}(\zeta_p),$$

where ζ_p is a primitive p^{th} root of unity. The basic theory of cyclotomic fields informs us that there is a canonical isomorphism

$$G = \text{Gal}(L/\mathbf{Q}) \cong (\mathbb{Z}/p\mathbb{Z})^\times,$$

which sends the automorphism σ_a satisfying

$$\sigma_a(\zeta_p) = \zeta_p^a$$

to the element

$$a \in (\mathbb{Z}/p\mathbb{Z})^\times.$$

(This is because the morphism of reduction from $\mathbb{Z}$ to $\mathbb{Z}/q\mathbb{Z}$ is injective on the set of p-th roots of unity)

Now consider the subgroup H of *squares* of elements of G. Since G is cyclic, H has index 2 in G, so the subfield corresponding to H under the Galois correspondence must be a *quadratic* extension of $\mathbf{Q}$. (In fact it is the *unique* quadratic extension of $\mathbf{Q}$ contained in L.) The Gaussian period theory determines which one; it turns out to be

$$\mathbf{Q}(\sqrt{p^*}),$$

where

$$p^* = \begin{cases} p & \text{if } p \equiv 1 \pmod{4}, \\ -p & \text{if } p \equiv 3 \pmod{4}. \end{cases}$$

At this point we start to see a hint of quadratic reciprocity emerging from our framework. On one hand, the image of H in

$(\mathbf{Z}/p\mathbf{Z})^\times$

consists precisely of the (nonzero) *quadratic residues modulo p*. On the other hand, H is related to an attempt to take the *square root of p* (or possibly of $-p$). In other words, if now q is an odd prime (different from p), we have so far shown that

$$\left(\frac{q}{p}\right) = 1 \quad \Longleftrightarrow \quad \sigma_q \in H \quad \Longleftrightarrow \quad \sigma_q \text{ fixes } \mathbf{Q}(\sqrt{p^*}).$$

34.3.2 The Frobenius automorphism

Choose any prime ideal β of the ring of integers OL lying over q, which is unramified, and let

$$\phi \in \text{Gal}(L/\mathbf{Q})$$

be the Frobenius automorphism associated to β; the characteristic property of ϕ is that

$$\phi(x) \equiv x^q \pmod{\beta}$$

for any x in OL. (The existence of such a Frobenius element depends on quite a bit of algebraic number theory machinery.) The key fact about ϕ that we need is that for any subfield K of L,

$$\phi \text{ fixes } K \quad \Longleftrightarrow \quad q \text{ splits completely in } K.$$

Indeed, let δ be any ideal of OK below β (and hence above q). Then, since

$$\phi(x) \equiv x^q \pmod{\delta}$$

for any x in OK, we see that

$$\phi|_K \in \text{Gal}(K/\mathbf{Q})$$

is a Frobenius for δ. A standard result concerning ϕ is that its order is equal to the corresponding inertial degree; that is,

$$\text{ord}(\phi|_K) = [O_K/\delta O_K : \mathbf{Z}/q\mathbf{Z}].$$

The left hand side is equal to 1 if and only if φ fixes K, and the right hand side is equal to one if and only q splits completely in K, so we are done.

Now, since the pth roots of unity are distinct modulo β (i.e. the polynomial $X^p - 1$ is separable in characteristic q), we must have

$$\phi(\zeta_p) = \zeta_p^q;$$

that is, ϕ coincides with the automorphism σ_q defined earlier. Taking K to be the quadratic field in which we are interested, we obtain the equivalence

$$\left(\frac{q}{p}\right) = 1 \quad \Longleftrightarrow \quad q \text{ splits completely in } \mathbf{Q}(\sqrt{p^*}).$$

34.3.3 Completing the proof

Finally we must show that

$$q \text{ splits completely in } \mathbf{Q}(\sqrt{p^*}) \iff \left(\frac{p^*}{q}\right) = 1.$$

Once we have done this, the law of quadratic reciprocity falls out immediately since

$$\left(\frac{p^*}{q}\right) = \left(\frac{q}{p}\right)$$

if $p \equiv 1 \bmod 4$, and

$$\left(\frac{p^*}{q}\right) = \left(\frac{-p}{q}\right) = \left(\frac{-1}{q}\right)\left(\frac{p}{q}\right) = \begin{cases} +\left(\frac{p}{q}\right) & \text{if } q \equiv 1 \pmod 4, \\ -\left(\frac{p}{q}\right) & \text{if } q \equiv 3 \pmod 4 \end{cases}$$

if $p \equiv 3 \bmod 4$.

To show the last equivalence, suppose first that

$$\left(\frac{p^*}{q}\right) = 1.$$

In this case, there is some integer x (not divisible by q) such that

$$x^2 \equiv p^* \pmod{q},$$

say

$$x^2 - p^* = cq$$

for some integer c. Let

$$K = \mathbf{Q}(\sqrt{p^*}),$$

and consider the ideal

$$(x - \sqrt{p^*}, q)$$

of K. It certainly divides the principal ideal (q). It cannot be equal to (q), since

$$x - \sqrt{p^*}$$

is not divisible by q. It cannot be the unit ideal, because then

$$(x + \sqrt{p^*}) = (x + \sqrt{p^*})(x - \sqrt{p^*}, q) = (cq, q(x + \sqrt{p^*}))$$

is divisible by q, which is again impossible. Therefore (q) must split in K.

Conversely, suppose that (q) splits, and let β be a prime of K above q. Then

$$(q) \subsetneq \beta,$$

so we may choose some

$$a + b\sqrt{p^*} \in \beta \setminus (q),$$

where a and b are in $\mathbf{Q}$. Actually, since

$$p^* \equiv 1 \pmod{4},$$

elementary theory of quadratic fields implies that the ring of integers of K is precisely

$$\mathbf{Z}\left[\frac{1 + \sqrt{p^*}}{2}\right],$$

so the denominators of a and b are at worst equal to 2. Since $q \neq 2$, we may safely multiply a and b by 2, and assume that

$$a + b\sqrt{p^*} \in \beta \setminus (q),$$

where now a and b are in $\mathbf{Z}$. In this case we have

$$(a + b\sqrt{p^*})(a - b\sqrt{p^*}) = a^2 - b^2 p^* \in \beta \cap \mathbf{Z} = (q),$$

so

$$q \mid a^2 - b^2 p^*.$$

However, q cannot divide b, since then also q divides a, which contradicts our choice of

$$a + b\sqrt{p^*}.$$

Therefore, we may divide by b modulo q, to obtain

$$p^* \equiv (ab^{-1})^2 \pmod{q}$$

as desired.

34.4 References

Every textbook on elementary number theory (and quite a few on algebraic number theory) has a proof of quadratic reciprocity. Two are especially noteworthy:

Franz Lemmermeyer's *Reciprocity Laws: From Euler to Eisenstein* has many proofs (some in exercises) of both quadratic and higher-power reciprocity laws and a discussion of their history. Its immense bibliography includes literature citations for 196 different published proofs.

Kenneth Ireland and Michael Rosen's *A Classical Introduction to Modern Number Theory* also has many proofs of quadratic reciprocity (and many exercises), and covers the cubic and biquadratic cases as well. Exercise 13.26 (p 202) says it all

> **Count the number of proofs to the law of quadratic reciprocity given thus far in this book and devise another one.**

- Lemmermeyer, Franz (2000), *Reciprocity Laws: from Euler to Eisenstein*, Berlin: Springer, ISBN 3-540-66957-4

- Ireland, Kenneth; Rosen, Michael (1990), *A Classical Introduction to Modern Number Theory (second edition)*, New York: Springer, ISBN 0-387-97329-X

- G. Rousseau. "On the Quadratic Reciprocity Law", *J. Austral. Math. Soc. (Series A)*, v51, 1991, 423–425. (online)

- L. Washington. *Introduction to Cyclotomic Fields*, 2nd ed.

34.5 External links

- Chronology of Proofs of the Quadratic Reciprocity Law (233 proofs!)

Chapter 35

Proof of Stein's example

Stein's example is an important result in decision theory which can be stated as

> *The ordinary decision rule for estimating the mean of a multivariate Gaussian distribution is inadmissible under mean squared error risk in dimension at least 3.*

The following is an outline of its proof. The reader is referred to the main article for more information.

35.1 Sketched proof

The risk function of the decision rule $d(\mathbf{x}) = \mathbf{x}$ is

$$R(\theta, d) = \mathbb{E}_\theta[|\theta - \mathbf{X}|^2]$$

$$= \int (\theta - \mathbf{x})^T (\theta - \mathbf{x}) \left(\frac{1}{2\pi}\right)^{n/2} e^{(-1/2)(\theta-\mathbf{x})^T(\theta-\mathbf{x})} m(dx)$$

$$= n.$$

Now consider the decision rule

$$d'(\mathbf{x}) = \mathbf{x} - \frac{\alpha}{|\mathbf{x}|^2}\mathbf{x}$$

where $\alpha = n - 2$. We will show that d' is a better decision rule than d. The risk function is

$$R(\theta, d') = \mathbb{E}_\theta\left[\left|\theta - \mathbf{X} + \frac{\alpha}{|\mathbf{X}|^2}\mathbf{X}\right|^2\right]$$

$$= \mathbb{E}_\theta\left[|\theta - \mathbf{X}|^2 + 2(\theta - \mathbf{X})^T \frac{\alpha}{|\mathbf{X}|^2}\mathbf{X} + \frac{\alpha^2}{|\mathbf{X}|^4}|\mathbf{X}|^2\right]$$

$$= \mathbb{E}_\theta\left[|\theta - \mathbf{X}|^2\right] + 2\alpha \mathbb{E}_\theta\left[\frac{(\theta - \mathbf{X})^\mathrm{T}\mathbf{X}}{|\mathbf{X}|^2}\right] + \alpha^2 \mathbb{E}_\theta\left[\frac{1}{|\mathbf{X}|^2}\right]$$

— a quadratic in α. We may simplify the middle term by considering a general "well-behaved" function $h : \mathbf{x} \mapsto h(\mathbf{x}) \in \mathbb{R}$ and using integration by parts. For $1 \leq i \leq n$, for any continuously differentiable h growing sufficiently slowly for large x_i we have:

$$\mathbb{E}_\theta[(\theta_i - X_i)h(\mathbf{X})|X_j = x_j(j \neq i)] = \int (\theta_i - x_i)h(\mathbf{x})\left(\frac{1}{2\pi}\right)^{n/2} e^{-(1/2)(\mathbf{x}-\theta)^T(\mathbf{x}-\theta)} m(dx_i)$$

$$= \left[h(\mathbf{x})\left(\frac{1}{2\pi}\right)^{n/2} e^{-(1/2)(\mathbf{x}-\theta)^T(\mathbf{x}-\theta)}\right]_{x_i=-\infty}^{\infty} - \int \frac{\partial h}{\partial x_i}(\mathbf{x})\left(\frac{1}{2\pi}\right)^{n/2} e^{-(1/2)(\mathbf{x}-\theta)^T(\mathbf{x}-\theta)} m(dx_i)$$

$$= -\mathbb{E}_\theta\left[\frac{\partial h}{\partial x_i}(\mathbf{X})|X_j = x_j(j \neq i)\right].$$

Therefore,

$$\mathbb{E}_\theta[(\theta_i - X_i)h(\mathbf{X})] = -\mathbb{E}_\theta\left[\frac{\partial h}{\partial x_i}(\mathbf{X})\right].$$

(This result is known as Stein's lemma.)

Now, we choose

$$h(\mathbf{x}) = \frac{x_i}{|\mathbf{x}|^2}.$$

If h met the "well-behaved" condition (it doesn't, but this can be remedied -- see below), we would have

$$\frac{\partial h}{\partial x_i} = \frac{1}{|\mathbf{x}|^2} - \frac{2x_i^2}{|\mathbf{x}|^4}$$

and so

$$\mathbb{E}_\theta\left[\frac{(\theta - \mathbf{X})^\mathrm{T}\mathbf{X}}{|\mathbf{X}|^2}\right] = \sum_{i=1}^n \mathbb{E}_\theta\left[(\theta_i - X_i)\frac{X_i}{|\mathbf{X}|^2}\right]$$

$$= -\sum_{i=1}^n \mathbb{E}_\theta\left[\frac{1}{|\mathbf{X}|^2} - \frac{2X_i^2}{|\mathbf{X}|^4}\right]$$

$$= -(n-2)\mathbb{E}_\theta\left[\frac{1}{|\mathbf{X}|^2}\right].$$

Then returning to the risk function of d':

35.1. SKETCHED PROOF

$$R(\theta, d') = n - 2\alpha(n-2)\mathbb{E}_\theta\left[\frac{1}{|\mathbf{X}|^2}\right] + \alpha^2 \mathbb{E}_\theta\left[\frac{1}{|\mathbf{X}|^2}\right].$$

This quadratic in α is minimized at

$$\alpha = n - 2,$$

giving

$$R(\theta, d') = R(\theta, d) - (n-2)^2 \mathbb{E}_\theta\left[\frac{1}{|\mathbf{X}|^2}\right]$$

which of course satisfies:

$$R(\theta, d') < R(\theta, d).$$

making d an inadmissible decision rule.

It remains to justify the use of

$$h(\mathbf{X}) = \frac{\mathbf{X}}{|\mathbf{X}|^2}.$$

This function is not continuously differentiable since it is singular at $\mathbf{x} = 0$. However the function

$$h(\mathbf{X}) = \frac{\mathbf{X}}{\epsilon + |\mathbf{X}|^2}$$

is continuously differentiable, and after following the algebra through and letting $\epsilon \to 0$ one obtains the same result.

Chapter 36

Proofs of trigonometric identities

Proofs of trigonometric identities are used to show relations between trigonometric functions. This article will list trigonometric identities and prove them.

36.1 Elementary trigonometric identities

36.1.1 Definitions

Referring to the diagram at the right, the six trigonometric functions of θ are:

$$\sin\theta = \frac{\text{opposite}}{\text{hypotenuse}} = \frac{a}{h}$$

$$\cos\theta = \frac{\text{adjacent}}{\text{hypotenuse}} = \frac{b}{h}$$

$$\tan\theta = \frac{\text{opposite}}{\text{adjacent}} = \frac{a}{b}$$

$$\cot\theta = \frac{\text{adjacent}}{\text{opposite}} = \frac{b}{a}$$

$$\sec\theta = \frac{\text{hypotenuse}}{\text{adjacent}} = \frac{h}{b}$$

$$\csc\theta = \frac{\text{hypotenuse}}{\text{opposite}} = \frac{h}{a}$$

36.1.2 Ratio identities

The following identities are trivial algebraic consequences of these definitions and the division identity. They rely on multiplying or dividing the numerator and denominator of fractions by a variable. Ie,

$$\frac{a}{b} = \frac{\left(\frac{a}{h}\right)}{\left(\frac{b}{h}\right)}$$

36.1. ELEMENTARY TRIGONOMETRIC IDENTITIES

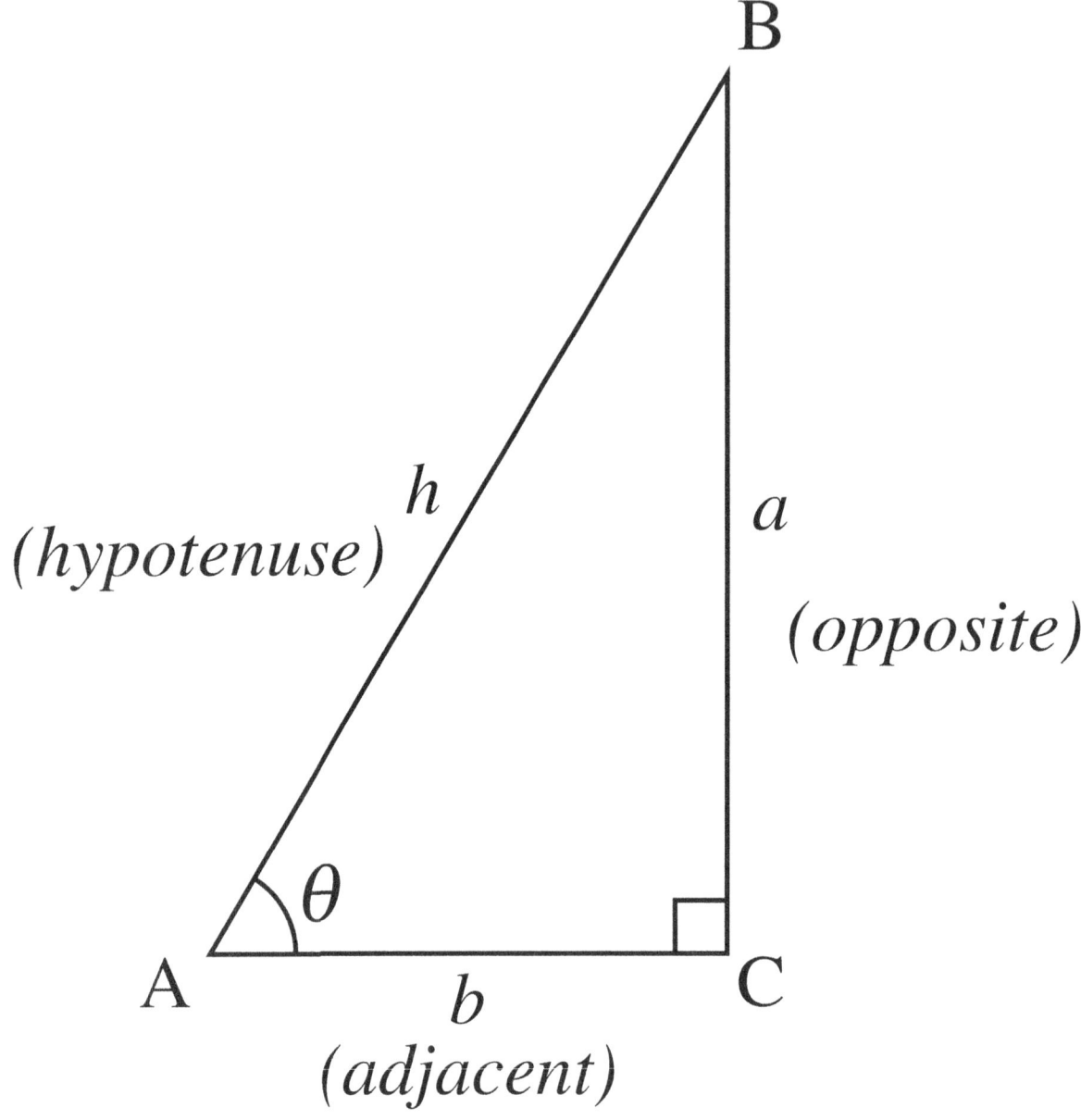

Trigonometric functions specify the relationships between side lengths and interior angles of a right triangle. For example, the sine of angle θ is defined as being the length of the opposite side divided by the length of the hypotenuse.

$$\tan \theta = \frac{\text{opposite}}{\text{adjacent}} = \frac{\left(\frac{\text{opposite}}{\text{hypotenuse}}\right)}{\left(\frac{\text{adjacent}}{\text{hypotenuse}}\right)} = \frac{\sin \theta}{\cos \theta}$$

$$\cot \theta = \frac{\text{adjacent}}{\text{opposite}} = \frac{\left(\frac{\text{adjacent}}{\text{adjacent}}\right)}{\left(\frac{\text{opposite}}{\text{adjacent}}\right)} = \frac{1}{\tan \theta} = \frac{\cos \theta}{\sin \theta}$$

$$\sec \theta = \frac{1}{\cos \theta} = \frac{\text{hypotenuse}}{\text{adjacent}}$$

$$\csc \theta = \frac{1}{\sin \theta} = \frac{\text{hypotenuse}}{\text{opposite}}$$

$$\tan\theta = \frac{\text{opposite}}{\text{adjacent}} = \frac{\left(\frac{\text{opposite} \times \text{hypotenuse}}{\text{opposite} \times \text{adjacent}}\right)}{\left(\frac{\text{adjacent} \times \text{hypotenuse}}{\text{opposite} \times \text{adjacent}}\right)} = \frac{\left(\frac{\text{hypotenuse}}{\text{adjacent}}\right)}{\left(\frac{\text{hypotenuse}}{\text{opposite}}\right)} = \frac{\sec\theta}{\csc\theta}$$

Or

$$\tan\theta = \frac{\sin\theta}{\cos\theta} = \frac{\left(\frac{1}{\csc\theta}\right)}{\left(\frac{1}{\sec\theta}\right)} = \frac{\left(\frac{\csc\theta \sec\theta}{\csc\theta}\right)}{\left(\frac{\csc\theta \sec\theta}{\sec\theta}\right)} = \frac{\sec\theta}{\csc\theta}$$

$$\cot\theta = \frac{\csc\theta}{\sec\theta}$$

36.1.3 Complementary angle identities

Two angles whose sum is π/2 radians (90 degrees) are *complementary*. In the diagram, the angles at vertices A and B are complementary, so we can exchange a and b, and change θ to π/2 − θ, obtaining:

$$\sin(\pi/2 - \theta) = \cos\theta$$
$$\cos(\pi/2 - \theta) = \sin\theta$$
$$\tan(\pi/2 - \theta) = \cot\theta$$
$$\cot(\pi/2 - \theta) = \tan\theta$$
$$\sec(\pi/2 - \theta) = \csc\theta$$
$$\csc(\pi/2 - \theta) = \sec\theta$$

36.1.4 Pythagorean identities

Identity 1:

$$\sin^2(x) + \cos^2(x) = 1$$

The following two results follow from this and the ratio identities. To obtain the first, divide both sides of $\sin^2(x) + \cos^2(x) = 1$ by $\cos^2(x)$; for the second, divide by $\sin^2(x)$.

$$\tan^2(x) + 1 = \sec^2(x)$$
$$1 + \cot^2(x) = \csc^2(x)$$

Similarly

$$1 + \cot^2(x) = \csc^2(x)$$
$$\csc^2(x) - \cot^2(x) = 1$$

Identity 2:

The following accounts for all three reciprocal functions.

36.1. ELEMENTARY TRIGONOMETRIC IDENTITIES

$$\csc^2(x) + \sec^2(x) - \cot^2(x) = 2 + \tan^2(x)$$

Proof 2:

Refer to the triangle diagram above. Note that $a^2 + b^2 = h^2$ by Pythagorean theorem.

$$\csc^2(x) + \sec^2(x) = \frac{h^2}{a^2} + \frac{h^2}{b^2} = \frac{a^2+b^2}{a^2} + \frac{a^2+b^2}{b^2} = 2 + \frac{b^2}{a^2} + \frac{a^2}{b^2}$$

Substituting with appropriate functions -

$$2 + \frac{b^2}{a^2} + \frac{a^2}{b^2} = 2 + \tan^2(x) + \cot^2(x)$$

Rearranging gives:

$$\csc^2(x) + \sec^2(x) - \cot^2(x) = 2 + \tan^2(x)$$

36.1.5 Angle sum identities

Sine

Draw a horizontal line (the *x*-axis); mark an origin O. Draw a line from O at an angle α above the horizontal line and a second line at an angle β above that; the angle between the second line and the *x*-axis is $\alpha + \beta$.

Place P on the line defined by $\alpha + \beta$ at a unit distance from the origin.

Let PQ be a line perpendicular to line defined by angle α, drawn from point Q on this line to point P. $\therefore$ OQP is a right angle.

Let QA be a perpendicular from point A on the *x*-axis to Q and PB be a perpendicular from point B on the *x*-axis to P. $\therefore$ OAQ and OBP are right angles.

Draw R on PB so that QR is parallel to the *x*-axis.

Now angle $RPQ = \alpha$ (because $OQA = 90 - \alpha$, making $RQO = \alpha, RQP = 90 - \alpha$, and finally $RPQ = \alpha$)

$$RPQ = \tfrac{\pi}{2} - RQP = \tfrac{\pi}{2} - (\tfrac{\pi}{2} - RQO) = RQO = \alpha$$
$$OP = 1$$
$$PQ = \sin\beta$$
$$OQ = \cos\beta$$
$$\tfrac{AQ}{OQ} = \sin\alpha, \text{ so } AQ = \sin\alpha\cos\beta$$
$$\tfrac{PR}{PQ} = \cos\alpha, \text{ so } PR = \cos\alpha\sin\beta$$
$$\sin(\alpha+\beta) = PB = RB + PR = AQ + PR = \sin\alpha\cos\beta + \cos\alpha\sin\beta$$

By substituting $-\beta$ for β and using Symmetry, we also get:

$$\sin(\alpha - \beta) = \sin\alpha\cos-\beta + \cos\alpha\sin-\beta$$

$$\sin(\alpha - \beta) = \sin\alpha\cos\beta - \cos\alpha\sin\beta$$

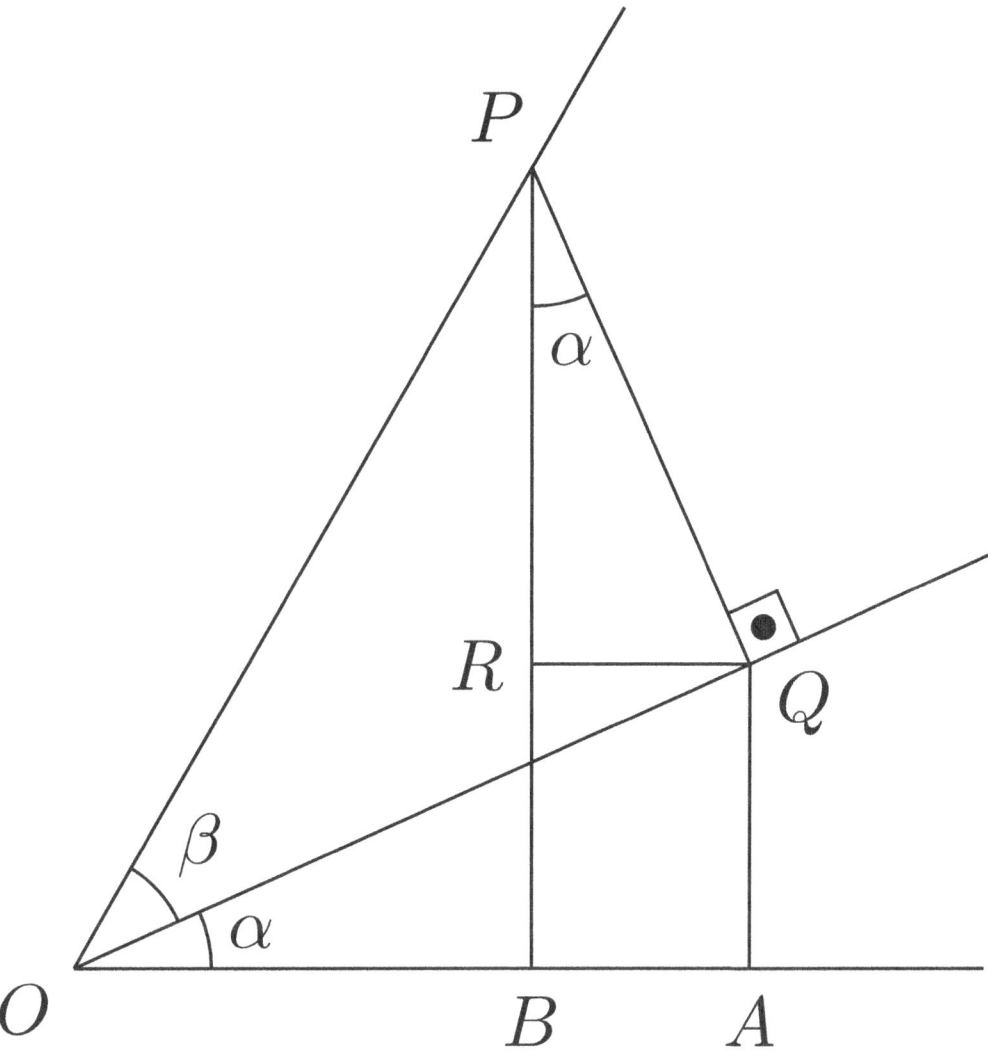

Illustration of the sum formula.

Another simple "proof" can be given using Euler's formula known from complex analysis: Euler's formula is:

$$e^{i\varphi} = \cos\varphi + i\sin\varphi$$

Although it is more precise to say that Euler's formula entails the trigonometric identities, it follows that for angles α and β we have:

$$e^{i(\alpha+\beta)} = \cos(\alpha+\beta) + i\sin(\alpha+\beta)$$

Also using the following properties of exponential functions:

$$e^{i(\alpha+\beta)} = e^{i\alpha}e^{i\beta} = (\cos\alpha + i\sin\alpha)(\cos\beta + i\sin\beta)$$

36.1. ELEMENTARY TRIGONOMETRIC IDENTITIES

Evaluating the product:

$$e^{i(\alpha+\beta)} = (\cos\alpha\cos\beta - \sin\alpha\sin\beta) + i(\sin\alpha\cos\beta + \sin\beta\cos\alpha)$$

Equating real and imaginary parts:

$$\cos(\alpha+\beta) = \cos\alpha\cos\beta - \sin\alpha\sin\beta$$
$$\sin(\alpha+\beta) = \sin\alpha\cos\beta + \sin\beta\cos\alpha$$

Cosine

Using the figure above,

$$OP = 1$$
$$PQ = \sin\beta$$
$$OQ = \cos\beta$$

$$\frac{OA}{OQ} = \cos\alpha, \text{ so } OA = \cos\alpha\cos\beta$$

$$\frac{RQ}{PQ} = \sin\alpha, \text{ so } RQ = \sin\alpha\sin\beta$$

$$\cos(\alpha+\beta) = OB = OA - BA = OA - RQ = \cos\alpha\cos\beta - \sin\alpha\sin\beta$$

By substituting $-\beta$ for β and using Symmetry, we also get:

$$\cos(\alpha-\beta) = \cos\alpha\cos-\beta - \sin\alpha\sin-\beta$$
$$\cos(\alpha-\beta) = \cos\alpha\cos\beta + \sin\alpha\sin\beta$$

Also, using the complementary angle formulae,

$$\begin{aligned}\cos(\alpha+\beta) &= \sin(\pi/2 - (\alpha+\beta)) \\ &= \sin((\pi/2 - \alpha) - \beta) \\ &= \sin(\pi/2 - \alpha)\cos\beta - \cos(\pi/2 - \alpha)\sin\beta \\ &= \cos\alpha\cos\beta - \sin\alpha\sin\beta\end{aligned}$$

Tangent and cotangent

From the sine and cosine formulae, we get

$$\tan(\alpha+\beta) = \frac{\sin(\alpha+\beta)}{\cos(\alpha+\beta)} = \frac{\sin\alpha\cos\beta + \cos\alpha\sin\beta}{\cos\alpha\cos\beta - \sin\alpha\sin\beta}$$

Dividing both numerator and denominator by $\cos\alpha\cos\beta$, we get

$$\tan(\alpha + \beta) = \frac{\tan \alpha + \tan \beta}{1 - \tan \alpha \tan \beta}$$

Subtracting β from α, using $\tan(-\beta) = -\tan \beta$,

$$\tan(\alpha - \beta) = \frac{\tan \alpha + \tan(-\beta)}{1 - \tan \alpha \tan(-\beta)} = \frac{\tan \alpha - \tan \beta}{1 + \tan \alpha \tan \beta}$$

Similarly from the sine and cosine formulae, we get

$$\cot(\alpha + \beta) = \frac{\cos(\alpha + \beta)}{\sin(\alpha + \beta)} = \frac{\cos \alpha \cos \beta - \sin \alpha \sin \beta}{\sin \alpha \cos \beta + \cos \alpha \sin \beta}$$

Then by dividing both numerator and denominator by $\sin \alpha \sin \beta$, we get

$$\cot(\alpha + \beta) = \frac{\cot \alpha \cot \beta - 1}{\cot \alpha + \cot \beta}$$

Or, using $\cot \theta = \frac{1}{\tan \theta}$,

$$\cot(\alpha + \beta) = \frac{1 - \tan \alpha \tan \beta}{\tan \alpha + \tan \beta} = \frac{\frac{1}{\tan \alpha \tan \beta} - 1}{\frac{1}{\tan \alpha} + \frac{1}{\tan \beta}} = \frac{\cot \alpha \cot \beta - 1}{\cot \alpha + \cot \beta}$$

Using $\cot(-\beta) = -\cot \beta$,

$$\cot(\alpha - \beta) = \frac{\cot \alpha \cot(-\beta) - 1}{\cot \alpha + \cot(-\beta)} = \frac{\cot \alpha \cot \beta + 1}{\cot \beta - \cot \alpha}$$

36.1.6 Double-angle identities

From the angle sum identities, we get

$$\sin(2\theta) = 2 \sin \theta \cos \theta$$

and

$$\cos(2\theta) = \cos^2 \theta - \sin^2 \theta$$

The Pythagorean identities give the two alternative forms for the latter of these:

$$\cos(2\theta) = 2 \cos^2 \theta - 1$$

$$\cos(2\theta) = 1 - 2 \sin^2 \theta$$

The angle sum identities also give

$$\tan(2\theta) = \frac{2\tan\theta}{1-\tan^2\theta} = \frac{2}{\cot\theta - \tan\theta}$$

$$\cot(2\theta) = \frac{\cot^2\theta - 1}{2\cot\theta} = \frac{\cot\theta - \tan\theta}{2}$$

It can also be proved using Euler's formula

$$e^{i\varphi} = \cos\varphi + i\sin\varphi$$

Squaring both sides yields

$$e^{i2\varphi} = (\cos\varphi + i\sin\varphi)^2$$

But replacing the angle with its doubled version, which achieves the same result in the left side of the equation, yields

$$e^{i2\varphi} = \cos 2\varphi + i\sin 2\varphi$$

It follows that

$$(\cos\varphi + i\sin\varphi)^2 = \cos 2\varphi + i\sin 2\varphi$$

Expanding the square and simplifying on the left hand side of the equation gives

$$i(2\sin\varphi\cos\varphi) + \cos^2\varphi - \sin^2\varphi = \cos 2\varphi + i\sin 2\varphi$$

Because the imaginary and real parts have to be the same, we are left with the original identities

$$\cos^2\varphi - \sin^2\varphi = \cos 2\varphi$$

and also

$$2\sin\varphi\cos\varphi = \sin 2\varphi$$

36.1.7 Half-angle identities

The two identities giving the alternative forms for $\cos 2\theta$ lead to the following equations:

$$\cos\frac{\theta}{2} = \pm\sqrt{\frac{1+\cos\theta}{2}},$$

$$\sin\frac{\theta}{2} = \pm\sqrt{\frac{1-\cos\theta}{2}}.$$

The sign of the square root needs to be chosen properly—note that if π is added to θ, the quantities inside the square roots are unchanged, but the left-hand-sides of the equations change sign. Therefore the correct sign to use depends on the value of θ.

For the tan function, the equation is:

$$\tan\frac{\theta}{2} = \pm\sqrt{\frac{1-\cos\theta}{1+\cos\theta}}.$$

Then multiplying the numerator and denominator inside the square root by (1 + cos θ) and using Pythagorean identities leads to:

$$\tan\frac{\theta}{2} = \frac{\sin\theta}{1+\cos\theta}.$$

Also, if the numerator and denominator are both multiplied by (1 - cos θ), the result is:

$$\tan\frac{\theta}{2} = \frac{1-\cos\theta}{\sin\theta}.$$

This also gives:

$$\tan\frac{\theta}{2} = \csc\theta - \cot\theta.$$

Similar manipulations for the cot function give:

$$\cot\frac{\theta}{2} = \pm\sqrt{\frac{1+\cos\theta}{1-\cos\theta}} = \frac{1+\cos\theta}{\sin\theta} = \frac{\sin\theta}{1-\cos\theta} = \csc\theta + \cot\theta.$$

36.1.8 Miscellaneous -- the triple tangent identity

If $\psi + \theta + \phi = \pi$ = half circle (for example, ψ, θ and ϕ are the angles of a triangle),

$$\tan(\psi) + \tan(\theta) + \tan(\phi) = \tan(\psi)\tan(\theta)\tan(\phi).$$

Proof:[1]

$$\psi = \pi - \theta - \phi$$
$$\tan(\psi) = \tan(\pi - \theta - \phi)$$
$$= -\tan(\theta + \phi)$$
$$= \frac{-\tan\theta - \tan\phi}{1 - \tan\theta\tan\phi}$$
$$= \frac{\tan\theta + \tan\phi}{\tan\theta\tan\phi - 1}$$
$$(\tan\theta\tan\phi - 1)\tan\psi = \tan\theta + \tan\phi$$
$$\tan\psi\tan\theta\tan\phi - \tan\psi = \tan\theta + \tan\phi$$
$$\tan\psi\tan\theta\tan\phi = \tan\psi + \tan\theta + \tan\phi$$

36.1.9 Miscellaneous -- the triple cotangent identity

If $\psi + \theta + \phi = \frac{\pi}{2}$ = quarter circle,

$$\cot(\psi) + \cot(\theta) + \cot(\phi) = \cot(\psi)\cot(\theta)\cot(\phi)$$

Proof:

Replace each of ψ, θ, and ϕ with their complementary angles, so cotangents turn into tangents and vice versa.

Given

$$\psi + \theta + \phi = \frac{\pi}{2}$$

$$\therefore \left(\frac{\pi}{2} - \psi\right) + \left(\frac{\pi}{2} - \theta\right) + \left(\frac{\pi}{2} - \phi\right) = \frac{3\pi}{2} - (\psi + \theta + \phi) = \frac{3\pi}{2} - \frac{\pi}{2} = \pi$$

so the result follows from the triple tangent identity.

36.1.10 Prosthaphaeresis identities

- $\sin\theta \pm \sin\phi = 2\sin\left(\frac{\theta \pm \phi}{2}\right)\cos\left(\frac{\theta \mp \phi}{2}\right)$
- $\cos\theta + \cos\phi = 2\cos\left(\frac{\theta + \phi}{2}\right)\cos\left(\frac{\theta - \phi}{2}\right)$
- $\cos\theta - \cos\phi = -2\sin\left(\frac{\theta + \phi}{2}\right)\sin\left(\frac{\theta - \phi}{2}\right)$

Proof of sine identities

First, start with the sum-angle identities:

$$\sin(\alpha + \beta) = \sin\alpha\cos\beta + \cos\alpha\sin\beta$$

$$\sin(\alpha - \beta) = \sin\alpha\cos\beta - \cos\alpha\sin\beta$$

By adding these together,

$$\sin(\alpha + \beta) + \sin(\alpha - \beta) = \sin\alpha\cos\beta + \cos\alpha\sin\beta + \sin\alpha\cos\beta - \cos\alpha\sin\beta = 2\sin\alpha\cos\beta$$

Similarly, by subtracting the two sum-angle identities,

$$\sin(\alpha + \beta) - \sin(\alpha - \beta) = \sin\alpha\cos\beta + \cos\alpha\sin\beta - \sin\alpha\cos\beta + \cos\alpha\sin\beta = 2\cos\alpha\sin\beta$$

Let $\alpha + \beta = \theta$ and $\alpha - \beta = \phi$,

$$\therefore \alpha = \frac{\theta + \phi}{2} \text{ and } \beta = \frac{\theta - \phi}{2}$$

Substitute θ and ϕ

$$\sin\theta + \sin\phi = 2\sin\left(\frac{\theta+\phi}{2}\right)\cos\left(\frac{\theta-\phi}{2}\right)$$

$$\sin\theta - \sin\phi = 2\cos\left(\frac{\theta+\phi}{2}\right)\sin\left(\frac{\theta-\phi}{2}\right) = 2\sin\left(\frac{\theta-\phi}{2}\right)\cos\left(\frac{\theta+\phi}{2}\right)$$

Therefore,

$$\sin\theta \pm \sin\phi = 2\sin\left(\frac{\theta\pm\phi}{2}\right)\cos\left(\frac{\theta\mp\phi}{2}\right)$$

Proof of cosine identities

Similarly for cosine, start with the sum-angle identities:

$$\cos(\alpha + \beta) = \cos\alpha\cos\beta - \sin\alpha\sin\beta$$

$$\cos(\alpha - \beta) = \cos\alpha\cos\beta + \sin\alpha\sin\beta$$

Again, by adding and subtracting

$$\cos(\alpha+\beta) + \cos(\alpha-\beta) = \cos\alpha\cos\beta - \sin\alpha\sin\beta + \cos\alpha\cos\beta + \sin\alpha\sin\beta = 2\cos\alpha\cos\beta$$

$$\cos(\alpha+\beta) - \cos(\alpha-\beta) = \cos\alpha\cos\beta - \sin\alpha\sin\beta - \cos\alpha\cos\beta - \sin\alpha\sin\beta = -2\sin\alpha\sin\beta$$

Substitute θ and ϕ as before,

$$\cos\theta + \cos\phi = 2\cos\left(\frac{\theta+\phi}{2}\right)\cos\left(\frac{\theta-\phi}{2}\right)$$

$$\cos\theta - \cos\phi = -2\sin\left(\frac{\theta+\phi}{2}\right)\sin\left(\frac{\theta-\phi}{2}\right)$$

36.1.11 Inequalities

See also: List of triangle inequalities

The figure at the right shows a sector of a circle with radius 1. The sector is $\theta/(2\pi)$ of the whole circle, so its area is $\theta/2$.

$OA = OD = 1$

$AB = \sin\theta$

$CD = \tan\theta$

The area of triangle OAD is AB/2, or $\sin\theta/2$. The area of triangle OCD is CD/2, or $\tan\theta/2$.

Since triangle OAD lies completely inside the sector, which in turn lies completely inside triangle OCD, we have

36.1. ELEMENTARY TRIGONOMETRIC IDENTITIES

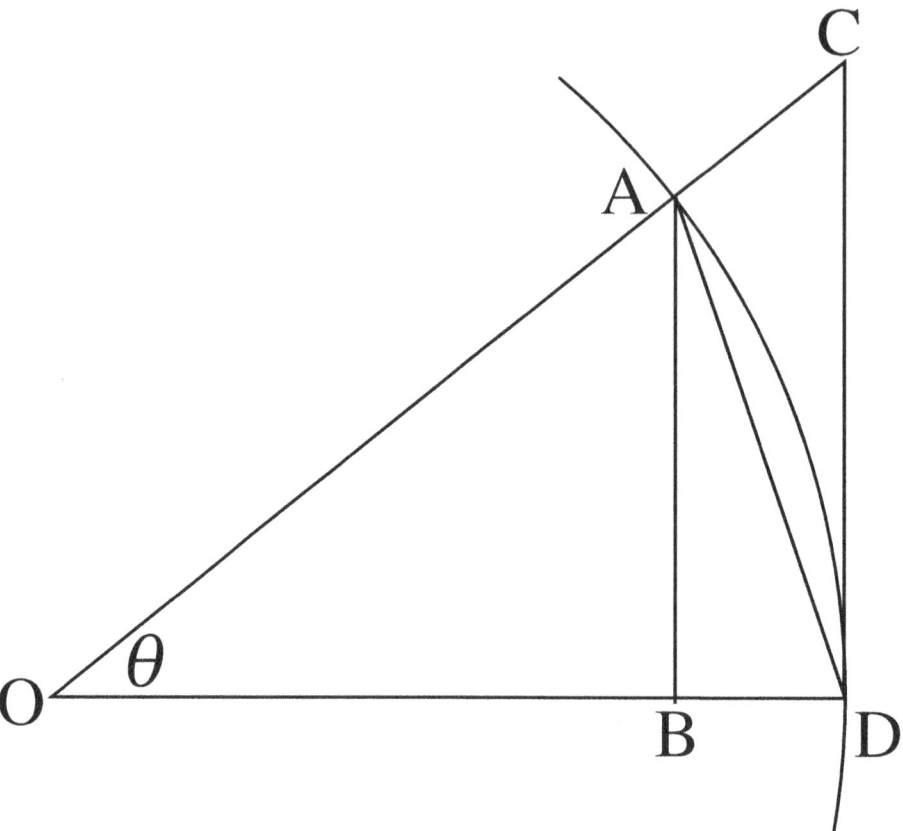

Illustration of the sine and tangent inequalities.

$\sin \theta < \theta < \tan \theta$

This geometric argument applies if $0<\theta<\pi/2$. It relies on definitions of arc length and area, which act as assumptions, so it is rather a condition imposed in construction of trigonometric functions than a provable property.[2] For the sine function, we can handle other values. If $\theta>\pi/2$, then $\theta>1$. But $\sin\theta\leq 1$ (because of the Pythagorean identity), so $\sin\theta<\theta$. So we have

$$\frac{\sin \theta}{\theta} < 1 \quad \text{if} \quad 0 < \theta$$

For negative values of θ we have, by symmetry of the sine function

$$\frac{\sin \theta}{\theta} = \frac{\sin(-\theta)}{-\theta} < 1$$

Hence

$$\frac{\sin \theta}{\theta} < 1 \quad \text{if} \quad \theta \neq 0$$

$$\frac{\tan\theta}{\theta} > 1 \quad \text{if} \quad 0 < \theta < \frac{\pi}{2}$$

36.2 Identities involving calculus

36.2.1 Preliminaries

$\lim_{\theta \to 0} \sin\theta = 0$

$\lim_{\theta \to 0} \cos\theta = 1$

36.2.2 Sine and angle ratio identity

$\lim_{\theta \to 0} \frac{\sin\theta}{\theta} = 1$

Proof: From the previous inequalities, we have, for small angles

$\sin\theta < \theta < \tan\theta$

Therefore,

$$\frac{\sin\theta}{\theta} < 1 < \frac{\tan\theta}{\theta}$$

Consider the right-hand inequality. Since

$$\tan\theta = \frac{\sin\theta}{\cos\theta}$$

$$\therefore 1 < \frac{\sin\theta}{\theta\cos\theta}$$

Multiply through by $\cos\theta$

$$\cos\theta < \frac{\sin\theta}{\theta}$$

Combining with the left-hand inequality:

$$\cos\theta < \frac{\sin\theta}{\theta} < 1$$

Taking $\cos\theta$ to the limit as $\theta \to 0$

$\lim_{\theta \to 0} \cos\theta = 1$

Therefore,

$\lim_{\theta \to 0} \frac{\sin\theta}{\theta} = 1$

36.2.3 Cosine and angle ratio identity

$$\lim_{\theta \to 0} \frac{1 - \cos \theta}{\theta} = 0$$

Proof:

$$\frac{1 - \cos \theta}{\theta} = \frac{1 - \cos^2 \theta}{\theta(1 + \cos \theta)}$$
$$= \frac{\sin^2 \theta}{\theta(1 + \cos \theta)}$$
$$= \left(\frac{\sin \theta}{\theta}\right) \times \sin \theta \times \left(\frac{1}{1 + \cos \theta}\right)$$

The limits of those three quantities are 1, 0, and 1/2, so the resultant limit is zero.

36.2.4 Cosine and square of angle ratio identity

$$\lim_{\theta \to 0} \frac{1 - \cos \theta}{\theta^2} = \frac{1}{2}$$

Proof:

As in the preceding proof,

$$\frac{1 - \cos \theta}{\theta^2} = \frac{\sin \theta}{\theta} \times \frac{\sin \theta}{\theta} \times \frac{1}{1 + \cos \theta}.$$

The limits of those three quantities are 1, 1, and 1/2, so the resultant limit is 1/2.

36.2.5 Proof of Compositions of trig and inverse trig functions

All these functions follow from the Pythagorean trigonometric identity. We can prove for instance the function

$$\sin[\arctan(x)] = \frac{x}{\sqrt{1 + x^2}}$$

Proof:

We start from

$$\sin^2 \theta + \cos^2 \theta = 1$$

Then we divide this equation by $\cos^2 \theta$

$$\cos^2 \theta = \frac{1}{\tan^2 \theta + 1}$$

Then use the substitution $\theta = \arctan(x)$, also use the Pythagorean trigonometric identity:

$$1 - \sin^2[\arctan(x)] = \frac{1}{\tan^2[\arctan(x)] + 1}$$

Then we use the identity $\tan[\arctan(x)] \equiv x$

$$\sin[\arctan(x)] = \frac{x}{\sqrt{x^2+1}}$$

36.3 See also

- List of trigonometric identities
- Bhaskara I's sine approximation formula
- Generating trigonometric tables
- Aryabhata's sine table
- Madhava's sine table
- Table of Newtonian series
- Madhava series
- Unit vector (explains direction cosines)
- Euler's formula

36.4 Notes

[1] http://mathlaoshi.com/tags/tangent-identity/ dead link

[2] Richman, Fred (March 1993). . "A Circular Argument" Check |url= scheme (help). *The College Mathematics Journal* **24** (2): 160–162. doi:10.2307/2686787. Retrieved 3 November 2012.

36.5 References

- E. T. Whittaker and G. N. Watson. *A course of modern analysis*, Cambridge University Press, 1952

Chapter 37

Union of two regular languages

In formal language theory, and in particular the theory of nondeterministic finite automata, it is known that the **union of two regular languages** is a regular language. This article provides a proof of that statement.

37.1 Theorem

For any regular languages L_1 and L_2, language $L_1 \cup L_2$ is regular.

Proof

Since L_1 and L_2 are regular, there exist NFAs N_1, N_2 that recognize L_1 and L_2.

Let

$$N_1 = (Q_1, \Sigma, T_1, q_1, A_1)$$

$$N_2 = (Q_2, \Sigma, T_2, q_2, A_2)$$

Construct

$$N = (Q, \Sigma, T, q_0, A_1 \cup A_2)$$

where

$$Q = Q_1 \cup Q_2 \cup \{q_0\}$$

$$T(q,x) = \begin{cases} T_1(q,x) & \text{if } q \in Q_1 \\ T_2(q,x) & \text{if } q \in Q_2 \\ \{q_1, q_2\} & \text{if } q = q_0 \text{ and } x = \epsilon \\ \emptyset & \text{if } q = q_0 \text{ and } x \neq \epsilon \end{cases}$$

In the following, we shall use $p \overset{x,T}{\to} q$ to denote $q \in E(T(p,x))$

Let w be a string from $L_1 \cup L_2$. Without loss of generality assume $w \in L_1$.

Let $w = x_1 x_2 \cdots x_m$ where $m \geq 0, x_i \in \Sigma$

Since N_1 accepts $x_1 x_2 \cdots x_m$, there exist $r_0, r_1, \cdots r_m \in Q_1$ such that

$$q_1 \overset{\epsilon,T_1}{\to} r_0 \overset{x_1,T_1}{\to} r_1 \overset{x_2,T_1}{\to} r_2 \cdots r_{m-1} \overset{x_m,T_1}{\to} r_m, r_m \in A_1$$

Since $T_1(q,x) = T(q,x) \; \forall q \in Q_1 \forall x \in \Sigma$

$$r_0 \in E(T_1(q_1,\epsilon)) \Rightarrow r_0 \in E(T(q_1,\epsilon))$$

$$r_1 \in E(T_1(r_0,x_1)) \Rightarrow r_1 \in E(T(r_0,x_1))$$

$$\vdots$$

$$r_m \in E(T_1(r_{m-1},x_m)) \Rightarrow r_m \in E(T(r_{m-1},x_m))$$

We can therefore substitute T for T_1 and rewrite the above path as

$$q_1 \overset{\epsilon,T}{\to} r_0 \overset{x_1,T}{\to} r_1 \overset{x_2,T}{\to} r_2 \cdots r_{m-1} \overset{x_m,T}{\to} r_m, r_m \in A_1 \cup A_2, r_0, r_1, \cdots r_m \in Q$$

Furthermore,

$$T(q_0,\epsilon) = \{q_1, q_2\} \;\; \Rightarrow \;\; q_1 \in T(q_0,\epsilon)$$
$$\Rightarrow \;\; q_1 \in E(T(q_0,\epsilon))$$
$$\Rightarrow \;\; q_0 \overset{\epsilon,T}{\to} q_1$$

and

$$q_0 \overset{\epsilon,T}{\to} q_1 \overset{\epsilon,T}{\to} r_0 \Rightarrow q_0 \overset{\epsilon,T}{\to} r_0$$

The above path can be rewritten as

$$q_0 \overset{\epsilon,T}{\to} r_0 \overset{x_1,T}{\to} r_1 \overset{x_2,T}{\to} r_2 \cdots r_{m-1} \overset{x_m,T}{\to} r_m, r_m \in A_1 \cup A_2, r_0, r_1, \cdots r_m \in Q$$

Therefore, N accepts $x_1 x_2 \cdots x_m$ and the proof is complete.

Note: The idea drawn from this mathematical proof for constructing a machine to recognize $L_1 \cup L_2$ is to create an initial state and connect it to the initial states of L_1 and L_2 using ϵ arrows.

37.2 References

- Michael Sipser, *Introduction to the Theory of Computation* ISBN 0-534-94728-X. *(See . Theorem 1.22, section 1.2, pg. 59.)*

37.3 Text and image sources, contributors, and licenses

37.3.1 Text

- **Mathematical proof** *Source:* https://en.wikipedia.org/wiki/Mathematical_proof?oldid=682829198 *Contributors:* AxelBoldt, Robert Merkel, The Anome, Toby Bartels, Youandme, Edward, TeunSpaans, Michael Hardy, Spazlink, TakuyaMurata, LittleDan, Pratyeka, Glenn, Tim Retout, Hashar, Revolver, Charles Matthews, Dysprosia, Jitse Niesen, Bjh21, Markhurd, Booya, Bcorr, Denelson83, Fredrik, Romanm, Lowellian, Gandalf61, Paul G, Hadal, Giftlite, Markus Krötzsch, Yekrats, CSTAR, Karl-Henner, Tyler McHenry, Hkpawn~enwiki, Petershank, Peter Kwok, Andreas Kaufmann, PhotoBox, Jason Carreiro, Leibniz, Paul August, Robertbowerman, Syp, El C, Skeppy, Stesmo, Jojit fb, Nk, ACW, Jumbuck, Kanie, Arthena, Itsmine, Oleg Alexandrov, Jslu, Mindmatrix, Rodrigo Rocha~enwiki, Bkkbrad, Pol098, Jeff3000, Tygar, Stevey7788, Seyon, Rjwilmsi, Salix alba, R.e.b., Alexb@cut-the-knot.com, KarlFrei, Jrtayloriv, BMF81, Chobot, Jersey Devil, DVdm, Borgx, Trovatore, Cheeser1, Boivie, FF2010, PTSE, Josh3580, HereToHelp, Willtron, Curpsbot-unicodify, Brentt, robot, SmackBot, KnowledgeOf-Self, Bomac, Jagged 85, Diegotorquemada, Gracenotes, Can't sleep, clown will eat me, Rantingsteve, Rrburke, Drackap, MichaelBillington, Occultations, Luke Gustafson, Byelf2007, Lambiam, Eliyak, Scottie 000, Heimstern, Loadmaster, Mets501, ZodoJats, Stephen B Streater, Newone, Mrdthree, Braddodson, CRGreathouse, CBM, Larrywcusick, Dgw, MarsRover, Myasuda, Gregbard, Cydebot, NotQuiteEXPComplete,Thijs!bot,Epbr123,Nnn9245,Escarbot,AntiVandalBot,Joasiak,Nikolas Karalis,JAnDbot,Leuko,Hut8.5,Four Dog Night,VoABot II,Soulbot, Singularity, Craw-daddy, Johnbibby, Seberle, JJ Harrison, David Eppstein, Bernard Hurley, Rohan Ghatak, JaimeLesMaths, Naveedahmad 14382,Wiki Raja,J.delanoy,Leon math,Numbo3,Maurice Carbonaro,Sefog,Vvitor,Smeira,He is a shithead,Lupussy,M.M.S.,Krish-nachandranvn, Tparameter, Policron, Jonjesbuzz, Dessources, Vanished user 39948282, Idioma-bot, VolkovBot, JohnBlackburne, Jimmaths,Greatwalk, TXiKiBoT,Gentlemath,Anonymous Dissident,Ocolon,AlleborgoBot,Symane,EmxBot,SieBot,YonaBot,Phe-bot,RiskAverse,RJaguar3, Triwbe, X-Fi6, LeadSongDog, 360 Degree, Barliner, Radon210, Byrialbot, Thehotelambush, DesolateReality, Nusumareta, Ran-domblue,Myrvin, Tautologist,Athenean,ClueBot,Justin W Smith,Dobermanji,Lozersk,Adrianwn,ChandlerMapBot,I am a violinist,Wiki-imee, DragonBot, Alexbot , Alejandrocaro35, La Pianista, Humanengr, Pichpich, Stickee, Gerhardvalentin, Kwjbot, WikiDao, EEng, Addbot,Betterusername,Glane23, AndersBot,Roux,ChenzwBot,Jaydec,Numbo3-bot,BOOLE1847,Legobot,Luckas-bot,Yobot,Kan8eDie,Jorge-Fierro,Texhausballa,AnomieBOT, JRB-Europe,Materialscientist,Are you ready for IPv6?,La comadreja,ArthurBot,LilHelpa,MauritsBot,Xqbot, Sourceholder, Almabot, J04n, Geometryfan, FrescoBot, LucienBOT, Pinethicket, FriedrickMILBarbarossa, Foodimentary, Di1000,Jauhienij,Leasnam,FoxBot,TobeBot, DixonDBot,Robert hoffman,EmausBot,AmigoDoPaulo,Josve05a,Sven Manguard,Jijo925,ClueBotNG, Wikigold96, CocuBot, Melville88, Widr, ساجد امجد ساجد, Keetanii, Helpful Pixie Bot, Joolsa123, Wasbeer, Solomon7968, Khonkhor-tisan, Kiewbra, Hebert Peró, Was123ification, Dexbot, Mogism, Brirush, Epicgenius, Gdaniel111, Jamesmcmahon0, Purnendu Karmakar,Schopenhauerswille,Blackbombchu,Coz7, Seansmoove27,Leegrc,EggyEggPercent,TheCoffeeAddict and Anonymous: 190

- **List of mathematical proofs** *Source:* https://en.wikipedia.org/wiki/List_of_mathematical_proofs?oldid=651793994 *Contributors:* Manning Bartlett, Edward, Michael Hardy, Dominus, Revolver, Pfortuny, Marc Venot, Tosha, Giftlite, Dbenbenn, Dissident, Gro-Tsen, Golbez, Zarvok, Peter Kwok, Shahab, Paul August, ZeroOne, ABCD, Oleg Alexandrov, Jacobolus, Salix alba, R.e.b., Mathbot, YurikBot, IanManka, Grubber, Buster79, Figaro, Googl, Zvika, Silly rabbit, Syrcatbot, Cydebot, Neko244, Weixifan, Leon math, JohnBlackburne, Jimmaths, Synthebot, Dmcq, OlEnglish, RJGray, Set theorist, Wbm1058, Joemkhan and Anonymous: 9

- **Bertrand's postulate** *Source:* https://en.wikipedia.org/wiki/Bertrand{ }s_postulate?oldid=684738304 *Contributors:* AxelBoldt, XJaM, Michael Hardy, COLETTE~enwiki, Dcoetzee, Bjh21, .mau., Chuunen Baka, Lowellian, Henrygb, Giftlite, Dbenbenn, Dissident, Porges, Poccil, Paul August, Djordjes, EmilJ, .:Ajvol:., Burn, Samohyl Jan, Shreevatsa, TheGoblin, Reddwarf2956, Mekong Bluesman, Rjwilmsi, FlaBot, Ysangkok, Sodin, Haonhien, Chobot, Lenthe, DYLAN LENNON~enwiki, Froth, Arthur Rubin, SmackBot, BeteNoir, JCSantos, PrimeHunter, Lhf, DMacks, SashatoBot, Richard L. Peterson, Wstomv, 'Ff'lo, Dajackstah, CRGreathouse, Pedro Fonini, JLISP, NERIUM, WinBot, JAnDbot, Txomin, Vanish2, Albmont, David Eppstein, Yecril, Philip Trueman, Gentlemath, Jobu0101, Nxavar, Spinningspark, Vongoiva~enwiki, SieBot, Portalian, Rdmabry, JLKrause, PhiEaglesfan712, Vikasatkin, Jsondow, Little Mountain 5, MystBot, Addbot, DOI bot, Ka Faraq Gatri, Luckas-bot, Yobot, Rubinbot, Citation bot, Xqbot, Motomuku, Citation bot 1, Trappist the monk, EmausBot, ZéroBot, TruthThruDigits, Sapphorain, ChuispastonBot, Mineallmine12, Dainiak, BG19bot, Anrnusna, TakuyaMishiba and Anonymous: 38

- **Proof of Bertrand's postulate** *Source:* https://en.wikipedia.org/wiki/Proof_of_Bertrand{ }s_postulate?oldid=684738416 *Contributors:* AxelBoldt, XJaM, Michael Hardy, Eric119, Mydogategodshat, Charles Matthews, Fredrik, Dbenbenn, Dissident, Porges, Supercoop, Wikiacc, Paul August, Gauge, EmilJ, Marco Polo, Gene Nygaard, Cgibbard, Woohookitty, Linas, Ryan Reich, Salix alba, FlaBot, Ysangkok, Bgwhite, Dmharvey, Buster79, Hv, Zvika, SmackBot, Eskimbot, PrimeHunter, Richard L. Peterson, Hiiiiiiiiiiiiiiiiiii, CRGreathouse, Gihanuk, Icek~enwiki, Jasperdoomen, Magioladitis, Vernanimalcula, David Eppstein, PMajer, Addbot, Touchatou, FrescoBot, Motomuku, SpaceFlight89, PleaseStand, Goldfruits, Danieljohngoldstein, Kodip, Kannan Nambiar th3an0maly, TakuyaMishiba and Anonymous: 29

- **Estimation of covariance matrices** *Source:* https://en.wikipedia.org/wiki/Estimation_of_covariance_matrices?oldid=680312980 *Contributors:* Michael Hardy, Zeno Gantner, Den fjättrade ankan~enwiki, Charles Matthews, Selket, Robinh, Bender235, O18, Landroni, Cburnett, Aaronh, Biomenne, Btyner, Stsmith, Ogrisel, Entropeneur, Bo Jacoby, Zvika, Jonas August, Ctacmo, Pagh, Lavaka, TNeloms, Jonathan A Jones, Johnbibby, David Eppstein, Algebraic, Ged.R, Jmath666, YohanN7, KoenDelaere, Melcombe, Skbkekas, Prax54, DOI bot, Fgnievinski, Wikomidia, Yobot, AnomieBOT, LilHelpa, J04n, FrescoBot, Citation bot 1, Aobha, Kiefer.Wolfowitz, Zfeinst, Mrtnmcc, Helpful Pixie Bot, BG19bot, Neil Frazer, Monkbot, Tsivo, Duncan8 and Anonymous: 36

- **Fermat's little theorem** *Source:* https://en.wikipedia.org/wiki/Fermat{ }s_little_theorem?oldid=682764660 *Contributors:* AxelBoldt, Mav, Zundark, Taral, XJaM, Michael Hardy, Blueshade, Ixfd64, Chinju, Nikai, Rotem Dan, Ideyal, Dysprosia, Sabbut, Robbot, Fredrik, Gandalf61, Bethenco, Wikibot, Tosha, Giftlite, Lupin, Peruvianllama, Dratman, Python eggs, Gubbubu, CryptoDerk, Profvk, Almit39, Sam Hocevar, Discospinster, Kooo, Ericamick, DonDiego, PittBill, Steveha, Obradovic Goran, Jumbuck, Neonumbers, Burn, Wtmitchell, Ling Kah Jai, Oleg Alexandrov, OwenX, Guardian of Light, JFG, Mensanator, Graham87, BD2412, Pako, Erkcan, Mathbot, Maxal, Chobot, Siddhant, Dmharvey, Vecter, RussBot, Arthur Rubin, SmackBot, RDBury, Mmernex, BeteNoir, Wzhao553, Hmains, Kurykh, PrimeHunter, TrogdorPolitiks, Bigmantonyd, Andrei Stroe, Rohit math, Karakal, Mr. Lefty, Hvn0413, Gauravk, Otac0n, Jingwang, Blackcloak, ChrisCork, Nutster, CRGreathouse, Misof, MC10, Fmalinkevich, Navigatr85, Heder~enwiki, Salgueiro~enwiki, JAnDbot, BlendsInWell, Connor Behan, Nthitz, David Callan, N4nojohn, Ttwo, Boris Allen, Andareed, Policron, Vcpandya, Pleasantville, Mrh30, BotKung, SieBot, OKBot, JMOprof, Sphilbrick, ClueBot, DragonBot, NuclearWarfare, Silas Maxfield, Johnuniq, Rangergordon, Virginia-American, Addbot, Download, Loupeter,

37.3. TEXT AND IMAGE SOURCES, CONTRIBUTORS, AND LICENSES 237

Legobot, Luckas-bot, Yobot, Charleswallingford, KamikazeBot, AnomieBOT, Rubinbot, Xqbot, Quintus314, Adrionwells, X7q, Sebastiangarth, AstaBOTh15, BasvanPelt, EmausBot, Vanished user bc8e8hwkjaflhw8tijwfiu, Tdupu10000, Zarboublian, Tttfffkkk, Toshio Yamaguchi, Donner60, Llightex, Anita5192, Wcherowi, Adityasinghhhhh, Adityasinghhhhhh, Gaurav178, Alpha1337Saint, Joel B. Lewis, Cenkner, Armadillopteryx, Mjbuoni, GKFX, M hariprasad, Dexbot, Razibot, Epicgenius, Purnendu Karmakar, José Carlos Matemáticas, Gavnerfsk and Anonymous: 102

- **Proofs of Fermat's little theorem** *Source:* https://en.wikipedia.org/wiki/Proofs_of_Fermat'{}s_little_theorem?oldid=661439778 *Contributors:* AxelBoldt, The Anome, Andre Engels, FvdP, Michael Hardy, Dominus, Cole Kitchen, Chinju, Tristanb, Charles Matthews, Timwi, Dysprosia, Schutz, Bethenco, Moink, Giftlite, Bob.v.R, Almit39, Rich Farmbrough, MaxPower, Mtaub, HasharBot~enwiki, Sligocki, Ling Kah Jai, Oleg Alexandrov, Woohookitty, Guardian of Light, Salix alba, Mathbot, YurikBot, Dmharvey, KSmrq, Gaius Cornelius, Mikeblas, GrinBot~enwiki, Zvika, Wutchamacallit27, Lupercus~enwiki, JRSpriggs, Anakata, Stannered, AntiVandalBot, David Eppstein, Error792, Yoni, S.riccardelli, Melsaran, Synthebot, BSoD, Horoball, DumZiBoT, SteveJothen, Addbot, Ytbau, Yobot, Armbrust, Eli Fogel, ??, Fox Wilson, Oneballeddie, Gwen-chan, Snotbot, Dunedubby, OskariVirtanen, ChrisGualtieri, Cerellon, Alkauskas, Somepeople4 and Anonymous: 57

- **Gödel's completeness theorem** *Source:* https://en.wikipedia.org/wiki/G%C3%B6del'{}s_completeness_theorem?oldid=656456908 *Contributors:* AxelBoldt, Mav, Michael Hardy, Modster, Mdebets, Tim Retout, Charles Matthews, Timwi, Dysprosia, Hyacinth, Aleph4, Gandalf61, Giftlite, Lethe, Lupin, Siroxo, Spoirier~enwiki, Mike Rosoft, Rich Farmbrough, Guanabot, Maksym Ye., Ben Standeven, El C, Chalst, Teorth, 3mta3, Gene Nygaard, Drbreznjev, Eric Qel-Droma, Oleg Alexandrov, Ott, Pdn~enwiki, Dzordzm, Flamingspinach, Dionyziz, Tim!, Mathbot, NavarroJ, Reetep, Algebraist, YurikBot, Tony1, Scope creep, SmackBot, RDBury, BeteNoir, Betacommand, Mhss, Dr. de Seis, Taggart Transcontinental, SashatoBot, Dan Gluck, Joseph Solis in Australia, Hilverd, Zero sharp, JRSpriggs, CBM, Myasuda, Gregbard, Spewin, Thijs!bot, Headbomb, Jbaranao, Morphriz, Meeples, TXiKiBoT, EuTuga, Da Joe, IsleLaMotte, Martarius, Hans Adler, El bot de la dieta, Thingg, Hugo Herbelin, Marc van Leeuwen, MrOllie, Lightbot, Legobot, OrgasGirl, Omnipaedista, BrideOfKripkenstein, Citation bot 1, WikitanvirBot, Jaydiem, Brad7777, ChrisGualtieri, Jochen Burghardt, Epicgenius, Salspaugh, 2.71828182845904523austen, Whiterray and Anonymous: 33

- **Original proof of Gödel's completeness theorem** *Source:* https://en.wikipedia.org/wiki/Original_proof_of_G%C3%B6del'{}s_completeness_theorem?oldid=679095433 *Contributors:* AxelBoldt, General Wesc, Mav, Ktsquare, Mjb, Michael Hardy, Eric119, Dysprosia, Giftlite, Mboverload, Rich Farmbrough, Guanabot, AlanBarrett, Ben Standeven, Pt, Nickj, Grue, Randall Holmes, 3mta3, Oleg Alexandrov, Woohookitty, Rjwilmsi, Tim!, Salix alba, Sodin, Trovatore, Zvika, Charles Moss, Dan Gluck, Zero sharp, CBM, Gregbard, Cydebot, Headbomb, Jjaazz, R'n'B, Mad7777, SieBot, Hans Adler, Lightbot, Legobot, Yobot, Citation bot 1, Bbbbbbbbba, Bahersabry and Anonymous: 25

- **Mathematical induction** *Source:* https://en.wikipedia.org/wiki/Mathematical_induction?oldid=687022745 *Contributors:* AxelBoldt, Bryan Derksen, Zundark, Tarquin, Jan Hidders, XJaM, Mjb, Youandme, Bdesham, Michael Hardy, Wshun, David Martland, Dominus, Kku, Meekohi, Sannse, TakuyaMurata, Ahoerstemeier, Snoyes, Александър, Charles Matthews, Dcoetzee, Sbloch, Dysprosia, Jitse Niesen, Selket, Markhurd, Furrykef, Sabbut, Bloodshedder, .mau., Aenar, Robbot, Iwpg, Altenmann, Gandalf61, Bruceq, Aetheling, Ruakh, Tobias Bergemann, Tosha, Giftlite, Nickptar, Fintor, Peter Kwok, Kousu, PhotoBox, Lucidish, Shipmaster, Ericamick, Paul August, Bender235, Elwikipedista~enwiki, Zenohockey, CXI, Spoon!, Aplusbi, NetBot, Obradovic Goran, Polylerus, Jumbuck, Msh210, Arthena, Dirac1933, Bsadowski1, Oleg Alexandrov, Zntrip, Mindmatrix, StradivariusTV, Acone, Drostie, Ruud Koot, Mpatel, Sartak, M412k, Ryan Reich, Palica, Mandarax, Chun-hian, JIP, Rjwilmsi, SMC, FlaBot, Maitch, Mathbot, Chobot, YurikBot, Borgx, Hairy Dude, RussBot, Hyad, KSmrq, Debroglie, TEB728, Trovatore, Nick, Jstrater, Schmock, Pyroclastic, Froth, Gadget850, Bota47, Addps4cat, Mgnbar, Brisvegas, Jwissick, Reyk, SmackBot, RDBury, Selfworm, Melchoir, Pgk, Bomac, Pokipsy76, Jagged 85, PJTraill, Chris the speller, Fintler, JCSantos, Trebor, Nbarth, Nixeagle, ConMan, Radagast83, Acepectif, Lambiam, Wvbailey, J. Finkelstein, Pliny, Mets501, MissingNOOO, Dreftymac, Blackhawk charlie2003, Zero sharp, JRSpriggs, CRGreathouse, Wafulz, CBM, WeggeBot, Gregbard, Cydebot, Zahlentheorie, JFreeman, BillWeiss, Headbomb, Pcbene, AntiVandalBot, Vic226, Gramby, Sekky, Alphachimpbot, MER-C, Wisnuops, David Eppstein, DerHexer, SquidSK, Stephenchou0722, MartinBot, JCraw, Maurice Carbonaro, TomS TDotO, Paulecoyote, OttoMäkelä, It Is Me Here, LordAnubisBOT, Tparameter, Policron, DavidCBryant, Omegamormegil, Fbarton, VolkovBot, AlnoktaBOT, Jimmaths, TXiKiBoT, Anonymous Dissident, Philogo, PaulTanenbaum, Gilisa, Pboulus, SieBot, Paradoctor, Alexsmail, MiNombreDeGuerra, Ellamosi, Ngriffeth, DesolateReality, Altzinn, Amahoney, AutoFire, Classicalecon, Gjakovit, ClueBot, Justin W Smith, The Thing That Should Not Be, Jdgilbey, Mathemajor, J8079s, DragonBot, Muro Bot, Tired time, SoxBot III, Dyhan81, Ost316, Brentsmith101, Kbdankbot, Addbot, DOI bot, Riyuky, CanadianLinuxUser, NjardarBot, MrVanBot, Chamal N, CarsracBot, LinkFA-Bot, Lightbot, Dminkovsky, Jarble, Yobot, Estudiarme, Jasonschock, AnomieBOT, 1exec1, MattTait, Citation bot, ArthurBot, TaySpace, Poetaris, DSisyphBot, J04n, RibotBOT, D'ohBot, Citation bot 1, Tkuvho, Kiefer.Wolfowitz, Unnachamois, Bozo the bear, Gabrielgmendonca, Mrhota, Vrenator, EdEveridge, Duoduoduo, Onel5969, Jowa fan, Whywhenwhohow, EmausBot, Vasanthloganathan, Slawekb, Ebrambot, Quondum, Noodleki, ClueBot NG, Moneysorter, Tideflat, B.dyck, Meingbg, Helpful Pixie Bot, DBigXray, BG19bot, FutureTrillionaire, Alpert1, CitationCleanerBot, Glacialfox, Justincheng12345-bot, Dexbot, Jochen Burghardt, I am One of Many, Tl. Gracchus, Glasbys, Beneficii, Monkbot, FourViolas, Swashski, Dsacf1234567 and Anonymous: 232

- **0.999...** *Source:* https://en.wikipedia.org/wiki/0.999...?oldid=687139691 *Contributors:* Zundark, The Anome, XJaM, Tommy~enwiki, Oliverkroll, Montrealais, D, Michael Hardy, Booyabazooka, Ixfd64, Tango, GTBacchus, Karada, Eric119, CesarB, William M. Connolley, Theresa knott, Darkwind, Jouster, Charles Matthews, Timwi, Dcoetzee, Dfeuer, Jitse Niesen, Hydnjo, Doradus, DJ Clayworth, Grendelkhan, Shizhao, AnonMoos, Phil Boswell, Aleph4, Donarreiskoffer, Robbot, Astronautics~enwiki, Fredrik, Altenmann, Gandalf61, Ashley Y, Merovingian, Timrollpickering, Rasmus Faber, Bkell, Quadalpha, Tobias Bergemann, Pablo-flores, Giftlite, Christopher Parham, Philwelch, Lethe, Bfinn, Rick Block, Sundar, Macrakis, Matt Crypto, Chowbok, Geni, Gdr, Slowking Man, Noe, MarkSweep, MisfitToys, Daniel.levine, Supadawg, Kevin B12, Ary29, Pmanderson, Gscshoyru, Kelson, M1ss1ontomars2k4, Davidstrauss, Lacrimosus, Porges, Guppyfinsoup, Orange Goblin, Sysy, Rich Farmbrough, Mhowkins, Lemontea, Paul August, Bender235, ESkog, ZeroOne, Kaisershatner, Jnestorius, Ben Standeven, BACbKA, Tompw, MisterSheik, Zenohockey, Kwamikagami, Triona, Phiwum, Chuaryw2000, Jpgordon, Causa sui, Bobo192, Touriste, Mike Schwartz, Reinyday, Dreish, .:Ajvol:., Malafaya, Of~enwiki, Slicky, Blotwell, Larry V, Hesperian, Crust, Nsaa, Jakew, Alansohn, Anthony Appleyard, Nsd, Diego Moya, Ronline, Riana, AzaToth, Sligocki, Yummifruitbat, Bart133, Blobglob, Schapel, Isaac, ReyBrujo, BlastOButter42, Computerjoe, Kusma, Shutranm, Mosesofmason, Pediddle, Ling Kah Jai, Saxifrage, Oleg Alexandrov, Zntrip, Thryduulf, Simetrical, Mindmatrix, Shreevatsa, Georgia guy, StradivariusTV, Kzollman, Pol098, Fbv65edel, Tylerni7, Mangojuice, Waldir, Gerbrant, CronoDAS, Dysepsion, Mandarax, SixWingedSeraph, JIP, Josh Parris, Rjwilmsi, Koavf, MarSch, Staecker, Seraphimblade, Bruce1ee, Salix alba, Tawker, SMC, Nneonneo, Boccobrock, Durin, Brighterorange, DoubleBlue, Dar-Ape, Algebra, Titoxd, VKokielov, RobertG,

Thebeatlesanthol, Kasama~enwiki, Kmorozov, Gurch, Ayla, TheDJ, Jrtayloriv, Fresheneesz, Argyrios Saccopoulos, BradBeattie, Mallocks, Mstroeck, MoRsE, Dougk, DVdm, Volunteer Marek, Bgwhite, Roboto de Ajvol, JPD, YurikBot, Wavelength, Maelin, PiAndWhippedCream, Hairy Dude, X42bn6, Brandmeister (old), Red Slash, Loom91, KSmrq, WikidSmaht, Ansell, Toffile, Stephenb, Gaius Cornelius, Ihope127, Cryptic, NawlinWiki, Hawkeye7, Msikma, Borbrav, Trovatore, Zarel, Długosz, Ttogreh, Dureo, SCZenz, Irishguy, Nick, HeroicJay, Scs, Paul.h, Misza13, Aaron Schulz, Jessemerriman, Jeremy Visser, BigMoosie, Jemebius, Occhanikov, WAS 4.250, FF2010, Tob~enwiki, Theda, Great Cthulhu, Arthur Rubin, Jesushaces, Red Jay, Kier07, HereToHelp, Anclation~enwiki, Katieh5584, Ajblue98, Arkon, Auroranorth, Zvika, Stumps, Lunch, Dupz, Itub, A bit iffy, SmackBot, RDBury, YellowMonkey, Nihonjoe, Incnis Mrsi, Maelwys, KnowledgeOfSelf, Melchoir, McGeddon, David.Mestel, Ccreitz, Flying Canuck, AnOddName, BiT, Trystan, KingRaptor, Aksi great, Gilliam, Smeggysmeg, Flewellyn, Adamstreed, Thumperward, Oli Filth, OrangeDog, Silly rabbit, Taxipom, Mdwh, JoeBlogsDord, Schi, Octahedron80, CMacMillan, Hengsheng120, Darth Panda, Calc rulz, Suicidalhamster, Magaroja, NYKevin, Can't sleep, clown will eat me, Timothy Clemans, OSborn, Rrburke, CorbinSimpson, Calbaer, Huon, King Vegita, ConMan, Flyguy649, NoIdeaNick, Cybercobra, Irish Souffle, Downwards, Khukri, Nibuod, Daqu, Savidan, Underbar dk, TedE, B jonas, Mjefm, Algr, Mini-Geek, Akriasas, Weregerbil, Astroview120mm, Meni Rosenfeld, Kendrick7, Metamagician3000, Pilotguy, Kukini, Homo sapiens, Nov ialiste, Luciand, Ceoil, TenPoundHammer, Ozhiker, SilverStar, Harryboyles, Xandi, Kuru, Euchiasmus, Freewol, BurnDownBabylon, Gizzakk, MvH, Disavian, Paladinwannabe2, JH-man, JoshuaZ, Tim Q. Wells, Mgiganteus1, IronGargoyle, Loadmaster, Grumpyyoungman01, Tyrrell McAllister, Androl, Yourmotherisanastronaut, Mikekelly, MrArt, Mets501, EdC~enwiki, Rubikfreak, Elb2000, Sasata, Inquisitus, Dl2000, Retio, Iridescent, WAREL, Wjejskenewr, PetaRZ, Amakuru, Lenoxus, Newyorkbrad, Rangi42, Civil Engineer III, Neurillon, Aliveboy, Tawkerbot2, Hamish2k, Michaelshull, Drlowell, Aherunar, CBM, Kylu, Matthew Auger, Outriggr (2006-2009), Moreschi, Alexignatiou~enwiki, FilipeS, Rudjek, Vectro, Doctormatt, Cydebot, Muahahahaha, Stebbins, Reywas92, MC10, Steel, Philbert2.71828, Badpazzword, Rishodi, He Who Is, DumbBOT, Chrislk02, T.A Shirakawa, Gionnico, Omicronpersei8, Malleus Fatuorum, Wandalstouring, Epbr123, King Bee, MatheMezzaMorphis163, Imthedragn, Headbomb, RevolverOcelotX, Jojan, Djdickmutt, Davkal, Ufwuct, Mentor1337, Tellyaddict, Poe Joe, Grayshi, Dugwiki, Flarity, Uruiamme, Mentifisto, Fildon, AntiVandalBot, RoyalAbidi, Majorly, Luna Santin, Aprogressivist, Mathisreallycool, Edokter, Dalassa, Exteray, Dylan Lake, Danger, Gdo01, Frustumator, Amarkov, Ap621, Husond, Barek, MER-C, Kprateek88, Staroftheshow86, Thenub314, Wlmh65, Tengfred, TAnthony, Gavia immer, Acroterion, Repku, Coffee2theorems, Casmith 789, Connormah, Bongwarrior, VoABot II, AlphaPhoenixDown, AuburnPilot, JamesBWatson, Appraiser, Mclay1, WODUP, YxSupermanxY, MetsBot, Alex Spade, Gustave the Steel, DerHexer, JaGa, Fantastic4boy~enwiki, King Dracula, Tkessler45, Jassi26, MartinBot, Spinafire, Rettetast, Mschel, AlexiusHoratius, Pbroks13, Pomte, Greenman248, N4nojohn, Brix., Quaternus~enwiki, Princemackenzie, Abecedare, DrKay, Trusilver, Intergr8, Shybunks, SiliconDioxide, Uncle Dick, TempestCA, Tdadamemd, P3net, M C Y 1008, Andareed, Ajmint, Brohacz, Dispenser, Tidaress, Screen111, Shhdez, Viper98, DarkFalls, Sgt alien, Scout1, Loasl, Cr2898, Vantucci, Krakowsdgdad, Hules002, Yuraty, Shinobi123, Paulmkgordon, Cayennecode, Rishartha the Pwninator, 888puma888, Tparameter, Craig Erb, Coolapi, Jorfer, ILoveFuturama, Sigmundur, Treisijs, Marxmanster, Bonadea, WinterSpw, Andrew Deans, Devindred, Ronbo76, Alcedias, CardinalDan, F.F.McGurk, 28bytes, VolkovBot, Leebo, JohnBlackburne, Nburden, LokiClock, Akwdb, Chango369w, VasilievVV, Nousernamesleft, Independentdependent, TXiKiBoT, Gentlemath, Ann Stouter, Anonymous Dissident, Retiono Virginian, Oxfordwang, Tantal-ja, Martin451, Digby Tantrum, Broadbot, Aaron Rotenberg, Abdullais4u, PDFbot, Wikiisawesome, Csdorman, Maxim, Rjgodoy, Homei, Jensen 198, Iapain wiki, Sylent, Asdfazerty, Brianga, Ceranthor, Tvinh, EverGreg, Katzmik, PokeYourHeadOff, SMC89, PlanetStar, Da Joe, Duplicity, Djuuss, Sunny910910, Wizzard2k, לריק, Paolo.dL, Pigpenguin666, Oxymoron83, Lightmouse, AuburnPiIot, Randomblue, Me, Hariva, Susan118, Mox83, Truthfulcynic, XDanielx, Extremecircuitz, Faithlessthewonderboy, ClueBot, The Thing That Should Not Be, Cliff, JuPitEer, PolarYukon, Blanchardb, TheSmuel, WestwoodMatt, Rockfang, Excirial, Eeekster, Urdead4g, Vivio Testarossa, Tlepp, NuclearWarfare, M.O.X, Swordstripe, I amm Beowulf!, Morel, BOTarate, Dsmurat, Aitias, Singularity42, Johnuniq, Indopug, Joopwiki, RMFan1, XLinkBot, Rangergordon, Marc van Leeuwen, Blehblehblehblehblehbleh, Jcbrd, Kojitakeo, Sero12, Addbot, DOI bot, Hellboy2hell, Ronhjones, TutterMouse, Wingspeed, CanadianLinuxUser, Flamergamer, Cst17, Ebaumsworlddotcom001, CarsracBot, Soberum, Numbo3-bot, Arbitrarily0, Luckas-bot, ZX81, Yobot, 2D, Ptbotgourou, Legobot II, Xqt, ArchonMagnus, Linket, MassimoAr, Tempodivalse, Szajci, Travswrong, DiverDave, AnomieBOT, Gpia7r, Jim1138, 9258fahsflkh917fas, AdjustShift, LlywelynII, Materialscientist, Citation bot, Felyza, Htrbabaseball, ArthurBot, Qorilla, JAK, Xqbot, Bozo9, Drilnoth, Mononomic, Gilo1969, Flying sheep, 78.26, Xdarkmagex123, Shadowjams, Spellage, SchnitzelMannGreek, Wowdie1, Fingerz, FrescoBot, Sławomir Biały, Alxeedo, Majopius, HJ Mitchell, Cannolis, Citation bot 1, Amplitude101, Intelligentsium, Tkuvho, Dr Marcus Hill, RandomDSdevel, Pinethicket, I dream of horses, Evileyelover22, Arctic Night, Abductive, JustifiedSkepticism, Jonesey95, Seryo93, Stpasha, Wikitanvir, Thesevenseas, Nameuserrandom1234, Lissajous, SkyMachine, Arbero, Ruyangyangcow, Alpir, Trappist the monk, Rentzepopoulos, Aoidh, Leonid 2, Dabawse, Julian Birdbath, Feffemannen, Suffusion of Yellow, DARTH SIDIOUS 2, Shiftnoise, EngineerFromVega, RjwilmsiBot, TjBot, Ripchip Bot, Eruditionfish, Grr the awesome, Calebmason2313, Otutusaus, DASHBot, Qwertyisbest, John of Reading, Kayrus, Racerx11, JaeDyWolf, Tommy2010, Josve05a, The Nut, Hanstheenforcer, Elio96, AManWithNoPlan, SeX-Eskimo, Staszek Lem, Mung Daal, Bulwersator, Iketsi, DASHBotAV, ClueBot NG, MelbourneStar, This lousy T-shirt, Satellizer, Indiansuperman24, BarrelProof, Joefromrandb, DokReggar, Wendidju, Infinite 9s, Bjacobs9743, MerlIwBot, SmartRamen, Calabe1992, Elijap29, DBigXray, Lowercase sigmabot, 1mathman, BG19bot, Interchangeable, MusikAnimal, Piguy101, Dentalplanlisa, Brad7777, That1guyme, Minsbot, Fighterf4u, Bonkers The Clown, Qetuth, NereusAJ, Mleeds12, Teammm, Cyberbot II, Erinmcleod, AliasThomasAnderson, Ducknish, Dexbot, Grapezxd, Jgrapez, Allen970307, Jamesx12345, Mjs1138, Parcheesidude, Cadillac000, Rltb, Georgeh109, AmericanLemming, EvergreenFir, DavidLeighEllis, Leland Prior, Pulvertaft, Panpog1, Blackbombchu, Dasintwist, Schwatzwutz, Seppi333, Ginsuloft, GreyWinterOwl, Stamptrader, EVDragon, Calculemus, CogitoErgoSum14, The f18hornet, Bilorv, LivinLikeLarry, Blain177, KillIrman, TheQ Editor, Dyott, Ydon205, Ffilozov, Wikpedian, Rezabehforooz, Tonyhungchihang123 and Anonymous: 742

- **Proof that 22/7 exceeds π** *Source:* https://en.wikipedia.org/wiki/Proof_that_22/7_exceeds_%CF%80?oldid=678187647 *Contributors:* Damian Yerrick, The Anome, Arvindn, Stevertigo, Bdesham, Michael Hardy, JakeVortex, GTBacchus, Paddu, Paul A, Tregoweth, LittleDan, Charles Matthews, Dcoetzee, Dysprosia, Fredrik, Lowellian, Sverdrup, Jachim69, Robinh, Giftlite, Dbenbenn, Smjg, BenFrantzDale, Python eggs, Tagishsimon, Anythingyouwant, Pmanderson, Neutrality, Damieng, Gazpacho, Mindspillage, Paul August, Night Gyr, Truthflux, Smalljim, Func, Reinyday, C S, Rbj, Neonumbers, Oleg Alexandrov, Woohookitty, Linas, Shreevatsa, Georgia guy, Jacobolus, Jerickson314, Rjwilmsi, Salix alba, AySz88, Diza, Glenn L, Bgwhite, WriterHound, Kauffner, Red Slash, KSmrq, Schmock, Ms2ger, Arthur Rubin, Kaicarver, Zvika, SmackBot, Xaosflux, Hmains, Jeffro77, EncMstr, Hgrosser, Sirgregmac, Lambiam, Jim.belk, Kirbytime, Dicklyon, TwistOfCain, Jafet, CRGreathouse, CBM, A876, Kaldosh, GadBeebe, Edokter, MER-C, Kaonslau~enwiki, VoABot II, Avicennasis, Tt 225, David Eppstein, JoergenB, Laurusnobilis, SJP, VolkovBot, Philip Trueman, Anonymous Dissident, Gerrish, Meters, Synthebot, Eronixpress, DonAByrd, ClueBot, Bold Clone, Johnuniq, DumZiBoT, RMFan1, Pistis888, Psycholian, Addbot, FokkerTISM, Lihaas, Yobot, Jim1138, Citation bot, Isheden, FrescoBot, DrilBot, PiRSquared17, Jhenderson777, Gotoquiz, Jowa fan, EmausBot, John Cline, Privatequelch, Thine Antique Pen, ClueBot NG, Wcherowi, The Master of Mayhem, Clearmaker, Helpful Pixie Bot, Th4n3r, Raumaan, Monkbot, PtolemyH, Sinterklaas99, Shreyanshjain16 and Anonymous: 81

- **Proof that e is irrational** *Source:* https://en.wikipedia.org/wiki/Proof_that_e_is_irrational?oldid=667547209 *Contributors:* Gareth Owen, Stevertigo, Michael Hardy, Paul A, Smack, Wooster, Revolver, Charles Matthews, Pmineault, Centrx, Giftlite, Ikitakoja, Icairns, Pt, EmilJ, R6MaY89, Woohookitty, Linas, Shreevatsa, Justinlebar, Rjwilmsi, Salix alba, Mathbot, Bgwhite, WriterHound, YurikBot, Lofty, Schmock, Arthur Rubin, Mordacil, HereToHelp, Zvika, Meshach, BeteNoir, Eskimbot, JCSantos, Oli Filth, Nbarth, AeroSpace, Cybercobra, Kirbytime, Tobias Pfanner~enwiki, Myasuda, RohanDhruva, Nyttend, David Eppstein, Jimothy 46, Synthebot, This, that and the other, Jsondow, Cmr08, Addbot, J.alzaili, 9258fahsflkh917fas, Ssola, Capricorn42, Citation bot 1, Jowa fan, GabKBel, Wackywill1001, Vaselli, Leostaley, Pirokiazuma, Kavigupta, SoSivr and Anonymous: 37

- **Proof that π is irrational** *Source:* https://en.wikipedia.org/wiki/Proof_that_%CF%80_is_irrational?oldid=686950348 *Contributors:* Michael Hardy, Giftlite, Dratman, SarekOfVulcan, Anythingyouwant, Dbachmann, Marco Polo, Hydriotaphia, Woohookitty, Shreevatsa, Rjwilmsi, Salix alba, R.e.b., Bgwhite, Kauffner, Trovatore, Schmock, Arthur Rubin, AndrewWTaylor, JCSantos, Nbarth, Vanished user v8n3489h3tkjnsd kq30u3f, Loadmaster, Dicklyon, Iridescent, Courcelles, CRGreathouse, Xanthoxyl, Stebulus, Thijs!bot, Headbomb, CobraWiki, Edokter, David Epp-stein, Kornfan71, Sallen2006, Plasticup, DavidCBryant, Jimothy 46, Kmhkmh, Thehotelambush, Egndgf, NuclearWarfare, Nicoguaro, Addbot, Giftiger wunsch, Favonian, Tide rolls, Yobot, AnomieBOT, Sterfried, 0x30114, KurtSchwitters, Johnnyaug, EmausBot, 999ers, George-Barnick, C lueBot NG, Wcherowi, O.Koslowski, Onvan25, BG19bot, Glevum, YFdyh-bot, Pirokiazuma, Escspeed, Tentinator, JAaron95, Lexikittygirl, Garfield Garfield, Zalet karake and Anonymous: 50

- **Divergence of the sum of the reciprocals of the primes** *Source:* https://en.wikipedia.org/wiki/Divergence_of_the_sum_of_the_reciprocals_of_the_primes?oldid=686022343 *Contributors:* Manning Bartlett, Michael Hardy, Eric119, Revolver, Charles Matthews, Timwi, Dcoetzee, Fredrik, Giftlite, Macrakis, Wmahan, Pmanderson, Wroscel, Topynate, Kevin Dorner, Pt, Msh210, Oleg Alexandrov, Woohookitty, Shreevatsa, Reddwarf2956, Ketiltrout, Dzzl, YurikBot, Schmock, DYLAN LENNON~enwiki, Jemebius, Arthur Rubin, Zvika, SmackBot, RDBury, BeteNoir, Incnis Mrsi, JCSantos, Thumperward, PrimeHunter, Kirbytime, Mathsci, WAREL, CRGreathouse, WinBot, Dricherby, Vanish2, Wlod, Stdazi, David Eppstein, Inhumandecency, J.delanoy, IKIZAMA, VolkovBot, Dmcq, Flyer22 Reborn, Capitalismojo, J.Gowers, Xodarap00, Belnumcree, Addbot, Jncraton, Lightbot, Yobot, Ptbotgourou, Xqbot, Sashikumarv, The Evil IP address, Edm1602a, Aliotra, Ripchip Bot, EmausBot, GoingBatty, Slawekb, Quondum, JordiGH, ClueBot NG, Helpful Pixie Bot, Wackywill1001, Adsrah, ChrisGualtieri, Pirokiazuma, Faizan, Seventeenman, Endohiraku, Kojinkara, MahbubAlam03, TakuyaMishiba, Loraof and Anonymous: 42

- **Gödel's ontological proof** *Source:* https://en.wikipedia.org/wiki/G%C3%B6del'{}s_ontological_proof?oldid=656751597 *Contributors:* DamianYerrick, AxelBoldt, Tobias Hoevekamp, Mav, The Anome, Michael Hardy, Brtkrbzhnv, Chinju, Eric119, Rossami, Big iron, Sethmahoney, Schneelocke, Rzach, Timwi, Charlesdarwin, Fairandbalanced, Shantavira, Sverdrup, Tobias Bergemann, Jorend, Karnan, Jonel, Rich Farm-brough, Guanabot, Smyth, Ben Standeven, Elipongo, Jemfinch, Guy Harris, Gene Nygaard, Madmardigan53, Pruss, Polyparadigm, Gra-ham87, Qwertyus, Dpv, Teque5, Rjwilmsi, Tim!, Koavf, Gareth McCaughan, Brighterorange, Winterstein, Bgwhite, Hairy Dude, Philopedia, KSchutte, Trovatore, Sarker112, Nfm, Newagelink, Infinity0, GrinBot~enwiki, SmackBot, Slashme, Eaglizard, Eskimbot, Squiddy, NYKevin, Miguel1626, Snowmanradio, CmdrObot, Ksoileau, Gregbard, X96lee15, Uvaphdman, Kaini, Magioladitis, Error792, Gwern, STBot, Djma12, NewEnglandYankee, Hersfold, Taraborn, Technopat, Ontoraul, LBehounek, BabyJonas, Csapajev, Jeffreykegler, Ceilican, SchreiberBike, Ed-itor2020, Wingspeed, Whelden, Legobot, Yobot, AnomieBOT, Cyan22, Mathonius, Steven Avraham Rosten, Larkusix, Mewulwe, Scott AHerbert, Argumzio, Machine Elf 1735, Dront, Edsu, Yunshui, ItsZippy, RjwilmsiBot, Kozation, EmausBot, AsceticRose, Cdhio, Ferek, Yetaz, Pandeist, Drpatten, ClueBot NG, Msanjelpie, Robertbak, Chester Markel, Helpful Pixie Bot, Absurdphilosopher, Tinynanorobots, Lord Mon-boddo, Lordgrenville, Sbaran7183 and Anonymous: 99

- **Proofs involving the addition of natural numbers** *Source:* https://en.wikipedia.org/wiki/Proofs_involving_the_addition_of_natural_numbers?oldid=647139965 *Contributors:* Toby Bartels, Aenar, Oleg Alexandrov, Woohookitty, Linas, Ruud Koot, GregorB, Salix alba, Jasonglchu, Zvika, NickelShoe, SmackBot, RDBury, Betacommand, Foxjwill, JRSpriggs, CBM, Skittleys, MER-C, Falcor84, Digitalr, Tassedethe, DrilBot, Jochen Burghardt and Anonymous: 6

- **Analyticity of holomorphic functions** *Source:* https://en.wikipedia.org/wiki/Analyticity_of_holomorphic_functions?oldid=676703493 *Contributors:* Michael Hardy, Giftlite, Dratman, Aaronbrick, Stephen Bain, Sligocki, Oleg Alexandrov, Tfz, Woohookitty, Linas, Shreevatsa, Tabletop, MarSch, Salix alba, Sodin, Crasshopper, Zvika, BeteNoir, Mct mht, Geometry guy, Wikiisawesome, SieBot, Mad2Physicist, Addbot, Jafeluv, 9258fahsflkh917fas, Xqbot, Netheril96, Nosuchforever, Brad7777, Tentinator and Anonymous: 15

- **Proofs involving covariant derivatives** *Source:* https://en.wikipedia.org/wiki/Proofs_involving_covariant_derivatives?oldid=613245630 *Contributors:* Michael Hardy, AugPi, Anthony Appleyard, Oleg Alexandrov, Woohookitty, Julyo, Salix alba, Dmharvey, SmackBot, RDBury, Ollivier, Magioladitis, David Eppstein, Lantonov, Erik9bot, Mgvongoeden, Brad7777, Monkbot and Anonymous: 1

- **Derivation of the Cartesian form for an ellipse** *Source:* https://en.wikipedia.org/wiki/Derivation_of_the_Cartesian_form_for_an_ellipse?oldid=627831734 *Contributors:* Delirium, Emperorbma, Charles Matthews, Premeditated Chaos, Peruvianllama, Oskar Sigvardsson, Water Bottle, Woohookitty, Linas, Rillian, Salix alba, SmackBot, RDBury, Ryan Roos, Aditya2504, Theminivann, Pjvpjv, Erik9bot, Dark Silver Crow, Loraof and Anonymous: 19

- **Derivation of the Routh array** *Source:* https://en.wikipedia.org/wiki/Derivation_of_the_Routh_array?oldid=673544099 *Contributors:* Michael Hardy, Giftlite, Rgdboer, Mdd, Bgwhite, Dspradau, SmackBot, Zaxxonal, CRGreathouse, Phatom87, Alaibot, Magioladitis, Mreiki, Maralia, Sdrtirs, Yobot, AnomieBOT, LilHelpa, Locobot, SPKirsch, BG19bot and Anonymous: 8

- **Deriving the Schwarzschild solution** *Source:* https://en.wikipedia.org/wiki/Deriving_the_Schwarzschild_solution?oldid=672134280 *Contributors:* Edward, Patrick, Charles Matthews, Robinh, Jason Quinn, Bender235, Mpatel, Christopher Thomas, Ems57fcva, Hillman, Widdma, Salsb, Colonies Chris, Georg-Johann, Radagast83, Turms, JanBielawski, JRSpriggs, Unown, Headbomb, Alphachimpbot, Setreset, Sun Creator, Terry0051, Addbot, Earthandmoon, EmausBot, Euty, Bibcode Bot, Hippokrateszholdacskai, IkamusumeFan, Jimthree60, Geométer, Elenceq and Anonymous: 21

- **Dual of BCH is an independent source** *Source:* https://en.wikipedia.org/wiki/Dual_of_BCH_is_an_independent_source?oldid=532048827 *Contributors:* HarryHenryGebel, DragonflySixtyseven, Uurtamo, Addbot and Greedyhalibut

- **Proofs of Fermat's theorem on sums of two squares** *Source:* https://en.wikipedia.org/wiki/Proofs_of_Fermat'{}s_theorem_on_sums_of_two_squares?oldid=686801545 *Contributors:* Michael Hardy, Chinju, Charles Matthews, Dcoetzee, Giftlite, D6, Rich Farmbrough, Zenohockey, Woohookitty, Karam.Anthony.K, Rjwilmsi, Salix alba, Magidin, Mathbot, Algebraist, Zvika, Gilliam, Psiphiorg, JCSantos, Colonies Chris, Daqu, Madmath789, LDH, CRGreathouse, Phauly, Limweizhong, Jaerik, Thom Tyrrell, JeffTowers, Strategist333, Arcfrk, GirasoleDE, NowhereDense, Marc van Leeuwen, Laudan08, AnomieBOT, Full-date unlinking bot, HighCrossRuff, Wisapi, Sapphorain, Tawarama, BG19bot, Chmarkine, ChrisGualtieri, Rhuaidhri, Teddyktchan and Anonymous: 29

- **Furstenberg's proof of the infinitude of primes** *Source:* https://en.wikipedia.org/wiki/Furstenberg'{}s_proof_of_the_infinitude_of_primes?oldid=662873580 *Contributors:* Michael Hardy, Giftlite, Bender235, Woohookitty, Shreevatsa, Salix alba, Sodin, PrimeHunter, RekishiEJ, CRGreathouse, Erzbischof, Sullivan.t.j, David Eppstein, Borat fan, VolkovBot, Lechatjaune, Thehotelambush, Malatinszky, Cenarium, MystBot, Addbot, DOI bot, Topology Expert, Favonian, Yobot, Citation bot, Point-set topologist, Elseif, Citation bot 1, Fly by Night, CitationCleanerBot and Anonymous: 9

- **Proofs involving the Moore–Penrose pseudoinverse** *Source:* https://en.wikipedia.org/wiki/Proofs_involving_the_Moore%E2%80%93Penrose_pseudoinverse?oldid=678965146 *Contributors:* Michael Hardy, Aenar, RainerBlome, Rich Farmbrough, Woohookitty, Ms2ger, Zvika, Smack-Bot, RDBury, Lavaka, HenningThielemann, Alaibot, MER-C, R'n'B, Skier Dude, Abelian, Mild Bill Hiccup, Morana, Qwfp, Addbot, Yobot, LilHelpa, Duoduoduo, Snotbot, Dschult, Helpful Pixie Bot, Neveritt, Monkbot and Anonymous: 12

- **Sharp-P-completeness of 01-permanent** *Source:* https://en.wikipedia.org/wiki/Sharp-P-completeness_of_01-permanent?oldid=685996269 *Contributors:* Michael Hardy, Charles Matthews, Mackensen, Altenmann, Giftlite, 4pq1injbok, Diego Moya, Kusma, Shreevatsa, Omnieiunium, Tznkai, Arthur Rubin, SmackBot, Karmastan, ForgeGod, Took, Aram.harrow, Bluebot, Ppadala, OrphanBot, Fuhghettaboutit, Mukadderat, Ylloh, CRGreathouse, Hardmath, Jayron32, Joachim Selke, David Eppstein, Keith D, Lantonov, Nsk92, Laudak, Kbdankbot, Deepmath, Twri, Thore Husfeldt, Citation bot 1, T. Canens, Torceval, Cobaltcigs, SporkBot, Macwhiz, Deltahedron and Anonymous: 17

- **Proof of Fermat's Last Theorem for specific exponents** *Source:* https://en.wikipedia.org/wiki/Proof_of_Fermat'{}s_Last_Theorem_for_specific_exponents?oldid=662410504 *Contributors:* Giftlite, Rich Farmbrough, Paul August, Georgia guy, Rjwilmsi, Magidin, Maxal, Sodin, Cydebot, Headgomb, KConWiki, David Eppstein, Ttwo, Proteins, GirasoleDE, Gamesguru2, MystBot, Addbot, Tassedethe, הרו ש, HerculeBot, Yobot, Kilom691, IskaralPust, AnomieBOT, Citation bot, Mcoupal, Citation bot 1, Double sharp, Duoduoduo, RjwilmsiBot, Neat maths, Howard L Kaplan, Bollyjeff, Vanished user fijw983kjaslkekfhj45, Toshio Yamaguchi, AUN4, Lanthanum-138, Joel B. Lewis, Helpful Pixie Bot, Brad7777, Spiros P. Andriopoulos, Monkbot and Anonymous: 8

- **Proof of the Euler product formula for the Riemann zeta function** *Source:* https://en.wikipedia.org/wiki/Proof_of_the_Euler_product_formula_for_the_Riemann_zeta_function?oldid=641462160 *Contributors:* Tobias Bergemann, Giftlite, Haham hanuka, Woohookitty, Shreevatsa, Jan Helge Salvesen, Salix alba, Mordacil, Zvika, Lambiam, Jim.belk, Loadmaster, Mon4, Asmeurer, CommonsDelinker, CSumit, Jjw, Iameukarya, Rosiestep, Cœlispex, Addbot, Vukini, Yobot, The Evil IP address, Gap9551, Balaonair, Kiravae, WikitanvirBot, Wmayner, Letsgoexploring, Helpful Pixie Bot, Alessandrofalconi, RichardMills65, Monkbot and Anonymous: 24

- **Proofs involving ordinary least squares** *Source:* https://en.wikipedia.org/wiki/Proofs_involving_ordinary_least_squares?oldid=679122919 *Contributors:* Michael Hardy, Ketiltrout, SmackBot, RDBury, Foofighter20x, Melcombe, EtudiantEco, Angry bee, Erik9bot, Kiefer.Wolfowitz, Stpasha, Vergnetp, Kchowdhary, Adam9007 and Anonymous: 13

- **Proofs involving the Laplace–Beltrami operator** *Source:* https://en.wikipedia.org/wiki/Proofs_involving_the_Laplace%E2%80%93Beltrami_operator?oldid=551265049 *Contributors:* Michael Hardy, Charles Matthews, Woohookitty, Linas, Salix alba, Gaius Cornelius, RDBury, Sillyrabbit, Jhausauer, Tesseran, Alaibot, RichardVeryard, MER-C, Geometry guy, Antixt, Stewy5714, Yobot and Anonymous: 6

- **Proofs of convergence of random variables** *Source:* https://en.wikipedia.org/wiki/Proofs_of_convergence_of_random_variables?oldid=679107359 *Contributors:* AlmostSurely, Bender235, Fram, SmackBot, Magioladitis, Jmath666, Melcombe, Eeekster, Bluemaster, Langmore, Jonesey95, Stpasha, RjwilmsiBot, Helpful Pixie Bot, Mynameisveryycool, ChrisGualtieri, DrWhitechalk, Lambda2012, Crispulop and Anonymous: 9

- **Proofs related to chi-squared distribution** *Source:* https://en.wikipedia.org/wiki/Proofs_related_to_chi-squared_distribution?oldid=685 *Contributors:* PAR, HannsEwald, Zvika, Mild Bill Hiccup, EtudiantEco, Kastchei, Toni parellada and Anonymous: 13

- **Proofs of quadratic reciprocity** *Source:* https://en.wikipedia.org/wiki/Proofs_of_quadratic_reciprocity?oldid=609838872 *Contributors:* CharlesMatthews, Lowellian, C S, Woohookitty, Salix alba, Holomorph, Dmharvey, RussBot, KSmrq, Zvika, -Ozone-, Digana, Jim.belk, CRGreathouse, ShelfSkewed, Stebulus, RobHar, David Eppstein, Chridd, Marc van Leeuwen, Virginia-American, Anticipation of a New Lover's Arrival, The, Helpful Pixie Bot, BG19bot and Anonymous: 10

- **Proof of Stein's example** *Source:* https://en.wikipedia.org/wiki/Proof_of_Stein'{}s_example?oldid=387216502 *Contributors:* Michael Hardy, Giftlite, Rich Farmbrough, Gary, Hq3473, Woohookitty, Salix alba, Zvika, Jwmillerusa, Alaibot, AlwynChen, Erik9bot and Anonymous: 3

- **Proofs of trigonometric identities** *Source:* https://en.wikipedia.org/wiki/Proofs_of_trigonometric_identities?oldid=685471069 *Contributors:* Michael Hardy, Charles Matthews, Tobias Bergemann, Giftlite, Edudobay, Gaussmarkov, Pearle, Rd232, RJFJR, Woohookitty, Linas, Shreevatsa, Prashanthns, Rjwilmsi, Salix alba, Sceptre, Grafen, MathMan64, Yahya Abdal-Aziz, Arthur Rubin, Chaleur, SmackBot, Richard B, Ikiroid, AeroSpace, William Ackerman, Kirbytime, Myasuda, Stannered, Kisfox, JamesBWatson, Indeed123, Nousernamesleft, Tomaxer, Derekjc, ClueBot, Alb31416, Mostargue, Estirabot, Floul1, Addbot, Yobot, DemocraticLuntz, Ipatrol, LilHelpa, Carrickdb, I dream of horses, Twistor96, John of Reading, SamHB, GabKBel, Wiwa1, Chewings72, Snotbot, Frietjes, Marechal Ney, Garygoh884, Mahdy Saffar, Solomon7968, Fritzmatias, Fylbecatulous, ! Biswas Amitava !, Adamb76, Alfazal, Jepsonr, IlFreccio, Dberard, Yut23, Monkbot, Magriteappleface, Loraof, JJMC89, Sgr ganesh, Shoil tanwer and Anonymous: 70

- **Union of two regular languages** *Source:* https://en.wikipedia.org/wiki/Union_of_two_regular_languages?oldid=687110531 *Contributors:* Charles Matthews, Rspeer, Gary, Ceyockey, Oleg Alexandrov, Qwertyus, Alaibot, CBKAtTopsails, Addbot, Luckas-bot, AnomieBOT, Ashutosh y0078, John Cline, Jochen Burghardt and Anonymous: 4

37.3.2 Images

- **File:0.999....ogg** *Source:* https://upload.wikimedia.org/wikipedia/commons/e/e5/0.999....ogg *License:* CC-BY-SA-3.0 *Contributors:*
- Derivative of 0.999... *Original artist:* **Speaker:** WODUP
  Authors of the article
- **File:4adic_333.svg** *Source:* https://upload.wikimedia.org/wikipedia/commons/5/59/4adic_333.svg *License:* CC-BY-SA-3.0 *Contributors:* ? *Original artist:* ?
- **File:999_Intervals_C.svg** *Source:* https://upload.wikimedia.org/wikipedia/commons/f/f4/999_Intervals_C.svg *License:* CC-BY-SA-3.0 *Contributors:* Own work *Original artist:* Melchoir
- **File:999_Perspective_Vector.svg** *Source:* https://upload.wikimedia.org/wikipedia/commons/e/e5/999_Perspective_Vector.svg *License:* CC BY-SA 3.0 *Contributors:* Original created by Melchoir in POV-Ray, edited by Sylvielmna in Inkscape *Original artist:* Melchoir, AzaToth, Mets501, Sopoforic Derivative Work: Sylvielmna
- **File:Ambox_important.svg** *Source:* https://upload.wikimedia.org/wikipedia/commons/b/b4/Ambox_important.svg *License:* Public domain *Contributors:* Own work, based off of Image:Ambox scales.svg *Original artist:* Dsmurat (talk · contribs)
- **File:Base4_333.svg** *Source:* https://upload.wikimedia.org/wikipedia/commons/9/9a/Base4_333.svg *License:* CC-BY-SA-3.0 *Contributors:* No machine-readable source provided. Own work assumed (based on copyright claims). *Original artist:* No machine-readable author provided. Melchoir assumed (based on copyright claims).
- **File:Cantor_base_3.svg** *Source:* https://upload.wikimedia.org/wikipedia/commons/f/f4/Cantor_base_3.svg *License:* CC-BY-SA-3.0 *Contributors:* ? *Original artist:* ?
- **File:Commons-logo.svg** *Source:* https://upload.wikimedia.org/wikipedia/en/4/4a/Commons-logo.svg *License:* ? *Contributors:* ? *Original artist:* ?
- **File:Cot(theta).svg** *Source:* https://upload.wikimedia.org/wikipedia/commons/f/f9/Cot%28theta%29.svg *License:* Public domain *Contributors:*
- -cot(theta).jpg *Original artist:*
- derivative work: Phatom87 (<a href='//commons.wikimedia.org/wiki/User_talk:Phatom87' title='User talk:Phatom87'>talk</a>)
- **File:Dominoeffect.png** *Source:* https://upload.wikimedia.org/wikipedia/commons/9/92/Dominoeffect.png *License:* CC-BY-SA-3.0 *Contributors:* ? *Original artist:* ?
- **File:Edit-clear.svg** *Source:* https://upload.wikimedia.org/wikipedia/en/f/f2/Edit-clear.svg *License:* Public domain *Contributors:* The *Tango! Desktop Project*. *Original artist:*
  The people from the Tango! project. And according to the meta-data in the file, specifically: "Andreas Nilsson, and Jakub Steiner (although minimally)."
- **File:Eisenstein-quadratic-reciprocity-1.svg** *Source:* https://upload.wikimedia.org/wikipedia/en/7/72/Eisenstein-quadratic-reciprocity-1.svg *License:* CC-BY-SA-2.5 *Contributors:* ? *Original artist:* ?
- **File:Eisenstein-quadratic-reciprocity-2.svg** *Source:* https://upload.wikimedia.org/wikipedia/en/8/8e/Eisenstein-quadratic-reciprocity-2. *License:* CC-BY-SA-2.5 *Contributors:* ? *Original artist:* ?
- **File:Eisenstein-quadratic-reciprocity-3.svg** *Source:* https://upload.wikimedia.org/wikipedia/en/6/62/Eisenstein-quadratic-reciprocity-3.svg *License:* CC-BY-SA-2.5 *Contributors:* ? *Original artist:* ?
- **File:Eisenstein-quadratic-reciprocity-4.svg** *Source:* https://upload.wikimedia.org/wikipedia/en/5/5c/Eisenstein-quadratic-reciprocity-4. *License:* CC-BY-SA-2.5 *Contributors:* ? *Original artist:* ?
- **File:Ellipse_derivation_1.svg** *Source:* https://upload.wikimedia.org/wikipedia/commons/6/6c/Ellipse_derivation_1.svg *License:* CC0 *Contributors:* Own work *Original artist:* Premeditated Chaos
- **File:Euler'{}s_formula.svg** *Source:* https://upload.wikimedia.org/wikipedia/commons/7/71/Euler%27s_formula.svg *License:* CC-BY-SA-3.0 *Contributors:* Drawn by en User:Gunther, modified by others. *Original artist:* Originally created by gunther using xfig, recreated in Inkscape by Wereon, italics fixed by lasindi.
- **File:Folder_Hexagonal_Icon.svg** *Source:* https://upload.wikimedia.org/wikipedia/en/4/48/Folder_Hexagonal_Icon.svg *License:* Cc-by-sa-3.0 *Contributors:* ? *Original artist:* ?
- **File:LambertContinuedFraction.JPG** *Source:* https://upload.wikimedia.org/wikipedia/commons/4/48/LambertContinuedFraction.JPG *License:* CC BY-SA 3.0 *Contributors:* Own work *Original artist:* KurtSchwitters
- **File:Legendre.jpg** *Source:* https://upload.wikimedia.org/wikipedia/commons/0/03/Legendre.jpg *License:* Public domain *Contributors:* http://www.numericana.com/answer/record.htm#legendre where it was cropped from here *Original artist:* Julien-Léopold Boilly
- **File:Leonhard_Euler.jpg** *Source:* https://upload.wikimedia.org/wikipedia/commons/d/d7/Leonhard_Euler.jpg *License:* Public domain *Contributors:*
  2. Kunstmuseum Basel
  Original artist: Jakob Emanuel Handmann
- **File:Nuvola_apps_edu_mathematics_blue-p.svg** *Source:* https://upload.wikimedia.org/wikipedia/commons/3/3e/Nuvola_apps_edu_blue-p.svg *License:* GPL *Contributors:* Derivative work from Image:Nuvola apps edu mathematics.png and Image:Nuvola apps edu mathematics-p.svg *Original artist:* David Vignoni (original icon); Flamurai (SVG convertion); bayo (color)
- **File:Oxyrhynchus_papyrus_with_Euclid'{}s_Elements.jpg** *Source:* https://upload.wikimedia.org/wikipedia/commons/8/8d/P._Oxy._I_29.jpg *License:* Public domain *Contributors:* http://www.math.ubc.ca/~{}cass/Euclid/papyrus/tha.jpg *Original artist:* Euclid

- **File:POV-Ray-Dodecahedron.svg** *Source:* https://upload.wikimedia.org/wikipedia/commons/a/a4/Dodecahedron.svg *License:* CC-BY-SA-3.0 *Contributors:* Vectorisation of Image:Dodecahedron.jpg *Original artist:* User:DTR
- **File:People_icon.svg** *Source:* https://upload.wikimedia.org/wikipedia/commons/3/37/People_icon.svg *License:* CC0 *Contributors:* OpenClipart *Original artist:* OpenClipart
- **File:Permanent-2powers01.png** *Source:* https://upload.wikimedia.org/wikipedia/en/f/f0/Permanent-2powers01.png *License:* PD *Contributors:* ? *Original artist:* ?
- **File:Permanent-Nonneg2Powers.png** *Source:* https://upload.wikimedia.org/wikipedia/en/4/4a/Permanent-Nonneg2Powers.png *License:* PD *Contributors:* ? *Original artist:* ?
- **File:Peter_Gustav_Lejeune_Dirichlet.jpg** *Source:* https://upload.wikimedia.org/wikipedia/commons/3/32/Peter_Gustav_Lejeune_Dirich jpg *License:* Public domain *Contributors:* Unknown *Original artist:* Unknown
- **File:Pierre_de_Fermat.jpg** *Source:* https://upload.wikimedia.org/wikipedia/commons/f/f3/Pierre_de_Fermat.jpg *License:* Public domain *Contributors:* http://www-groups.dcs.st-and.ac.uk/~{}history/PictDisplay/Fermat.html *Original artist:* ?
- **File:Pierre_de_Fermat.png** *Source:* https://upload.wikimedia.org/wikipedia/commons/4/4b/Pierre_de_Fermat.png *License:* Public domain *Contributors:* ? *Original artist:* ?
- **File:Portal-puzzle.svg** *Source:* https://upload.wikimedia.org/wikipedia/en/f/fd/Portal-puzzle.svg *License:* Public domain *Contributors:* ? *Original artist:* ?
- **File:Proofs-of-Fermats-Little-Theorem-bracelet1.svg** *Source:* https://upload.wikimedia.org/wikipedia/commons/8/80/Proofs-of-Fermat. svg *License:* Public domain *Contributors:* en:Image:Proofs-of-Fermats-Little-Theorem-bracelet1.png *Original artist:* en:User:Dmharvey, User:Stannered
- **File:Proofs-of-Fermats-Little-Theorem-bracelet2.svg** *Source:* https://upload.wikimedia.org/wikipedia/commons/0/08/Proofs-of-Ferma2. svg *License:* Public domain *Contributors:* en:Image:Proofs-of-Fermats-Little-Theorem-bracelet2.png *Original artist:* en:User:Dmharvey, User:Stannered
- **File:Proofs-of-Fermats-Little-Theorem-dynamic1.png** *Source:* https://upload.wikimedia.org/wikipedia/en/8/82/Proofs-of-Fermats- png *License:* PD *Contributors:* ? *Original artist:* ?
- **File:Proofs-of-Fermats-Little-Theorem-dynamic2.png** *Source:* https://upload.wikimedia.org/wikipedia/en/1/1d/Proofs-of-Fermats- png *License:* PD *Contributors:* ? *Original artist:* ?
- **File:Proofs-of-Fermats-Little-Theorem-dynamic3.png** *Source:* https://upload.wikimedia.org/wikipedia/en/f/f6/Proofs-of-Fermats- png *License:* PD *Contributors:* ? *Original artist:* ?
- **File:Question_book-new.svg** *Source:* https://upload.wikimedia.org/wikipedia/en/9/99/Question_book-new.svg *License:* Cc-by-sa-3.0 *Contributors:*
Created from scratch in Adobe Illustrator. Based on Image:Question book.png created by User:Equazcion *Original artist:*
Tkgd2007
- **File:Sieve_of_Eratosthenes_animation.gif** *Source:* https://upload.wikimedia.org/wikipedia/commons/b/b9/Sieve_of_Eratosthenes_ gif *License:* CC-BY-SA-3.0 *Contributors:* ? *Original artist:* ?
- **File:Sound-icon.svg** *Source:* https://upload.wikimedia.org/wikipedia/commons/4/47/Sound-icon.svg *License:* LGPL *Contributors:* Derivative work from Silsor's versio *Original artist:* Crystal SVG icon set
- **File:Sum_of_reciprocals_of_primes.svg** *Source:* https://upload.wikimedia.org/wikipedia/commons/f/f9/Sum_of_reciprocals_of_primes. svg *License:* CC0 *Contributors:* Own work *Original artist:* User:Dcoetzee
- **File:Tan(theta).jpg** *Source:* https://upload.wikimedia.org/wikipedia/commons/3/38/Tan%28theta%29.jpg *License:* Public domain *Contributors:* Transferred from en.wikipedia to Commons by Phatom87 using CommonsHelper. *Original artist:* Zaxxonal at English Wikipedia
- **File:Text_document_with_red_question_mark.svg** *Source:* https://upload.wikimedia.org/wikipedia/commons/a/a4/Text_document_with_ red_question_mark.svg *License:* Public domain *Contributors:* Created by bdesham with Inkscape; based upon Text-x-generic.svg from the Tango project. *Original artist:* Benjamin D. Esham (bdesham)
- **File:TrigInequality.svg** *Source:* https://upload.wikimedia.org/wikipedia/commons/e/eb/TrigInequality.svg *License:* CC-BY-SA-3.0 *Contributors:* Image:TrigInequality.png *Original artist:* Traced by User:Stannered
- **File:TrigSumFormula.svg** *Source:* https://upload.wikimedia.org/wikipedia/commons/0/03/TrigSumFormula.svg *License:* CC-BY-SA-3.0 *Contributors:* Image:TrigSumFormula.png *Original artist:* Remade by Edudobay using Eukleides.
- **File:Trigonometric_Triangle.svg** *Source:* https://upload.wikimedia.org/wikipedia/commons/4/48/Trigonometric_Triangle.svg *License:* CC-BY-SA-3.0 *Contributors:* en:Image:Trigonometric Triangle.PNG/Image:Trigonometry triangle.svg *Original artist:* Traced by User:Stannered
- **File:Twocolumnproof.png** *Source:* https://upload.wikimedia.org/wikipedia/commons/1/13/Twocolumnproof.png *License:* Public domain *Contributors:* ? *Original artist:* ?
- **File:Weak_field_approximation_diagram.svg** *Source:* https://upload.wikimedia.org/wikipedia/commons/5/5e/Weak_field_approximation_ diagram.svg *License:* CC BY-SA 4.0 *Contributors:* Own work *Original artist:* IkamusumeFan
- **File:Wiki_letter_w_cropped.svg** *Source:* https://upload.wikimedia.org/wikipedia/commons/1/1c/Wiki_letter_w_cropped.svg *License:* CC-BY-SA-3.0 *Contributors:*
- Wiki_letter_w.svg *Original artist:* Wiki_letter_w.svg: Jarkko Piiroinen
- **File:Wiktionary-logo-en.svg** *Source:* https://upload.wikimedia.org/wikipedia/commons/f/f8/Wiktionary-logo-en.svg *License:* Public domain *Contributors:* Vector version of Image:Wiktionary-logo-en.png. *Original artist:* Vectorized by Fvasconcellos (talk · contribs), based on original logo tossed together by Brion Vibber

37.3.3 Content license

- Creative Commons Attribution-Share Alike 3.0

Made in the USA
Monee, IL
05 February 2025

11584344R00142